The Final Countdown
Tribulation Rising
The AI Invasion
Volume 2

FIRST PRINTING

Billy Crone

Cover Design:
CHRIS TAYLOR

To my wife, Brandie.

Thank you for being so patient
with a man full of dreams.
You truly are my gift from God.
It is an honor to have you as my wife
and I'm still amazed that you willingly chose
to join me in this challenging yet exhilarating
roller coaster ride called the Christian life.
God has truly done exceedingly abundantly above all
that we could have ever asked or even thought of.
Who ever said that living for the Lord was boring?!
One day our ride together will be over here on earth,
yet it will continue on in eternity forever.
I love you.

Contents

Preface

Like many of us, I had been familiar with the technological term of AI or Artificial Intelligence for quite some time now, especially with my electronics background. However, it wasn't until a few years ago that I really began to dive into this eclectic topic in the research phase of our documentary entitled, *"Attack of the Drones: Skynet is Coming."* It was here that I was forced to go down this rabbit hole for the first time and not only was it a real eye opener to say the least, but it left a taste in my mouth for more. The more I chewed on it, the more I began to see how AI is a pivotal historical development needed to pull off the last days scenarios mentioned in the Book of Revelation and other prophetic texts. Thus I determined to chase the rabbit down the hole once again, and this time, exhausted every Biblical text I could think of proving just how important the rise of Artificial Intelligence really is to the rise of the Antichrist kingdom and global tyranny that is coming to this planet much sooner than people want to believe. I am now totally convinced that AI was not only prophesied to emerge across the planet nearly 2,600 years ago, but sure enough, as God warned in that same prophetic text, Artificial Intelligence will lead to the destruction of humanity. Not so surprisingly, even the secularists agree and give the same dire warning. Therefore, what you will read in the pages of this book, I believe, is the most current, exhaustive, up-to-date Biblical study on the topic of Artificial Intelligence you will find anywhere on the planet. And it is my prayer that you will be equipped with this last days timely information from God to quickly share with as many people as you can how to escape through Jesus Christ, the horrors that AI and the Antichrist will be bringing to the planet much sooner than we can imagine. Time is of the essence! One last piece of advice; when you are through reading this book, will you please READ YOUR BIBLE? I mean that in the nicest possible way. Enjoy, and I'm looking forward to seeing you someday!

Billy Crone
Las Vegas, Nevada
2021

Chapter Twelve

The Future of Media Conveniences with AI

The **3ʳᵈ way** AI is being pitched to take over our Entertainment is **AI Will Control Our Media**. Because we all know how hard it is to generate news, let alone turn the channel to find the news in the first place, let alone find a good movie on TV, or wait for somebody to come up with a decent one! Oh, the inhumanity of it all!

So hey, wouldn't it be great if someone could create all that media for us so we wouldn't have to interrupt our lazy cushy hedonistic lives and avoid all that suffering? Well hey, worry no more! Your wish is AI's command! That's right, even our media is now being pitched to be controlled by AI and it will not just lead to a life of convenience and lack of suffering, it will actually lead to a Big Brother nightmare and slaughter in the 7-year Tribulation! But don't take my word for it. Let's listen to God's.

Revelation 13:11-15 "Then I saw another beast, coming out of the earth. He had two horns like a lamb, but he spoke like a dragon. He exercised all the authority of the first beast on his behalf and made the earth and its inhabitants worship the first beast, whose fatal wound had been healed. And he performed great and miraculous signs, even causing fire to come down from heaven to earth in full view of men. Because of the signs he was given power to do on behalf of the first beast, he deceived the inhabitants of the earth. He ordered them to set up an image in honor of the beast who was wounded by the sword and yet lived. He was given power to give breath to the image of the first beast, so that it could speak and cause all who refused to worship the image to be killed."

Now as we saw before in this text, we saw how the Bible clearly says that the False Prophet, in the last days, is not only going to dupe the whole world into worshiping the Antichrist, but he's what? He's going to "make" them, he's going to "order" them, he's going to "cause" them to do whatever he says to do, otherwise they will what? They will die, right?

Now again, focus on the words, "make" "order" and "cause." This implies that we have some serious global enforcement going on here! Because again, that is the context. It's global. He forces the whole planet to do whatever in the world he wants them to do, or they are going to die! And as we saw before in our study on Modern Technology, this is actually Big Brother on steroids! The Antichrist and False Prophet are literally going to be pulling the trigger on the death of anyone on the whole planet because they are literally micro-managing the whole planet! You need Big Brother to pull that off!

Now I want you to focus on the phrase there, "All who refused to worship the image is to be killed." This tells us that not everyone is going to worship the beast, right? It is as plain as day, right there in the text. Some are actually going to refuse to obey this order. But the problem is, I can see the Antichrist killing and taking out the average Joe on the planet for refusing to obey, you know, the people nobody knows who they are in the first place. But wait a second, what about those who refuse to obey who are high up in positions around the world, or other important global positions of authority, or well-known celebrities, or maybe even other world leaders? How are you going to take them out? And that's not even counting the whole race of Jewish people who will also refuse to worship the Antichrist as god. They will never do that! So how are you going to justify killing all these people without having a revolt on your hands?

Can you say AI to the rescue? Folks, all you do is what evil dictators did in the past and have always done, like Hitler. You brainwash the people to go along with your murder spree, and then you won't have a revolt on your hands when you start knocking people off! And how do you do that? You get control of all the media and create a "propaganda machine" and use that media machine to convince millions of people to go along with your slaughter. And that's exactly what Hitler did, if you know your history. Many believe he was a type or a foreshadowing of the Antichrist, when he slaughtered the Jewish people and he used the media.

"The atrocities committed by the Nazis at the height of their power are staples of every lesson on World War II. But just how did a democracy like Germany fall under the sway of a tyrant in the first place? As Chancellor, Hitler wasted no time tightening his grip on every aspect of German life. He tripled the size of the military, violating the Treaty of Versailles. Under Hitler's orders rival parties were banned. While para-military groups cracked down on protests and executed military opponents. Anti-Semitic laws prohibited Jews from working, voting and occupying public spaces. A propaganda department produced art, films and books praising Hitler and embracing his vision of a better Germany."

Simon Whistler: *"In modern times we look back at the Nazi's Third Reich and wonder how on earth could anyone follow Adolph Hitler when you could clearly see he was an evil dictator. This was all thanks to Joseph Goebbels, the minister of propaganda. He knew exactly how to manipulate the minds of the nation with fake news, Nazi idealized books and propaganda films. Goebbels was single-handedly responsible for brainwashing the entire population and destroying people who were caught in his web of lies."*

"Inciting Hate Against the Jews" by Joseph Goebbels: *"Party Comrades, a good government can no more exist without propaganda than good propaganda without a good government. One needs to augment* *the other. And even if today's Jewish newspapers still believe they can intimidate the National Socialist movement with veiled threats, even if they believe they have permission to circumvent our emergency decrees, they should take care, for one day our patience will run dry, and then the Jews will have their brazen, lying mouths stuffed once and for all."*

Simon Whistler: *"Goebbels was desperate to regain Hitler's trust, so in 1939 he incited a mob by publishing a fake story in the newspaper that said a German man had been shot by a Jewish man in Paris. He spoke in the streets proclaiming that the Jews were destroying society. He led an angry mob to burn down a synagogue. That same year he traveled to Poland with a documentary film crew to show what life was like inside a Jewish ghetto. The people living in the ghetto with sunken faces, wore rags and were generally dirty and sick. We now know that this is because the Nazis forced them to live in inhumane conditions in the first place.*

However, Goebbels saw this as an opportunity to compare them to animals, saying they were choosing to live that way. He said they were less human, and they needed to be exterminated. The movie was called 'The Eternal Jew' and was introduced in 1940. In the movie he tried to say all Jews were criminals. If the Nazis were to lose the war, and if the Jews were to take over the world, they would force the German people to live in squalor. 'For that reason alone, the Jews hate us. They despise our culture and learning, which they perceive as towering over their nomadic worldview. They fear our economic and social standards, which leave no room for their parasitic drives.'"

This documentary generally frightened the German people, and they felt that concentration camps were absolutely justified. After visiting Auschwitz in person, Joseph Goebbels wrote that the death and torture of these prisoners was so gruesome that he did not even want to write the details in his private journal. He wrote: 'It may be barbaric, but they truly deserve it.'

Hitler declared himself Fuhrer and absolute dictator. With Germany under his total control Hitler would shift his focus to Global Domination, setting the stage for World War II."[1]

And it's going to happen again, except this time it's going to be WWIII or the global wars mentioned in the 7-year Tribulation, led by the actual Antichrist, which many again, believe Hitler was a type of. But notice how Hitler got the people to go along with his evil schemes, as evil and sick as they were. He took control of what? The media, arts, books, movies, news, you name it, to brainwash the people.

Well guess what? AI again, right now, is being pitched to do the same thing except this time, it is not just controlling the media of one country, it's the media around the whole planet. Just in time for the 7-year Tribulation, slaughter of the Jewish people and many more people unfortunately, and we see that here.

Zechariah 13:8-9 "In the whole land, declares the LORD, two-thirds will be struck down and perish; yet one-third will be left in it. This third I will bring into the fire; I will refine them like silver and test them like gold. They will call on my name and I will answer them; I will say, 'They are my people,' and they will say, 'The LORD is our God.'"

And again, this a good news-bad news, passage for the Jewish people. They finally wake up at this point, the midway point, and realize that Jesus is the One and Only Messiah, but it comes at a horrible price. Two-thirds perish! We also know that the Antichrist is going to seek to slaughter anyone who follows God during the 7-year Tribulation. We see that here.

Revelation 6:9-10 "When He opened the fifth seal, I saw under the altar the souls of those who had been slain because of the Word of God and the testimony they had maintained. They called out in a loud voice, 'How long, Sovereign Lord, holy and true, until You judge the inhabitants of the earth and avenge our blood?'"

And the picture in the text is it is bloody murder! A horrible slaughter! So, the question is, how is the Antichrist going to get the bulk of the planet, not just one country, to go along with this global plan to slaughter anyone who chooses to follow God instead of Him? AI to the rescue!

Right now, AI is being pitched to basically be the "new electronic Joseph Goebbels" around the whole planet controlling all the world's media on a scale that Hitler would have had a heyday with! It's going to brainwash people around the globe to help the Antichrist justify killing a ton of people. And I want to show you how it is being set up for that.

The **1st way** AI is taking over the media on a global scale is with **The News**. That's right, we all know how hard it is to come up with some sort of news-worthy related item for people to watch or read. I mean, whose got time for that, right? It's hard enough to watch it let alone having to come up with it!

So hey, wouldn't it be great if we just allow AI to create it all for us, huh? And as crazy as that sounds, it's already being done on a massive scale! First of all, we're so lazy, we can't even muster up the energy to find news in the first place and we are now having AI even do that! For instance, you can use **Jottr**, an AI APP that, "Learns what you like, filters

what you don't like and widens your horizons." Actually, it decreases your horizons because it tells you what to read.

Or there's also **News360**, another AI APP that, "Learns what you

like, don't like, and finds news articles, blog posts, stories from all across the Internet that are most relevant to you." Or relevant to what AI wants you to think is relevant to you.

And as lazy as that is, it gets even worse. Now we are so lazy that we not only want AI to find us the news in the first place, but we're now letting AI create the news for us without our help! I'm not joking! We are actually removing the human element completely when it comes to News!

"How to Create Effective Content with the Help of Artificial Intelligence."

"Artificial intelligence has become a real buzzword these last couple of years. No longer restricted to the realm of science fiction, AI is being improved upon each day in very real ways, and this includes its advanced widespread application of content creation."

That is a fancy way of saying, AI is writing the news for us! Another article puts it even more bluntly.

"AI may have written this article. But is that such a bad thing? 'Imagine how productive Woodward and Bernstein might have been if only they had robots to write their articles for the Washington Post. With AI on their side, they might have taken down the Nixon Administration in days instead of years.'"

In fact, *"A lot of people don't realize this, but a lot of the news stories you read right now are already written by Artificial Intelligence. You get these news releases about things that are happening in sports, for example, or in business. But people are not creating these pieces anymore. It's actually AI that's releasing this information."*

"Lots of us spend hours on our mobile phones reading updates about events and news flashes never realizing it is AI that's generating this stuff now."

Isn't that crazy? And who are some of the people pushing for this?

"In 2013 Amazon founder Jeff Bezos, now the richest man on earth, bought The Washington Post for $250 million. And taking a cue from the profitable AI-based algorithms Amazon uses to sell its books and everything else on the planet, Bezos' Post began using AI to write articles four years later."

So, this has been going on at least since 2017. AI is writing the news! I didn't know that did you? And it's not just Jeff Bezos who's getting in on this, so are the other big AI Tech moguls like Google! Shocker! And they want to go worldwide!

"Artificial Intelligence to power news media? Google partners for a new 'Journalism AI' project."

"Part of the Google News Initiative (GNI) the 'Journalism AI' project will focus on research and training for newsrooms on the intersection of AI and journalism, to help the news industry to use Artificial Intelligence in more innovative ways."

"We'll also collaborate with newsrooms and academic institutions to create the best practices and produce free online training, in a multitude of languages, on how to use AI in the newsroom for journalists worldwide."

So, there it is, they admitted it! They want AI to take over the news on the planet! The question is, "Why would Google, or anyone else for that matter, want to control all the news content worldwide?" Because then you could control the "mind" of the people worldwide! As we already saw that they stated in public, "We want to be like the mind of god." Remember that? Sounds like they are making good on their pledge! But it gets even worse! They also want to, "Use AI to gauge the accuracy of the news." Really? Yeah, listen to their rationale.

"As the impact of fake news has grown, so too have attempts to detect and remove it. But this is unreliable, and we need an AI-driven approach to accurately spot fake news stories."

For instance, *"If a website has published fake news before, there's a good chance they'll do it again," the researchers explain. "By automatically scraping data about these sites, the hope is that our system can help figure out which ones are likely to do it in the first place."*

Wait a second, so now you're going to predict when somebody's planning to say something you don't like, that you consider Fake News, in the future, and prevent them from even doing that before you ever get a chance! You'll never get the News out there! This is nuts! In fact, *"Researchers have already created an Artificial Intelligence with the ability to identify Fake News in 2 minutes."* Who says? So much for freedom of the press. And most people don't realize it's already being done!

Narrator: *"The goal of the online hate index is to help Tech platforms to better understand the growing amount of hate on social media and to use that information to help address the problem. By combining Artificial Intelligence and machine learning and social science, the Online Hate Index will ultimately uncover and identify trends and patterns in hate speech across different platforms. The next stage of our project, we will look at specific targeted populations in a more detailed manner. We will examine content on multiple SM sites, and we will identify strategies to deploy the model more broadly."*

News Commentator: *"Forty percent of Americans under the age of thirty-five tell pollsters they think that the first amendment is dangerous because you might use your freedom to say something that might hurt someone else's feelings. Guess what?! There are some really passionately held views about the abortion issue on this panel today. Can you imagine a world where you might decide that pro-lifers are prohibited from speaking about their abortion views on your platform?"*

Mark Zuckerberg*: "I do generally agree with the point you are making, which is as we are able to technologically shift towards AI, proactively look at content, I think that is going to create massive questions for society."[2]*

Yeah, like whatever happened to the First Amendment?! But you might be thinking, "Oh come on, you're being an alarmist here Pastor Billy! I'm sure we can trust this AI technology to deliver accurate news and information for us. What are you talking about? There's nothing to worry about." Really? And I quote, I wish I was making this up.

"Technologists taught AI to lie and create fakes." I'm not making this up!

"Open AI introduced a new version of AI that is able to mislead." That's their words, not mine! *"This upgrade will allow Artificial Intelligence to create so-called fake news and even write and send abusive spam."*

So much for accurate news! And now you are going to let it control the whole information structure on the planet? It gets even worse!

"Experts are rightly sounding the alarm because it's not difficult to mislead the population anywhere around the globe and now it's not known for what purpose and for who will be using this AI update that creates fake news."

Uh, I got a name, it rhymes with the Antichrist! Talk about a dream come true for him! Now he can create any narrative he wants at any time on the whole planet! And again, they want to take this technology across the globe! *"This Technology is needed to make the world a more truthful place."* Really? You're actually controlling the news on a global scale and dictating what you want the whole planet to hear.

This is dangerous folks! A total control of all the information people get from around the planet! I mean, Hitler burned the books to control the news and information people were getting back in his day. But now these Tech moguls and others around the planet want to use AI basically to "burn anything online" anywhere on the planet, that they don't want people to know about. And I wonder why? I think it has everything to do with the 7-year Tribulation slaughter!

We will make sure people go along with the Antichrist's wicked murder of the Jews and anyone who follows God during that time! I don't care how high up of a position you might have, even a celebrity, no matter who you are or what status you have, you name it! We'll make sure the whole planet goes along with our dastardly deed because we'll use the media, controlled now by AI, to spin a narrative that will brainwash the whole planet to go along with it! A total repeat of what Hitler did, except this time it is being done on a global scale just in time for the 7-year Tribulation!

But that's not all, speaking of history, we all know that some of the people in Germany didn't go along with Hitler back in the day, controlling all the news and information people got to see, they did speak out. So, what's to stop people from doing it again? Well, can you say AI to the rescue again?

Folks, I kid you not, I wish I was making this up too! It's like they thought of every conceivable way around this scenario! So much so are these Tech Moguls and others around the planet pushing for AI to control all the news content around the whole globe, that they are now even going to the next step of having AI robots report the news for us, so we humans don't have to. Isn't that convenient? Here is the transcript of that video.

Fox Business Reports: *"I think I might be in trouble. China has developed a virtual anchor to deliver the news. Watch this:"*

Chinese Narrator: *"The news anchor AI, whose voice and appearance are copied from the original anchor Qui Hao, whose name is Chin Hou."*

Robot Anchor: *"I am the Chinese AI anchor who just joined Xinhua news agency today. My appearance and voice are based on Xinhua news anchor Qiu Hao, but I never need to rest. I am created by face recognition, face modeling, speech synthesis, deep learning and other intelligence techniques."*

Chinese Narrator: *"Another very lifelike female AI Chinese anchor, whose name is Xin Xiaoment. And she is considered to be world's first AI female news anchor from China."*

Female Robot Anchor: *"My name is Xin Xiaoment. My image and voice are modeled after Qu Meng, a news anchor at Xinhua's New Media Center. I will serve as your broadcaster during this year's 'Two Sessions.' My partner and I will present you with a better news broadcast."*

Chinese Narrator: *"And the last one is another Chinese male AI news anchor from the same news agency, Xinhua's News."*

Second Male Robot Anchor: *"Hello, everyone, I am an English artificial intelligence anchor. This is my very first day at Xinhua's News Agency.*

My voice and appearance are modeled from the real anchor at Xinhua. The development of the media industry calls for continuous innovation and deep integration with the national advanced technologies."

Fox Business: *"Well that was exciting, what do you think?"*

"The point is, they can work tirelessly, 24 hours a day. There would be no difficult contract negotiations to get through as well. No vacations. As they say with technology, it continues to develop in advance so who knows, in five years, what that could maybe translate to."[3]

Well, I'll translate it for you. The human element has been totally removed from the news and all information sharing on a global scale, and if Hitler were alive today, he would go nuts with this. You can see where it is headed! If you can control the news on a global scale, then you can control the minds of the people on a global scale! And this is exactly what the Antichrist needs to do if he is going to brainwash the planet into going along with his dastardly deeds and his last days holocaust in the 7-year Tribulation! All this is current technology that is not coming, but already being put into place, just in time for the 7-year Tribulation!

The **2nd way** AI is taking over the media on a global scale is with the **Arts**. Believe it or not, even the Arts are being taking over by AI, and it is like these Tech guys are taking a page out of Hitler's playbook. Why? Because as we saw, this is what Hitler did. He not only took over the news, information, books, media, but even art was taken over by him to brainwash the people into going along with his evil schemes!

And believe it or not, AI is right now being pitched to do the same thing! *"AI Art is on the Rise."* Also known as "computer generated art" it is not only catching on, but it's able to paint portraits on its own using robotics or it can even generate them with a computer and they're fetching some pretty hefty prices at some pretty famous art galleries!

Cloudpainter.com: *"Cloudpainter.com is a network connecting computers and robots, using them to replicate artistic creativity. And in* *doing so, experiment with artificial intelligence. We try to have it make as many of its own creative decisions as possible. And it is getting pretty good at it. So here are many paintings made by Obvious, several of my family, a couple of reproductions, and these abstract portraits here. Similar to how humans learn to paint, Obvious got to this point with a lot of practice. Over the past ten years it has painted over one thousand canvases with the combination of AI and human directions."*

CBS This Morning: *"This will be the first auction house to offer artwork created by Artificial Intelligence. We are seeing the portrait of Edmond Bellamy, right here in studio 57, only on CBS This Morning. It was created by the Paris based art collective, Obvious. It used an algorithm to analyze data, set from 15,000 portraits, and then created a very unique image. Christies International and Multiples, Richard Lloyd, is here with the piece. Explain to us how it works."*

Richard Lloyd: *"Obvious used a piece of software. They uploaded thousands of images into the computer and then at that point it actually splits itself in two. One half is called the generator and that analyzes those thousands of portraits and learns what a portrait is. So, it has two kind of round things here and it sort of passes through the images, then it thinks now I am going to start my own version of those. The second half of the computer, the discriminator tries to spot that. Then every time the cycle is run, if the discriminator is able to say, 'wait a minute', then the generator runs it again. And the cycle finishes when the discriminator says, 'I give up, I can't tell the difference between the computer-generated version and the human version.' And that is what is popped out."*

CBS This Morning: *"This looks good and how much do you think it would sell for?"*

Richard Lloyd: *"Well, we have estimated it at seven to ten-thousand dollars."*

And how much did it actually go for?

Christies: *"For the first time ever, a piece of art created by Artificial Intelligence program, sold at an art auction for almost forty-five times its estimated value. The piece entitled Edmond Bellamy, was made by a Paris art collective named Obvious and sold for $432,500."[4]*

That is crazy! You thought it would go for $7,000-$10,000 but it actually goes for 45 times that amount, $432,500, nearly a half a million! Crazy! In fact, speaking of crazy, as wild as that is, I kid you not, recently, "AI painted a self-portrait" as you can see here. This is what it came up with after reading, "3,000 New York Times articles on itself, AI then generated that image. Which if you know art, certainly classical art, it looks a whole lot like this piece.

That's the classical piece depicting "The Hand of God" reaching down to "The hand of man." Who do you think you are AI, God? That's your impression of yourself! That is what it drew of itself. But speaking of drawings, another propaganda tool that's been used traditionally to influence people and how they view the world, and the people of the world, is not just Classical Art, but cartoons. Believe it or not, many people don't realize how important cartoons were during WWII, either in print or on the screen, to "paint" your enemies as evil. Remember these cartoons even here in the U.S.?

And of course, even in print, cartoons were used in WWII to make the Jewish People, of all people, look really bad, as you can see here.

Well, believe it or not, even this realm, cartoons, is being taken over by AI as well. It can now generate cartoons from scratch, including this classic one we used to all watch!

AI creates 'Flintstones' cartoons from text description.

Nicole Lee, Senior Editor, Engadget: *"The days of painstakingly hand drawn animation might be limited. Researchers have developed Artificial Intelligence that can create short Flintstones cartoons using just a short text description. The scientists trained their AI named Craft by having it examine over 25,000 clips from the classic TV series. Each of them annotated to describe the actions and characters in each scene. With that knowledge under its belt the AI only needed a brief summary of a new scene to insert character's backgrounds and props all by itself. Need Betty and Wilma having a conversation on the couch, just ask. Studios could finish cartoons that much sooner. It might not be all good news. AI automation could lead to job cuts with animators whose works are no longer needed. Either way there comes a time when many animations have little input from humans."[5]*

And that time appears to be here now with AI controlling all the cartoons even as well as all the other art forms. Why? Because that is what you need to do to convince people do go along with your dastardly deeds. Even a Holocaust. You have to control the Art Industry just like Hitler did! All current technology, all here now, just in time for the last days slaughter in the 7-year Tribulation! Not by chance folks!

The **3rd way** AI is taking over the media on a global scale is with **The Movies**. What? Yeah, because again, remember what Hitler did. He got control of the news, the information, the media, books, and even the art. But you have to control the movies if you're going to mold the minds of people and influence them to go along with your dastardly deeds! Movies are one of the most powerful tools for doing just that historically! And believe it or not, AI is already doing that as well! I kid you not, they have left no stone unturned!

First of all, concerning visual media, we've already been conditioned to allow AI to manipulate our photos for us, with all kinds of AI apps. We see that with those filters people apply to their photos making them look younger or even something freaky like a squirrel. But AI Visual manipulation has advanced way beyond just spicing up your photos to now even being able to randomly generate photos of people who do not even exist. It totally makes them up!

"This video shows several people, all dressed differently and male and female. But none of these people are real. As the clip continues, each person is changing outfits and gender. They look very real, but they are generated randomly and automatically by AI. Researchers in Japan developed a new algorithm to create the virtual models. They will be used in the advertising and fashion industries." [6]

Wow! So now you can randomly generate any person doing anything to create any narrative you want, without the danger of an original person saying it's not true! There's nobody to object that's not a real person! Virtual models is right! You can virtually create anything you want! Now you can visualize people doing anything you want them to, to create a riot or some other response like make a certain genre look bad in the public eye. Hitler would have loved this kind of stuff!

In fact, AI can now even imitate your voice in case it wanted to attach that to a photo or video of you or something!

"Terrifying AI learns to mimic your voice in under 60 seconds."

"When it comes to personal privacy and overall security, we often think of passwords, fingerprints, and even our own faces as being the keys that unlock our world, but what about your voice?"

"If someone could perfectly mimic your voice, what kind of damage could they do? If they contacted people you know, could they lie their way into gaining private information about you?"

"Unfortunately, we live in a world where such a danger is real, thanks to extremely powerful Deep Learning (AI) technology that can mimic your voice using just 60 seconds of your speech."

And if you think about it, we have already given these Tech Moguls tons of our voice samples to use for these kinds of purposes via speaking to Siri on our phone or Alexa in our home. They have got way more than 60 seconds!

But speaking of manipulating, as we have seen before in our Modern Technology study, all this technology is leading to another concern called "deep fakes" where the AI technology can actually create a fake video of you or anyone doing anything, they want you to, as well as saying whatever they want you to. It is ripe for abuse as this guy shares.

"Artificial Intelligence has the power to change society. This algorithm is able to detect terrorist propaganda online with a 94% success rate. Perhaps a vital tool in countering radicalization. Despite positive breakthroughs there are potential dangers with AI according to a new report. Automated hacking is identified as one of the most imminent applications of AI, especially so-called fishing attacks."

Shahar Avin, Report Co-author, University of Cambridge: *"So you can model someone's topics of interest or preferences, their writing style, the writing style of a close friend and have the machine automatically create a message that looks a lot like something they would click on."*

"It's been less than a week. Programmers from the University of Washington, last year built an AI algorithm to create a fake video of Barack Obama, allowing them to put any words into the former President's mouth."

Shahar Avin: *"You create videos and audio recordings that are pixel to pixel, indistinguishable from real videos and real audios of people."*[7]

Wow! Now you can have anybody doing anything, and they can't deny it! And it gets even worse! AI has now moved beyond the photos and voices and video samples that we give it to create deep fakes. To, "Being able to literally randomly generate videos from scratch!" I kid you not! This is wild!

Burger King's new AI ad, "Burger King to Unveil Ad Campaign Created by Artificial Intelligence." That was two years ago by the way!

And "AI has managed to make a slightly more appetizing cheeseburger."

It even created a photo of a cheeseburger that looks way better than the original, who cares if it's not real! It's totally manipulated!

In fact, AI is even being looked at to create not just commercials and advertisements from scratch, but whole movies from scratch! From script writing, to marketing, to film editing, you name it! AI is poised to, "Radically alter the movie industry." In fact, here's one of the new and improved movie trailers it came up with recently!

This is a trailer from the movie "Morgan" from 20th Century Fox.

Female character speaking to Morgan: *"It's nice to meet you, Morgan."* Morgan is a robot, standing behind a glass barrier.

Morgan: *"It's nice to meet you."*

Bold Business: *"In 2016, AI came to Hollywood. 20th Century Fox and IBM Watson produced a movie trailer using AI for the first time. Using visual and audio analysis, Watson identified the most powerful and tension packed scenes from the sci-fi thriller, 'Morgan' and created a trailer like no other."*

Female character speaking to Morgan: *"Don't be afraid."*

Bold Business: *"Technology is changing the face of film making. Companies from around the world are leveraging on AI to improve film production, from script writing, all the way to marketing."*[8]

In fact, speaking of AI writing your scripts for you, one of the guys who invented an AI Automatic Script Writer program called "Benjamin" was…

"A political ghostwriter who also worked on the Obama campaign and then later at the White House during the Obama years writing everything from letters to proclamations to speeches."

And this guy later decides to develop an AI program that can do all that script writing automatically. I'm sure they wouldn't use that for political purposes, would they, for some nefarious reason? And I quote… *"With Benjamin, I can throw out the maddest notion, and it will start deconstructing and figuring out how it might be achieved. I think it's just a matter of time when AI will be creating a blockbuster like no one has ever seen!"*

Yeah, to influence people to certain political outcomes and make certain political figures look better than others! Isn't this crazy? Hitler would have had a heyday with all this technology!

In fact, speaking of manipulation, you can now not only use AI to generate writing, music, news, information, books, media, art, cartoons, photos, videos and even whole movies to boot, to mold the minds of people, but in case all that doesn't convince you to vote or even worship a future political person that rhymes with the Antichrist, now there's even talk of developing an AI app that will make sure you choose him!

One scenario was called the "Pick Me" app and it would act as an "AI election advisor" to "rate and recommend the politicians you should vote for." You know, so you don't have to do all the research. "Let the Machine do it." How lazy can you be? All under the guise of convenience, AI is poised to control all our media to manipulate the minds of people all around the planet to obey, worship and even follow a future political figure in the 7-year Tribulation. I wonder who that might be?

How much more proof do we need? The AI invasion has already begun, and it's a huge sign that we're living in the last days! And that's precisely why, out of love, God has given us this update on *The Final Countdown: Tribulation Rising* concerning the AI invasion to show us that the Tribulation is near, and the 2nd Coming of Jesus Christ is rapidly approaching. And that's why Jesus Himself said:

Luke 21:28 "When these things begin to take place, stand up and lift up your heads, because your redemption is drawing near."

People of God, like it or not, we are headed for *The Final Countdown*. The signs of the 7-year Tribulation are Rising! Wake up! And so, the point is this. If you are a Christian, and you're not doing anything for the Lord, shame on you! Get busy doing something for Jesus now! Stop wasting your life! We need you! Don't sit on the sidelines! Get on the front lines and help us! Let's get busy working together doing something splendid for Jesus with what time is left and get busy saving souls! Amen?

But if you're not a Christian, then I beg you, please, heed these signs, heed these warnings, give your life to Jesus now! Because this AI technology is not going to lead to a life of wonderful dreams and a modern-day utopia, but a nightmare beyond your wildest imagination in the 7-year Tribulation! Do not go there! Get saved now through Jesus! Amen?

Chapter Thirteen

The Future of Gaming Conveniences with AI

The **4th way** AI is being pitched to take over, is our entertainment. AI **Will Control Our Gaming**. That's right, because who's got time to play sports anymore or get into shape for sports in the first place, along with other forms of recreational activities that we rely upon to entertain ourselves to death, right? I mean, aren't you tired of getting Carpal Tunnel Syndrome from the remote control or throwing your back out trying to play sports of any kind?

Well hey, wouldn't it be great if someone could control all of our gaming options for us, so we don't have to do any of the work? We can just sit back and relax, and not interrupt our lazy, hedonistic lives and avoid all that suffering? Well hey, worry no more! Your wish is AI's command! That's right, even our games are now being pitched to be controlled by AI as well and it will not lead to a life of convenience and lack of suffering, it will actually lead to a Mark of the Beast scenario in the 7-year Tribulation! But don't take my word for it. Let's listen to God's.

Revelation 13:11-18 "Then I saw another beast, coming out of the earth. He had two horns like a lamb, but he spoke like a dragon. He exercised all the authority of the first beast on his behalf and made the earth and its inhabitants worship the first beast, whose fatal wound had been healed. And he performed great and miraculous signs, even causing fire to come down from Heaven to earth in full view of men. Because of the signs he was given power to do on behalf of the first beast, he deceived the inhabitants of the earth. He ordered them to set up an image in honor of the beast who was wounded by the sword and yet lived. He was given power to give breath to the image of the first beast, so that it could speak and cause all who refused to worship the image to be killed. He also forced everyone, small and great, rich and poor, free and slave, to receive a mark on his right hand or on his forehead, so that no one could buy or sell unless he had the mark, which is the name of the beast or the number of his name. This calls for wisdom. If anyone has insight, let him calculate the number of the beast, for it is man's number. His number is 666."

Now, as we have seen before in this text, the Bible clearly says there is coming a day when all the inhabitants of the earth are not only under the authority of the Antichrist and economy of the Antichrist, but what? They are going to receive the mark of the Antichrist. They will actually be deceived into receiving some sort of mark on their bodies, either in or on the right hand or forehead, to connect themselves to the Antichrist system so they can "buy and sell" and escape the threat of death!

But again, as always, "Could this really happen?" Could the entire world really be deceived into receiving the Mark of the Beast, and is there any evidence that it's really going to take place, just like the Bible said? And more importantly, do we have the technology for this to happen in our lifetime? Uh, yeah! In fact, it is already happening as we sit here!

And once again, all under the guise of convenience, this Mark of the Beast Technology is not only here, but it's all the rage in the Gaming Industry and people are lining up to get a mark out of pure laziness and selfish convenience. And wonder of wonder, guess who is being pitched to control it all! That's right! AI to the rescue! And I want to demonstrate that to you.

The **1ˢᵗ way** AI is taking over our gaming, on a global scale, is with **Sports**. You see, as it was already stated, we all know how hard it is to get into shape for sports, let alone have to push the button on the remote control to turn the channel on the TV to find some sports to watch, right? I mean, if I get another callous on my thumb, I don't know what I'm going to do, how about you?

So again, wouldn't it be great if we just allowed AI to monitor and play all our sports for us, so we don't have to, huh? And as crazy as that sounds, it is already being done on a massive scale! For instance, in order to "enhance" our so-called Sporting Experiences, because we always have to be doing something different to please the viewers so they won't change the channel, believe it or not, they are already putting microchips into all kinds of sports-related items, including the players, so we can have better stats and up to date minute by minute factoids that we all just have to have. So, you don't change that channel and they keep you interested!

For instance, how about Golf as you can see here. *"Callaway Golf Reveals New Artificial Intelligence-Enhanced Driver."* Because it's just

too hard to figure out how to make one yourself!

"Callaway Golf released a new product line for 2019, unveiling a line of golf clubs engineered with the help of AI to determine the optimum design."

"It will result in greater ball speed and more distance for golfers."

That's right, your life as a golfer is now complete, thanks to AI. And that's just the equipment. They are also looking at AI to even coach you into a better game! What? Yes, I know, it is true!

From Your SwingShot 8/12/18
Annotated by Joe Instructor

"AI is Transforming Coaching in Golf." "New technology could transform how amateurs play the game and lower their handicap."

Could it be true? Yes, it is! Please don't cry men!

"Golf.com has launched a new platform that uses AI to analyze a person's swing and automatically identify issues with their action. The 'Swing AI' system detects the issues and develops an interactive lesson plan and then "tracks" the student's progress."

Notice the word "track" there.

Then, *"The PGA partnered with Microsoft using Microsoft's (AI) to review all the data given to broadcasters to "enhance our viewing of Golf," with Doppler radar "tracking" club and golfer's movement, while the shot distance, power, and trajectory of your golf ball is also measured."*

And then, *"Other companies are using "sensors" embedded into the clubs and the balls to create data where it's transmitted to the cloud so AI can become, "The World's Smartest Caddie."*

"Just like a human caddie, AI will know the individual's golf game and help players react and refine their game." "With sensors embedded into the balls and clubs, AI will improve the games of golfers everywhere."

In fact, speaking of which:

"We now have the technology to tell us what's wrong with our golf game."

Moment of silence please. But that is not all! Thanks to AI, you can now:

"Play some of the world's most famous courses, like St Andrews, without ever having to go there."

Boy, talk about convenient! But still that's not all! Thanks to AI never again will you have to play on a junky golf course! Why? Because, AI technology is now being combined with satellite and drone imagery to monitor growing conditions and report problem areas on golf courses before they ever happen.

"We can count the beetles on a leaf!"

Isn't that convenient? I mean, who does not want AI controlling all of our golfing experiences? They will know everything about our game, including us and our intimate whereabouts! But I'm sure that's for good, right? Oh, but that's still not all. All kinds of sports are getting in on the AI action! Including gymnastics.

"By using Lidar and Artificial Intelligence, we can now improve accuracy with judging and awarding points."

Why? Because *"Human eyes can't accurately measure the bending angle of joints"* but guess who can? That's right AI!" And it is already being put into play!

"Montreal's 2017 World Championships were the first to use the AI

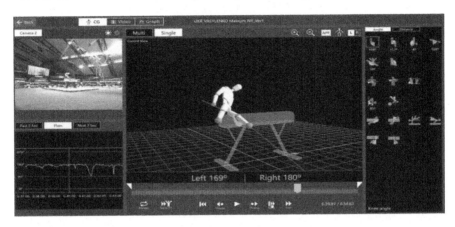

system and then it was phased into the World Championship competition and now they're looking at going fully automated by 2020."

But hey, maybe golfing and gymnastics isn't your thing, maybe it's running. That's right, runners, you're next in line! *"Artificial Intelligence*

helps runners go the extra mile." "AI technology knows no boundaries and it has encroached the sports arena with a wealth of data on athletes being collected that could help lift and upgrade their experience in marathons. AI solutions now recognize marathon runners and analyze their body motions."

"We can show professionals the angles of the foot streak, the gait, etc. and give feedback to the athletes such as, 'You may want to see this position, or you may want to adjust your gear."

Wow! That brings a whole new meaning to, *"You can run but you cannot hide!"* But that's still not all! Maybe golfing, gymnastics, or even running isn't your cup of tea! Speaking of "cup of tea" maybe it is those European Sports. You know what I'm saying? Well worry no more! Cricket is getting in on the action as you can see here.

"Artificial Intelligence Cricket Bat to be a Game Changer. It will change the sport."

"Microsoft's Azure Cloud platform powers the equipment and it uses AI and the Internet of Things (IoT) to bring sports closer to fans and real-time sports analytics."

In other words, they are monitoring everything about it!

"The technology also gives a more accurate idea of how the player should improve, and it even helps find new batting talent when looking at shot power." "This is just the beginning of what's possible for not only Cricket, but all sports."

Oh, but that's not the only European Sport that's experiencing this AI Invasion, so is Soccer!

"Soccer team turns to Artificial Intelligence for tactics."

"Teams are asking Alexa to help make their tactical decisions and to try to get an edge over their opponents," They're using "AI for Smarter Coaching."

"The best soccer players aren't necessarily the ones with the best physical skills. The difference lies in the ability to make the right split-second

decisions on the field about where to run and when to tackle, pass or shoot."

But man can't always spot these things so what are we to do?

"AI tracks players and the ball throughout each game and then analyzes all the data to improve performance and decision-making skills."

AI to the rescue again! Including, I saved the best for last, football! That's right folks, no sport is left untouched by AI. Unless of course you are into Olympic Crocheting, or something like that, I can't help you with

that! But seriously, even football in America is being invaded with AI.

"Footballs with Tracking Chips" will *"Allow officials to make exact first-down calls and determine with certainty whether a ball crosses the goal line."*

Oh, come on now, who does not need that? Don't you hate it when the officials keep getting it wrong? And it gets worse every year, right? Well, AI to the rescue! And that's still not all! Even the players can get in on the microchipping action as well.

"*In 2013, Zebra Technologies began tracking the movements of players using RFID tags in shoulder pads in the NFL.*" "*To determine the players' position and speed.*"

"*We work with teams who used to have to employ people who literally had to follow the quarterback around everywhere he went and count every throw.*" "*Now they don't have to anymore.*"

AI can do it all! Now folks, here is my point! This is all under the guise of convenience and an improved sporting experience, anybody starting to see a pattern here? First, they start by monitoring the sporting equipment with AI and microchipping technology, then it eventually moves to the players themselves, right? Do you think this is possibly a conditioning process, via sports, to get us all to, one day, accept a microchip ourselves so that AI can monitor, track, and analyze all our behaviors and improve our performance as well? Yeah, that's exactly what it is, but don't take my word for it, let's ask the so-called experts. They are saying that this is exactly where it's headed real soon!

PBS News Hour Reports: *"Human microchips, they are already being used to identify ranch animals and pets. But the process of implanting chips in people is also on the rise."*

Malcolm Brabant, Special Correspondent: *"These microchips they use are called Nearfield Communication. It is exactly the same sort of technology used in contactless debit and credit cards. Developers believe that consumers will soon be free to spend with the implant. On Sweden's high speed trailway, for the past eighteen months, they have implanted passengers to put their travel payment on the chip."*

Carla Grelsson: *"I think it is really good, with the chip ticket. I love all companies that use the chip technology."*

Malcolm Brabant: *"IT student, Kieran Anderson, is being chipped by Dr. Geoff Watson, a consultant anesthetist, whose driving research into medical benefits."*

Dr. Geoff Watson: *"The microchip is actually inside here (the needle) and this piece of plastic will push it out (of the syringe) once it's in under the skin."*

He put the needle under the skin and pushes on the syringe.

Dr. Watson: *"This is the implant coming out. You will feel a little bit of pressure, and then a clunk."*

He pushes the plastic and it pops into the students' hand.

Dr. Watson: *"You are now a cyborg."* And laughs. *"Kieran has a Nearfield Communication microchip in his hand, and that can be used to store data, store information, his name, address, contact details, anything he chooses to put on there. Blood type, allergies, basic medical information."*

Kieran Anderson: *"The idea of everything being put together is great, makes it easier for your automatic payments, without you having to do anything."*

Channel 8 News: *"It happens in seconds. Your child wanders off and panic sets in. But what if you had an extra layer of security inside your child?"*

A Mother: *"If it will save my kid, there is nothing too extreme."*

Keith Cate, Chanel 8 News: *"The technology is here, but just how far would you go to keep your kids safe? Our expert insists putting this technology in place is just a matter of time."*

"It would save a life, reunite a family, find their missing Alzheimer's patient."

RT News Reports: *"Israel is putting into practice its plan to get back to normality as its schools and kindergartens are beginning to open their doors. But concerns have been raised by the country's Prime Minister, proposing a controversial measure to track children's social distancing by microchipping them."*

Prime Minister Netanyahu: *"I spoke to the heads of our technology in order to find measures that Israel is good at, such as sensors. For instance, every person and every kid, I want it on kids first, would have a sensor that would sound an alarm when you get too close. Like the ones on cars."*[1]

USA Today: *"A small firm recently embedded microchips into their employees as a way to bypass company badges and corporate log-ons and for the cafeteria kiosks which are available on a cashless payment plan like Apple pay. True proponents tell us that a chip in the hand is better than grabbing for the cell phone because you can never forget it and you can't lose it and you have the capability to communicate with machines."*

Talking Tech: *"You will be chipped, it's just a matter of time."*

PBS News Hour: *"Already, governments are keen on gathering your fingerprints. They are keen on capturing your iris scan. They want to be able to do this at a distance so they can identify you as you walk through areas. They are doing facial recognition. These are all fallible technologies. They are hard to do well. But the idea that they can actually just embed you is a modern form of tattooing bar codes on people. We have seen governments do that in the past. This is just the next generation."*

Nightly News: *"When Elias Goldberger goes to work, he doesn't need ID and he doesn't need money. In fact, much of what he needs to get through the day is hidden right there, just below the surface, in his hand."*

Elias Goldberger: *"You want to touch it?"*

Reporter: *"Yeah, it feels like a grain of rice."*

Reporter: *"Embedded in his hand is the microchip. It serves as his keys, his ID, and his wallet."*

Elias Goldberger: *"Yes, I use it like to get around the building, and buy snacks."*

Reporter: "Then let's buy some snacks." As she tries to open the door of the snack machine, it won't open. *"It won't open."*

Elias Goldberger: *"So what I need to do is scan my chip, and log in, and from there I get access to the refrigerator."*

Reporter: *"Popular TV shows like 'Black Mirror' has imagined chips as a utopian future."*

Scene from Black Mirror: *"Insert chip with a local antiseptic and you're good to go."*

Reporter: *"In Sweden the microchips are already here. The microchip implants us with the same technology that is in contactless credit cards which has made cash pretty much obsolete in Sweden. At this tech fair there is a chipping event for those on the cutting edge, merging their hands with this new technology."*

Event visitor: *"I thought it would be fun, right?"*

Reporter: *"In the future do you think everyone is going to be chipped?"*

Event visitor: *"I am certainly convinced that millions of people will find it very, very valuable to have a smart device under their skin."*

Reporter: *"Human microchipping may be our future, but in Sweden it is already a reality."*

On Assignment: *"Hannes has made it his mission to convince more of us to get microchipped. He is what is known as a biohacker. Someone that wants to improve their body with technology."* As they are sitting at the table for the interview. *"What's wrong with just having contactless paying cards, we've all got phones, we've all got a set of keys, what is the point?"*

Hannes Sjoblad, Bio-hacking Entrepreneur: *"The point is to reduce the hassle of exacting these things. I mean, in the morning, when you stand there, going out your front door, you check your purse or your pockets, to make sure you have your wallet, your charger, your keys, your phone, all* *your stuff. What if you could reduce that by half? It would declutter your life."*

Reporter: *"Isn't that being just the utmost laziness?"*

Hannes Sjoblad*: "No, it's convenience. And convenience is a pretty powerful force."[2]*

Well, that's reassuring! What's going on here? This is nuts! How did we get to this place where they are actually saying bluntly that we are all going to get micro-chipped, it is just a matter of time? You know, just like the Bible warned about nearly 2,000 years ago, including using a marking system to make payments, i.e. "buying and selling," like the people in the 7-year Tribulation?

Can you see the trend for lazy convenience? I'm telling you this is exactly the reason why the so-called "experts" are saying why we're in the shape we're in! It's because of the seduction tool of laziness to get us to go along with it! Yeah, you're right! It will even convince you to get a microchip in your hand to "buy and sell" stuff. Call it laziness or convenience, that's the tool they're using to get us to go along with this, in our selfish hedonistic culture!

And hey, it will not only "buy and sell" as we saw, but it too is coupled with AI to improve your daily performance and output of work, just like in the Sports Arena. Or it could even become our "Life Enhancement Coach" to monitor and analyze everything you do on a daily basis for your supposed good. And it'll even track you, your pet, your kid, your grandparent, Alzheimer's patient, wherever you go on the planet, so nobody'll ever get lost, meaning they will always know where you are at any given time, anywhere in the world! Isn't that convenient? Yeah, it's called a setup, just in time for the 7-year Tribulation!

The **2nd** **way** AI is taking over our gaming, on a global scale is with **Video Games.** You see, the scenario we just saw is with chipping in the "hand." But what about the other option mentioned there in our opening text?

Revelation 13:16-17 "He also forced everyone, small and great, rich and poor, free and slave, to receive a mark on his right hand or on his forehead, so that no one could buy or sell unless he had the mark, which is the name of the beast or the number of his name."

You see, it's not just getting a mark on your hand, to get connected to the Global Antichrist System, it's in the where? In the "head" as well. And to me personally, I don't know about you, but I can see the so-called "convenience" of getting the "hand" chipped. As mentioned, that is where we're already conditioned to carry our cell phones to make payments. So, the "hand" is a natural conversion, and it's better because you can't lose it or get it stolen, it's always with you. But the "head." I mean, who's going to get a chip in the "head? How is that "convenient?" I'm thinking it's an eye sore or a pain! Well not anymore! Thanks to AI in the Video Gaming Industry, you can now get a brain chip in your head to make your gaming experience much more convenient! Don't believe me?

"Elon Musk says his AI-brain-chip company will be putting implants in humans within a year."

And notice it's not just a chip in the head but one that's connected with, guess who? AI or Artificial Intelligence! He even went on to say,

"It's going to be as common as a Lasik Eye procedure."

And the rationale they have with a Brain Chip is,

"It will help treat brain injury and trauma, as well as restore eyesight, hearing, and limb movement."

And why would you say no to that? But if that doesn't get you, even the Video Gaming Industry is conditioning us to go in the same direction as well. Let me show you how. First of all, it started with "Virtual Reality Headsets," and they have been out for a while now.

And the rationale is, it not only saves you the hassle of trying to find a big enough screen to really get into your game, but it brings you so close to your game that you're "totally immersed" into it! It's like you're really there inside your head!

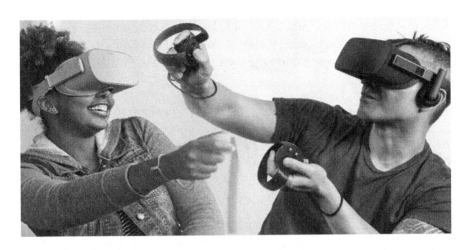

Then it went to "External Brain Headsets," as you can see here, because we all know how hard it is to push those buttons on those game controllers. Again, if I get another callous, I don't know what I'm going to do! But seriously, whether you realize it or not, the Gaming Industry has already developed headsets that read your "brain signals." Why? So you can control your gaming experience with "your thoughts" with your mind! And as crazy as it is, it's now the latest trend! Move over

Virtual Reality! But the problem is, it is leading to a new danger out there called, "Brain Hacking" where now, because of these headsets, they can actually "hack into your brain" and read your thoughts! I'm not making this up!

Lee Newton, Sourcefed: *"It's official. We are living in the future but not necessarily the 'all hover boards and robots future.' Oh no, we're living in the 'they can possibly read our minds and nothing is a secret kind of future.'"*

Joe Bereta: *"Oh, man, I knew we would get here eventually just hide your brains, kids."*

Lee Newton: *"Scientists demonstrated how they can hack someone's brain to find out things like in numbers, addresses, or whether or not someone is important to you. You know the stuff they would use against you in a CIA movie to make you talk."*

Joe Bereta: *"The craziest thing about it all is that the researchers from the Universities of Oxford, Geneva and California found a way to pluck this stuff right out of your head using a $300 off of the shelf headset. The Epoc headset, which is the headset that lets you control and interact with your computer through brainwaves. Computer games through brainwaves. The devices have access to your EEG which is essentially your electrical brain signal data that contains a neurological phenomenon triggered by subconscious activity."*

Lee Newton: *"Ooh, science. And if that isn't creepy enough the team found that they could find a person's home 60% of the time with a 1 in 10 chance and had a 40% chance of recognizing the first number of a pin number. Mind you that is only the first number. This is a computer gaming headset. I repeat, a computer gaming headset. And it seems promising when it comes to interrogation and detection of criminal details."*[3]

So, they will know whether you've been obeying like a good slave! Folks, this is nuts! How many of you had any idea this technology was available let alone already out there! It's wild! Most don't! But you talk about lazy! I'll run the risk of somebody "hacking into my actual brain and know my most intimate thoughts" all because it's just way too hard to use my "hands" to control the game anymore! Ouch!

And speaking of laziness and "reading your thoughts," I mean what's next, are you going to come out with some sort of "head device" I wear that will put a vacation into my mind so I don't have to go anywhere like in this movie?

Scene from Total Recall: *The main character is boarding the train. While standing at the aisle he notices the TV is advertising something very interesting.*

TV commentator: *"Are you dreaming of a vacation to the bottom of the ocean? But you can't float the bill. Have you always wanted to climb a mountain on Mars? But now you are over the hill? Then come to Recall Inc. Where you can buy the memory of your ideal vacation, cheaper, better and safer than the real thing. So, don't let life pass you by. Call Recall for the memory of a lifetime."*

As he looks at the screen, he starts to think. This may not be such a bad idea. In the background the music is playing, 'Recall, Recall, Recall.' He makes the arrangements to take this special dream vacation.

He is at the facility and is in the chair that is going to take him on his vacation. He lays back, and his head is in a circular enclosure. His arms are strapped down, and he is ready to go.

The researcher tosses the program to her assistant and as he looks at it, he mentions that this one is new. He hadn't seen it before. The researcher asks him how he prefers his women. Blond, brunette, red head.

Main Character: *"Brunette."*

Assistant: *"Boy, is he going to have a wild time. He's not going to want to come back. That's for sure."*

He closes his eyes and falls asleep. The researcher and her assistant just look at each other and grin. [4]

And there you go, a completely "Computer Generated Vacation" in your brain! But good thing that's just a science fiction movie! Actually, it's already here, thanks to the new and improved Brain Chips! You see there's just one drawback with these Computer Headset Gaming Devices. You can lose them, or somebody could steal them. And how inconvenient would that be to have to go back to using your hands! The agony of it all! But if you'd just get a Brain Chip in your head, you could not only put the computer screen directly behind your eyes, and get rid of those bulky Virtual Reality goggles for an immersive video game experience, but just like that movie, you could even pick out whatever "dreams" you want to dream at night in your head, just like you're scrolling through Netflix to find a movie to watch! Think I'm kidding? It's already here!

Julia Sieger, Tech24: *"Can we trust our own brain? And what can our dreams teach us? We are going to talk to Moran Cerf, a neuroscientist from Northwestern University. Thank you very much for being with us. Tell me, is it possible to hack the human brain and is it as easy as hacking a computer?"*

Moran Cerf, Neuroscience Professor: *"It's actually probably easier. The idea is that we are now starting to understand how to influence a person to change their mind and behave in a certain way. Now a lot of companies are interested in that, because this tackles the very idea of free will. Amazon cannot just know what you want but will actually start offering things that they know you want. Because they know you better than yourself. That is where we are entering into a world where your brain attracts the outside world behind your back.*

Julia Sieger: *"Now you have also studied dreams quite a bit. Can deciphering their meanings be a key to our brains?"*

Moran Cerf: *"Yes. Dreams have been fascinating people for a millennia. They mean something. If you don't believe me, you can just try to tell your boyfriend or your husband in the morning that you dreamt of your ex-boyfriend and see how they respond. They think that dreams mean something whether they do or not.*

For the first time in history, we can actually access their understanding. Up to now, or during the time of Freud, they had to ask you to tell them what you believe, what is the story. But now we can actually get access to your dreams with neuroscience while you are sleeping and understand what they are telling us.

They are telling us about who you are, what you want, what you think of the future, what you want to do next and what you think about this world and its possibilities. It helps to understand you better and to control the world around you better.

And we can start manipulating your dreams by creating dreams for you. So, you can go to sleep and dream what I have set up for you. Maybe a film maker, like Spielberg could make dreams for people and so on, so you could go to sleep after a nice evening and continue the evening in your sleep. The companies like Netflix, YouTube, Hulu, or Tech24 can put content in, and generally the ultimate reality is what our brain creates for us."[5]

See, no more Virtual Reality Headsets! They're bulky anyway! But if I just get a Brain Chip in my head, I can avoid all this! And then I can customize my thoughts or dreams just like picking out a movie on Netflix! I mean, isn't that convenient? Because you all know how hard it is to scroll down with your thumb on the controller! Ouch! I mean, what are they going to call this new service…Dreamflix? Probably!

But I wish I was making this whole thing up, but I'm not! Total Recall and Arnold Schwarzenegger eat your heart out! This is crazy! But it gets worse! Speaking of crazy, people are already starting to see where this is all headed. If you can monitor my thoughts and even now inject new made up thoughts into my head with this technology, then does that mean you can also erase my thoughts? And that's precisely what these people are asking!

Travis Stork, MD: *"When times get tough, do you ever wish you could just push that reset button and just start all over again? We all thought about it and wished we could do it from time to time but be careful of what you wish for because scientists are now working to make these wishes a reality."*

"It may sound like something out of a science fiction film but believe it or not, scientists are working on implanting microchips into human brains to delete memories. Prototypes of the chips are allegedly being tested on epilepsy patients in hospitals with good results. But the creator hopes to one day expand this chip, not just for the sick, but for the healthy population where users could remove experiences and get this, buy new ones."

"The chip is predicted to be as common as cell phones in the future. Are you ready?"

Dr. Andrew Ordon: *"Maybe there is a use to eliminate those bad thoughts, PTSD memories, traumas, this whole area of functional brain surgery, whether it be a chip or whether it be brain stimulation. I mean this is an area that is evolving at a crazy rate. I remember when Neal Martin, the neurosurgeon was on, he said just wait, the time is coming that we are going to able to control so many of these behaviors. Obesity, for example, he said we are going to be able to control it all through the brain."*

Travis Stork, MD: *"This is West Worldish. With someone in the control room somewhere, controlling your memories, controlling your thoughts, this is very creepy."*[6]

Yeah, it's creepy! But good thing it's never going to happen. Are you kidding me? Folks, it's already being done! You see it's not just the monitoring of your thoughts and the injection of memories or movies or dreams into your thoughts, or even, as you saw, the erasing of your thoughts, with this technology that they want to do. Believe it or not, they also want you to get a Brain Chip in your head so they can control your thoughts and in essence control you, by making you into a type of remote-controlled robot! I am not joking! In fact, some would say it's already been done and has been for a while!

RT Reports: *"It's the stuff out of the movies, but a group of veterans have filed a lawsuit against the CIA and the U.S. Army claiming the government planted remote devices in their brains. So, could this really be happening? Well, joining us to help discuss this is Dr. Colin Ross, President of the Colin Ross Institute for Psychological Trauma. Dr. Ross, is this really happening? Did the government really take part in mind control experiments on soldiers? What kind of stories have you heard from these survivors of these experiments? I know you have had access to thousands of documents from the CIA."*

Dr. Colin A. Ross: *"Well, there are all kinds of experiments where no real consent was given, where people really didn't know what was going on. And they were basically tricked. And I think in the brain-electrode experiments its kind of a combination of both. Some patients were told they were going to get an electrode put in their brain, but it was for some therapy purpose, when it was really for research. Others were told to go here, and volunteer and they didn't really have much choice. Then others were given a more exact story."*

RT Reports: *"So what exactly would the government do when they would control someone's mind? What would they make someone do when they manipulated their brain?"*

Dr. Colin A. Ross: *"Well, what is described in the documents by the public speakers is, there are actual photographs of a 16-year-old girl that has a series of electrodes in her brain. Depending on which buttons are pushed on the transmitter she is either strumming her guitar, pounding furiously on the wall, or staring off into space. With animals, they are* *actually directed to walk or swim to a target. So, you can control the actual physical motion and the mental state. How detailed or fine-tuned that has become since 1970, again I don't know because it is all classified. It must have gotten a lot more developed."*

RT Reports: *"How fast can a person's mind be taken over? Does it happen over a period of weeks or days?"*

Dr. Colin Ross: *"The electrodes are a little different. You just put the electrodes in, you push the button and it happens right away."[7]*

A remote-controlled person! Being worked on for decades now, now perfected with a simple Brain Chip, that is as simple as a Lasik eye procedure, isn't that convenient? Ready to make your life so much more comfortable! And what did Elon Musk say? *"We'll be putting these into humans within a year."*

All under the guise of convenience, just in time for the 7-year Tribulation! In fact, speaking of which, maybe this has something to do with the reason why the Bible says for those who receive the Mark of the Beast, in their right hand or forehead, it seals their own doom, forever!

Revelation 14:9-11 "A third angel followed them and said in a loud voice: 'If anyone worships the beast and his image and receives his mark on the forehead or on the hand, he, too, will drink of the wine of God's fury, which has been poured full strength into the cup of His wrath. He will be tormented with burning sulfur in the presence of the holy angels and of the Lamb. And the smoke of their torment rises for ever and ever. There is no rest day or night for those who worship the beast and his image, or for anyone who receives the mark of his name."

So why is it that anyone who receives the Mark of the Beast is doomed to the Lake of Fire in the last days? Well, stir all this together, and maybe it's because that Brain Chip will be controlling their thoughts, as well as their behavior, and the last thing the Antichrist and AI will ever allow them to think is about Jesus and getting right with God. That's one tape they'll never let you play in your head!

Now, I'm not saying, "Thus saith the Lord," but it sure makes you wonder, doesn't it? But it's all current technology, all here now, being pushed under the guise of convenience for a lazy, hedonistic, selfish, self-centered culture, with AI controlling it all on a global basis, just in time for the 7-year Tribulation!

How much more proof do we need? The AI Invasion has already begun, and it's a huge sign that we're living in the last days! And that's precisely why, out of love, God has given us this update on *The Final Countdown: Tribulation Rising* concerning the **AI Invasion** to show us that the Tribulation is near, and the 2nd Coming of Jesus Christ is rapidly approaching. And that's why Jesus Himself said:

Luke 21:28 "When these things begin to take place, stand up and lift up your heads, because your redemption is drawing near."

People of God, like it or not, we are headed for The Final Countdown. The signs of the 7-year Tribulation are Rising! Wake up! And so, the point is this. If you are a Christian and you're not doing anything for the Lord, shame on you! Get busy doing something for Jesus now! Stop wasting your life! We need you! Don't sit on the sidelines! Get on the front lines and help us! Let's get busy working together doing something splendid for Jesus with what time is left and get busy saving souls! Amen?

But if you're not a Christian, then I beg you, please, heed these signs, heed these warnings, give your life to Jesus now! Because this AI technology is not going to lead to a life of wonderful dreams and a modern-day utopia, but a nightmare beyond your wildest imagination in the 7-year Tribulation! Do not go there! Get saved now through Jesus! Amen?

Chapter Fourteen

The Future of
Agriculture with AI

The **4ᵗʰ area** AI is making an invasion, is in **Agriculture**. That's right. AI is not only already taking over all our finances and shopping experiences, of what we "buy and sell," including the delivery options, but for those of you who wanted to grow your own food or store your own water, and get off the grid, so to speak, and become self-sufficient to avoid this Antichrist regime, would that work out for you? I don't think so! As if they haven't really thought of that one!

Believe it or not, the Antichrist and AI are about to take over the planet, not only in businesses, finances, and conveniences, but even our food supply! Talk about total global control! I know it sounds crazy, but I am telling you, it is already being done and it's exactly what the Antichrist is going to do in the 7-year Tribulation. But don't take my word for it. Let's listen to God's:

Revelation 6:1-8 "I watched as the Lamb opened the first of the seven seals. Then I heard one of the four living creatures say in a voice like thunder, 'Come!' I looked, and there before me was a white horse! Its rider held a bow, and he was given a crown, and he rode out as a conqueror bent on conquest. When the Lamb opened the second seal, I heard the second living creature say, 'Come!' Then another horse came out, a fiery red one. Its rider was given power to take peace from the earth and to make men slay each other. To him was given a large sword. When the Lamb opened the third seal, I heard the third living creature say, 'Come!' I looked, and there before me was a black horse! Its rider was holding a pair of scales in his hand. Then I heard what sounded like a voice among the four living creatures, saying, 'A quart of wheat for a day's wages, and three quarts of barley for a day's wages, and do not damage the oil and the wine!' When the Lamb opened the fourth seal, I heard the voice of the fourth living creature say, 'Come!' I looked, and there before me was a pale horse! Its rider was named Death, and Hades was following close behind him. They were given power over a fourth of the earth to kill by the sword, famine and plague, and by the wild beasts of the earth."

Which in today's population would be about 2 billion people! But as we can see here in our text, dealing with the first part, the first half of the 7-year Tribulation, not only is one-fourth of the planet going to be annihilated as an act of God's Judgment upon this wicked world system, but due to the famine conditions mentioned there, our world, the whole planet, is actually going to be on some sort of "global food distribution program" just to stay alive, right? What did it say there? For one day's work you could get a quart of wheat for yourself, or you could opt for three quarts of barley, a less nutritional meal, literally animal feed, as we saw before, to feed you and your family of two. Now, here's the point. That is exactly what the globalists and elitists are pitching right now to do with AI! They not only want to use AI to control our whole world's finances, businesses, and conveniences, but they also want to use it to control our whole world's food supply! In fact, they are already doing it! We have already seen in some of our other prophecy studies that all the world's food supply, right now, is currently in the hands of just three entities. Cargill/Monsanto, ConAgra, and Novartis/ADM. They freely admit that if you want to really control the world, forget the fuel, forget gas and oil, you need to control food. And I quote...

Dwayne Andreas, former chairman of ADM: *"The food business is far and away the most important business in the world. Everything else is a luxury. Food is what you need to sustain life every day."*

"Food is fuel. You can't run a tractor without fuel, and you can't run a human being without it either. Food is the absolute beginning."

...of what?

And I quote, *"Those who control the global food system, have the ultimate in economic power."*

In other words, whoever controls the food supply controls the world! And that is exactly what the Bible says is going to happen in the last days, as we saw in our opening text! Think about it! In order to distribute the food on a global basis in famine conditions, in the 7-year Tribulation, what has got to happen? Somebody has got to be controlling the whole food supply on the planet! Otherwise, how could you pull it off? And believe it or not, it is not only already being done, but they want AI to run the whole system and manage all of our agriculture on the whole planet. It's sounds crazy but here's the term they have for it. It is called "Smart Agriculture." And again, as we saw before, anytime you see the word "smart" before something, it really means, "Big Brother" i.e. "Total Control." And this is what they want to do with all of our agriculture or food supply! Where...

"AI and robotics will augment the production, processing and packaging of food products in timed cycles, that limits human intervention."

In other words, AI will be controlling the whole thing, without humans involved! And lest you think this is some crazy conspiracy theory...

"AI in the agriculture market was valued at $600 million in 2018 but it is expected to reach 2.6 billion by 2025."

In other words, they are putting all their eggs into this basket, including the United States! And what they are calling it, is the "Fourth Revolution in Farming," and by doing this they will, "Be able to feed the world." Or literally, control the world because that is what it is going to lead to! But let me show you how they have been doing this, step by step, allowing AI to literally control our total food supply and how it is being put into place.

The **1st way** AI is taking over agriculture on a global basis is with **Automated Tractors**. That's right, who needs humans anymore, including farming? If I can have AI control all my finances, businesses, and convenience options, like delivery of food, what we buy and sell; then why can't I have it be in charge of, or controlling, all of that as well? The other important thing we buy and sell, includes the tractors that plant and harvest it. Well, guess what? It is already being done! Autonomous Tractor, Future of Farming: Commercial now being shown for the tractors for sale:

"Pure electric drive, zero emissions, wheels or tracks, ultra-compact power unit, highly maneuverable, 500KW total power, 5-15T variable ballast, better soil protection, fully integrated tractor/implement." This tractor has it all.[1]

And that's John Deere! One of the biggest manufacturers of tractors and farm equipment around the whole planet! And the rationale they're saying is that we need to make this switch to have AI control all the tractors in farming.

"AI in Agriculture will usher in a more efficient, easier manner of growing crops throughout the agricultural chain." i.e. the entire world.

"Virtually every aspect of agriculture will be impacted by artificial intelligence over the next 10 years. It will become more automated."

But that's not all, the other benefits of this is that they say, *"AI field robots can identify weeds and diseases immediately and deal with them, solving the problems themselves."*

In other words, they do not need humans anymore for any aspect of farming! And then they even go on to warn, *"In the long run AI will replace a farmer's knowledge entirely."* Let alone the farmer himself, as we'll see in a second.

The **2nd way** AI is taking over agriculture, on a global basis, is with **Automated Drones**. You see, you not only need AI to control the ground in farming, but you also need it to control the air! That is, if you are going to take over the whole planet's agriculture. And that is because anybody who knows anything about farming, knows that you have things in the air that you need, that has to do with farming, like crop dusting, spraying pesticides, monitoring the crops for diseases, etc. Well guess what? AI can do that as well! In fact, it is already being done! Here are the Top 5 benefits of allowing AI drones to monitor and control your farm.

David Plummer at Aerial Influence: *"Here are the top 5 reasons you should be using a drone on your farm in 2019. Here we go:*

5. Water & fertilizer efficiency; Drones like the Mavic II, enterprise dual or the Mitrice 200 with an XT sensor, they have thermal capabilities. Now what that means is that these thermal sensors can detect which portions of your field are over or under watered, allowing you to water only the areas that need to be watered, which lessens the amount of fertilizer runoff and wastes as well. Let's not forget drones like the DJI Enterprise Series which allows you to spot, spray fertilizer, water and even spread seed. We have that one right here at Ariel Influence as well as all the DJI Enterprise products. So, make sure you contact us to answer any of your questions about those drones.

4. Saving you time. Gone are the days of solely counting your field on foot, on a tractor or via airplane. Drones can quickly and easily get a quick snapshot of your crops even in hard to reach areas. As most of our drones have the ability to fly up to four miles away from the remote control, as long as it is still within a line of sight. So, imagine scouting your back 10 from your front porch. That sounds pretty good right?

3. Increase your yields. So, if you can scout your fields more frequently because you have a drone you will be able to find potential problems before they get out of hand. That is going to save your crops and increase your yields. Simple as that.

2. Mapping. Today's drones are so advanced that you can draw a map right on the screen of the remote control that you're using to fly the device and have the drone fly that precise path. If you need really precise maps or flight paths, you should step up to something that uses RTK, like the Agras series, and the Mitrice 210 RTK, with those drones you can get centimeter accuracy. It's like GPS on steroids. But it comes at a price, speaking of price, let's get to the number one reason that you should be using a drone on your farm in 2019.

1. To save you money. The purpose of all this is to help you save your hard-earned cash. Here's an example, on average it costs around $2.00 per acre for a visual inspection or aerial survey. By using a drone, the return on investment could be realized within a single season. That's right, you can pay for your drone in one season by taking advantage of the technology that drones are bringing to the agricultural world."[2]

Huh? You would have to be a fool to not want to use one of these drones in your farming procedures! And then eventually you can have them, along with the tractors on the ground, controlled by satellites to direct the mapping so you will not have to! Isn't that great? Yeah, if you want to work yourself out of a job! And we will get to that in a second!

The **3rd way** AI is taking over agriculture, on a global basis, is with **Automated Livestock**. Now, this is another important aspect of farming. You see, farmers not only grow crops of plants for food, but they also grow animals or livestock for food, right? And believe it or not, even that is also being automated with AI, including the use of microchip technology, to monitor the livestock's every move, for efficiency of course! For instance:

"Companies around the globe are leveraging on Artificial Intelligence to make more informed decisions and better run their businesses," on their farms. How?

"This week, Dairy Farmers of America announced an investment in SomaDetect, a dairy technology start-up that will help farmers utilize AI to more closely monitor the health of their herd and improve milk quality."

"It's a game-changing technology that allows our farmers to know the health of each cow and quality of milk in real time."

And they mean in real time because everything about the cow or livestock, not just the location, but its overall physical well-being and even when it's best to get it in for mating, will be monitored in this system. They're calling it, "The Internet of Cows" or "The Connected Cow" and all aspect of the cow is stored and monitored by AI in what's called, "The Cow Cloud" I kid you not! It is the new way of farming for Digital Farmers that creates what's called, "Precision Agriculture" where they know everything, as seen here.

Yasir Khokhar, Founder Connecterra: *"A few years ago I was staring outside my window watching the neighbors cow graze the land. I joked about strapping a sensor on a dairy cow and learn what it does all day. It turned out; productivity is a real issue for farmers. There are 1.5 billion cows on the planet and the milk they produce is a major staple of many diets. It is a fact that by 2050 there will be over 9 billion people on the planet and food production needs to increase by 60% to sustain that. So how are we going to feed everyone?*

I got working with an old friend of mine on a device that became the Fitbit for cows. Using complex mathematics that is the basis of artificial intelligence, we're teaching our algorithms the behavior of cows. For example, we can detect problems of eating and feeding disorders. We can know when it is the best time to get a cow pregnant. But farmers don't want data, they don't want to become analysts. They want to know what has gone wrong, why it happened and what they should do about it. I believe in the future world, where sensors and artificial intelligence will be ubiquitous.[3]

In other words, AI will be controlling, all over the planet, not just the farming, but the livestock as well! And these systems are not just for cows either! These AI controlled systems are not only, *"Good enough for the Internet of Cows, but sheep, trees, rivers, (i.e. their water supply), tractors, pumps,"* you name it.

"Very soon all of these things are going to be sensor equipped (i.e. a microchip in them) that will make them location aware, remote-controlled data generating devices."

And from there it will go into the cloud, the Cow Cloud or whatever cloud, where all this data is going to be processed by AI controlling the whole thing

The **4th way** AI is taking over agriculture, on a global basis, is with **Automated Apps.** You see just in case there are still a few human farmers left, they have figured out a way to have AI making the decisions or controlling them, in order to control the whole food supply. They are doing that with a ton of AI farming apps that are just like Alexa or Google Home as we saw before. And just like we got used to having AI Alexa and AI Google Home telling us what to do individually in the home, now farmers are doing the same with these AI controlled farming apps. Here is what they can do:

"AI is already transforming agriculture in important ways." And with sensors embedded in all aspects of farming, AI Farming apps will help farmers, 'Make smarter decisions about when to irrigate and how much fertilizer to apply.'"

For instance, "The AI Sowing App draws on more than 30 years of climate data, combined with real-time weather information, and sophisticated weather forecasting models powered by Azure AI (which is made by Microsoft) …

"To determine the optimal time to plant, ideal sowing depth, how much manure to apply, and much, much, more."

"It can also provide things like temperature, rain, wind speed and direction, and solar radiation."

"According to research, 90% of crop losses are due to weather related events and 25% of these losses could be prevented by using predictive AI weather modeling." In other words, their apps.

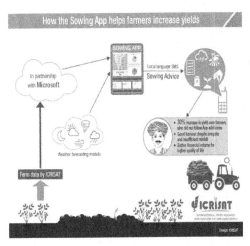

But that's not all. *"The information is shared with farmers via text messages and the average increase ranged upwards to 30 percent."* Wow! That is a big improvement!

And, *"With this level of precise AI knowledge, Farmers around the world are able to make pinpoint decisions about when to plant, when to water, when to harvest, that will save in labor, reducing overall costs, and improve output."* And who doesn't want to do that?

And believe it or not, many third world countries are getting in on these AI farming apps as well, to improve their crops, which means, it's going global and AI is controlling all the decisions for farming everywhere! I mean, what is next? You are going to get rid of the human element altogether? Uh huh! That is exactly where it is headed!

The **5th way** AI is taking over agriculture, on a global basis, is with **Automated Workers**. You see, all this AI farming technology has advanced so much, you no longer need people anymore for anything in the Agriculture Industry, including the workers themselves! AI can do it all.

"This autonomous bot roams greenhouses looking for ripe pepper to pick. It's called Sweeper. It is a collaboration between researchers from Europe and Israel. Every so often, Sweeper will stop and examine a pepper plant using LED lights and a camera that can recognize color and distance. With the help of computer vision, Sweeper decides if the pepper is ripe. If it is, the bot uses a razor to cut off the stem. To pick a single pepper, it takes 24 seconds. But the researchers say it can work 20 hours a day. Sweeper is not the only harvesting robot. Argobot is testing a machine that picks strawberries. Green Robot Machinery has a cotton-picking robot and Israel start-up, MetroMotion, is working on a tomato-picking bot. Experts expect the agricultural robots market will reach $75 billion by 2025.

Farmers, these are the droids you are looking for. They can inspect fruit trees or count individual fruit. This is the Ladybird. (It looks like a large solar panel attached to a tractor.) It scuttles along vegetable rows. It can tell the difference between vegetables and weeds and exterminate weeds with herbicide. Then there's Shrimp (It looks like a large tabletop attached to a small tractor), which can herd cows. These autonomous robots increase yields and minimize herbicide use. Where will robots go next?[4]

Yeah, good question! I think it is obvious! They are going to be controlling every aspect of the total global food supply, just like **Revelation 6** says that the Antichrist's going to do! And as crazy and obvious as that is, now they are getting people, around the world, to go along with this total takeover of our food supply. They are pitching that this AI Robotic Farm Worker will, not just save time and savings, but again, it's the perfect worker! Think about it. These AI robots can work non-stop, 24 hours a day on the farm, they don't need breaks, there's no need to offer them, let alone pay for, a healthcare plan, that's so expensive, including stock options, 401k plans, and it never whines or complains or gets sick, and it does what it's told every single time! Who wouldn't want an AI robot working on their farm? And if you think about it, they could even pitch this as a solution to fixing the proverbial Illegal Immigration crisis too, right? Think about it. With all these AI robot workers, then there are no illegal immigrants for the farmers to be tempted to hire as their labor force, just get a robot! Then there are no farm worker positions available for these illegal immigrants to be tempted with to cross the border as well! Those jobs are gone! AI robots will have taken that over! AI has even solved that the immigration crisis as well, without a wall! And we all know how controversial that is! But another huge thing they say the AI Robot Worker Force will do, is not only *"Minimize field labor, but improve food quality and safety."*

"Robots can pick food in a much more sanitized manner than humans, and they can also weed crops, milk cows, and help with the overall worker shortage that farmers are now experiencing."

CBS This Morning Reports: *"Agriculture in the United States is more than a 360 billion dollar annual business. But it is in trouble. American farmers are getting older. Their average age is just over 58 and farming in general is facing a major labor shortage. Harold Barnett met with some growers to see how they are trying to find a solution with the help of technology from Silicon Valley."*

Harold Barnett: *"Good Morning, we know that farming is tough work, getting up before dawn to take care of cows like these. You need to bring them to milking machines like these, twice a day. Generally speaking, though this is not the type of work people want to do anymore. It has become more difficult to find all sorts of agricultural labor. Farmers I recently met with told me that the key to fix that is Artificial Intelligence.*

For just about a hundred years, Gary Wisnosky's family has been running the Wish Fruit Farm business."

Gary: *"I think if my grandfather was still alive, he would be totally mesmerized to see what the future has turned into."*

Harold Barnett: *"In order to keep his crop healthy, he needs six hundred people to harvest six hundred acres every two to three days. But finding that amount of labor is becoming unsustainable."*

Gary: *"We have seen a shrinking labor force and an aging labor force. We actually had fields that we were abandoning early in the season which is a really painful thing for a grower to do."*

Harold Barnett: *"In an effort to avoid losing $20,000 per acre on abandoned fields, Gary is part of a team of engineers that are working to fully automate the process."*

Paul Bissett: *"We have separated all the things that a picker is doing in various pieces of the robot."*

He shows us how the tractor works with the blades that divide the plants.

Harold Barnett: *"Paul Bissett is the chief operating officer of the company behind it all.*

Paul Bissett: *"We are collecting between 50 and 100 images per plant and all those images are fed into our AI system in order to tell us 'this is a good berry and this is the one we want to go after.'"*

Harold Barnett: *"So you are telling me in real time, this machine is looking at the plant thinking what to do next, based on the imaging and executing that action?"*

Paul Bissett: *"Exactly."*

Harold Barnett: *"So is this as fast as a human right now?"*

Paul Bissett: "This machine, commercialized, will replace the thirty people you saw in the field, earlier today with Gary."

CBS Morning News: *"This is a global problem, the lack of farm laborers."[5]*

Ooh! And the global solution is what? AI is controlling the whole thing! From tractors, to drones, to livestock, to information of what to do, when to do it, harvesting, workers, you name it, no humans are needed anymore. AI can do it all! And what are we talking about? The total control of our food supply on a global basis, just in time for the 7-year Tribulation! Now, lest you think this total AI automation of farms is still light years away, it is already being done! Here are just two examples. A traditional farm, and a hydroponic farm.

Global News: *"Now when it comes to farming smart, not many do it better than Jonathan Gill."*

Jonathan Gill: *"We were the first to grow an entire crop without ever going into the field."*

Global News: *"Gill, the events keynote speaker, is showcasing his cutting-edge research of his automated farming."*

Jonathan Gill: *"Work an entire area of land with a Hactare without anyone sitting in the driver's seat or doing anything with the ground."*

Global News: *"They call it the 'hands free Hactare.' Gill and his team in the United Kingdom, with their automated farm machinery and drone systems grow and harvest a crop of spring barley. Now the whole idea of unmanned vehicles making their way through a field causes at least one farmer some concern."*

Craig Walsh: *"It's kind of scary too, when you think of Terminator, okay now it's running itself."*

CNBC News: *"This farm is run entirely with robots."*

The farmer has different flat containers with a different bunch of plants in each one. A robot arm is planting new plants in each container as it rolls past each container.

Brandon Alexander, Iron Ox Co-founder: *"If we are going to double the food production, over the next 30 years to feed the population there has to be a radical change.*

For Iron Ox, that change starts with its two robots. This 1,000-pound mobile robot navigates autonomously around the indoor, hydroponic farm in San Carlos, California. It moves trays of plants to the processing area. Then a robotic arm moves plants from one tray to another sorting them by life cycle to optimize space. Baby plants are close together, mature plants are farther apart. The robots also use machine learning and AI to detect pests and diseases.

Jon Binney, Iron Ox Co-Founder and CTO: *"It's not just that the robots can move things around very efficiently, it's also that they can help you avoid ever having a plant go bad."*

Iron Ox designed its farm around the robot's capabilities.

Brandon Alexander: *"We're able to do the equivalent of 30 acres of outdoor farming in just a single acre of our robotic farm."*

High tech indoor farms are not new. The company says its non-GMO and pesticide-free produce will be as cheap as traditional agriculture.

Jon Binney: *"One of the great things about the robot is that they don't really get tired, and they don't really care about what hours they work, so as long as they have juice in the batteries they can keep going."*[6]

And do what? Totally automate the whole world's food supply with no humans needed. AI can now do it all, for the first time in man's history! But you might be thinking, "Well hey, that's all nice and dandy, AI and robots controlling all these corporate farms and individual farms or even hydroponic farms, but I'm just going to grow my own food supply and eat whatever I want, when I want, and I'll be self-sufficient." No, you won't! Again, you really think they haven't thought of that? Step by step they are even preventing that! And that is why, slowly but surely, over the years, we have been conditioned to allow the governments around the world to dictate what we can and cannot eat. Whether or not we can grow food in the first place, or even store up water. Here is how they have been doing it! Fear! All under the guise of fear we have allowed the government to control our whole food supply! I mean, haven't you heard? We have a horrible food crisis on our hands! AHHH!

The global food crisis

"The world's 200 wealthiest people have as much money as about 40% of the global population, and yet 850 million people have to go to bed hungry every night."

We have got salmonella outbreaks in our vegetables; kids are allergic to peanut butter! No! Not the peanut butter! Yes folks, even the peanut butter! Think of the kids! What are they going to do? How will they survive? And then there is that E. Coli thing, the gluten related issues, lactose intolerance, the hoof and mouth disease! Oh no! Save us! How? Well hey, how about cry out for what they want you to cry out for, let the government control it all for our safety, right? Yeah right, and that is exactly what they started to do!

Back in the Obama administration, he said, *"We are not just designing laws that will keep the American people safe but enforcing them."* And he is talking about the food supply! In fact, if you do not follow these new food laws, you might end up losing your property. And that's because he launched the "Food Safety Modernization Act."

That called for, "The creation of a Food Safety Administration to "allow the government to regulate food production at all levels, and even mandates property seizure, fines of up to $1 million dollars per offense and criminal prosecution for those who fail to comply to regulations."

And then another Bill called the "Food Safety and Tracking Improvement Act" that's backed by Monsanto, Archer Daniels Midland, and Tyson states, *"We must ensure that the Federal Government has the ability and authority to protect the public,"* and it, *"Calls for the establishment of a national database of our food supply with electronic records to identify where the food was grown, prepared, handled, manufactured, processed, distributed, shipped, warehoused, imported, to ensure the safety of the food."* That is total control of the food! And we have already had SWAT team seizures of private property, with semi-automatic rifles of food cooperatives across America.

That involved families herded on to couches in the living room, keeping guns trained on the parents, children, infants and toddlers, for hours on end, without the ability to make a phone call, nor told what crime they were being charged with, nor read their rights. As well as tens of thousands of dollars worth of food being taken, including all their personal stockpile of food for that coming year, along with their cell phones, computers and all their contact records. So much for storing up your own food supply! But it gets even worse!

They also want to control our water supply, you know, the other thing that goes along with our food supply! And according to the Bible, that is going to become a huge issue during the 7-year Tribulation as well.

Revelation 8:6,8-11 "Then the seven angels who had the seven trumpets prepared to sound them. The second angel sounded his trumpet, and something like a huge mountain, all ablaze, was thrown into the sea. A third of the sea turned into blood, a third of the living creatures in the sea died, and a third of the ships were destroyed. The third angel sounded his trumpet, and a great star, blazing like a torch, fell from the sky on a third of the rivers and on the springs of water, the name of the star is Wormwood. A third of the waters turned bitter, and many people died from the waters that had become bitter."

So here we see the Bible tells us that during the 7-year Tribulation that not only does the saltwater, the sea, get judged by God, but even the fresh water gets judged by God, right? It said the springs! And so, what happened? Many people will die as a result, right? And so, this tells us whoever controls that water supply during the 7-year Tribulation, also gets to control a whole lot of people as well, right? Uh huh! And that is exactly what these same people who are pushing for this total control of our food supply are saying we need to do right now as well. Because haven't you heard? We have got a serious water crisis on our hands!

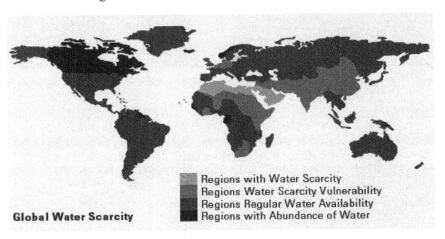

Global Water Scarcity

Regions with Water Scarcity
Regions Water Scarcity Vulnerability
Regions Regular Water Availability
Regions with Abundance of Water

There is your fear factor!

There are chemicals being dumped into our streams. There are poisons that are leaching into our water table, and there are those dreaded toxins being released into our municipal water plants, Oh no! What will we do? Save us Government! Save us! And because of that, that is what we've been told for years by the government that, *"It's illegal to collect rainwater in many states here in the U.S."*

And how the U.N. is seeking to control the whole planet's drinking water with a program they are pushing called, "The Blue Planet Project"

SUSTAINING THE
BLUE PLANET
GLOBAL WATER EDUCATION CONFERENCE

It's being called, "The Mother of All Power Grabs," but oddly enough the news media has totally ignored it, "What should be the biggest news story of the year." And if that wasn't bad enough, they've been pushing another so-called crisis to brainwash us into letting them dictate what we literally can or cannot eat or drink. And it is not just the Food Safety and Water Safety Crisis, but the Obesity Crisis.

Oh no! What will we do?! Now, this is where it gets down to the individual level. They not only want to control the growing of food, the storing of water, but the actual ingesting of food or water, including what kind and how much, down to the individual! You know, for your own safety, of course! And this is why they've been issuing Fat Report Cards to shame kids in school to "curb" this behavior.

WORLD OBESITY DAY

2. Prevalence of adult excess weight (overweight or obese) (2014)

If current trends continue, 2.7 billion adults worldwide will be overweight or obese by 2025 – up from 2.0 billion in 2014

© World Obesity Federation 2015

Overweight or obese is BMI > 25kg/m2
Estimates from World Obesity Federation and World Health Organization
Map © World Obesity Federation 2015

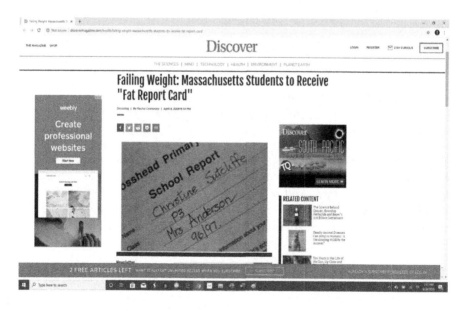

Discover

Failing Weight: Massachusetts Students to Receive "Fat Report Card"

And in certain areas they also launched "Food Police" that are secretly photographing children's lunches to analyze the contents whereupon they then contact the parents and encourage them to "improve" their nutrition. Then the FDA is looking at limiting the salt intake of Americans....

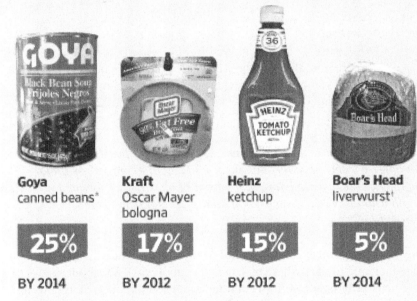

Hold the Salt

How some companies are planning to cut sodium in their products:

Goya	Kraft	Heinz	Boar's Head
canned beans*	Oscar Mayer bologna	ketchup	liverwurst†
25%	17%	15%	5%
BY 2014	BY 2012	BY 2012	BY 2014

Some other companies planning sodium reductions:
Au Bon Pain, FreshDirect, Hain Celestial, LiDestri Foods, Mars Food U.S., McCain Foods, Red Gold, Starbucks, Subway, Unilever, Uno Chicago Grill, White Rose

*Already offers some low-sodium products †Had already cut sodium 15% to meet 2012 target
Sources: the companies; New York City Department of Health and Mental Hygiene

…with the excuse that it is for the "protection" of the health of the public, which some people are now dying from because they are not getting enough salt! Then there is the so-called Fat tax, Twinkie tax, junk food tax, taxes on soda, chips you name it, and they keep adding to the list. It's getting bigger by the day. Punishing you for not eating or drinking what they say you should eat or drink. That is total control! I like what one person said, "What's left to eat?" Are you going to start feeding us "wheat" and "barley" now? Makes you wonder where all this is headed. It is almost like somebody is following a script or something! In fact, they even go on to say, even in our own country:

"The American public cannot be trusted to make their own choices, so we must do it for them."

What are you, the Antichrist? And for those of you who think you can still hide out somewhere, anywhere from this tyranny over our food supply, think again. They are now using, I kid you not, AI to spot obesity from space!

ESA via AP: *"According to Science Magazine, Artificial Intelligence can spot obesity from space through the use of satellite images. Artificial Intelligence can estimate a region's level of obesity by knowing the neighborhoods rate of over-weight adults. AI can help target intervention."*

In other words, if you get out of line, it will find you anywhere on the planet, and dictate what you can and cannot eat! Talk about shades of the 7-year Tribulation, it is all coming together! I am not making this up! In fact, they are now saying that AI can not only control and monitor and produce and enforce the total food supply on the planet, just in time for the 7-year Tribulation, but now they are saying AI can dictate whether you get to live or die in the Global Controlled Antichrist System.

AI for Good; Solving Humanity's Challenges With Artificial Intelligence

Pedro Domingos, Professor University of Washington: *"One of the biggest issues confronting the world today is the impact that we human beings are having on the planet. Human beings are flying the planet, but they are flying blind and flying blind is very dangerous."*

Jason Pontin: *"There is a poem by Phillip Larkin where he mourns the current state of the world and then he says, but at least somewhere there will be a clean ocean, untouched by humanity that will always exist. We are poisoning the last wild environment that we had. We receive a lot of our food from the ocean. It's the basis of the environmental system. We need a solution to creating or recreating the clean oceans that we had. Now there's some really interesting experiments that you might run. You could put algae back in the ocean. There are a bunch of things that we could do. No one wants to run these large geoengineering experiments on a real global scale, but what you want is to run them in simulations and AI's might do that in a way that you could at least begin to experiment with."*

Pedro Domingos: *"The main problem we have right now is that we don't understand how these systems work. Like, ecosystems are very complex. Everything interacts with everything, and we don't know what the impact is or what we do but with more sensors and better machine learning to build models based on those sensors we can actually understand how a whole ecosystem works. How the things we put into the atmosphere changes the climate, etc. Then we can actually do the following thing, which is, we can simulate an ecosystem on a computer and then we can see what are the impacts of doing different things and then we can do those things, that actually, for example, gives us the most gain for what we want, while minimally impacting the ecosystem. We also, for example, use it to understand which species really are important, and which species are less important. So that at the end of the day we often have to make hard trade-offs, and sometimes species that would be significant and not be very important because the whole ecosystem revolves around them and sometimes species that are very, you know, that we're very fond of because maybe they're cuddly animals that are not to be less important.*

So, what I can give us is the ability to actually see further down the road so that we can actually make better decisions."[8]

And what are those hard decisions? Whether or not AI considers you a valuable species, that's good for the environment, you know, save the planet, anywhere in the world, on a global basis! You better obey what it says and do what it says, including how you handle your food supply! Folks, do you see where this is going? As crazy as it is, right now, AI is being pitched on a global basis to dictate all aspects of life, including our food supply and whether or not we get to live or die in this new Antichrist system, just in time for the 7-year Tribulation! How much more proof do we need? The AI Invasion has already begun, and it is a huge sign that we're living in the last days! And that's precisely why, out of love, God has given us this update on *The Final Countdown: Tribulation Rising* concerning the AI Invasion to show us that the Tribulation is near, and the 2nd Coming of Jesus Christ is rapidly approaching. And that is why Jesus Himself said:

Luke 21:28 "When these things begin to take place, stand up and lift up your heads, because your redemption is drawing near."

People of God, like it or not, we are headed for The Final Countdown. The signs of the 7-year Tribulation are Rising! Wake up! And so, the point is this. If you are a Christian and you are not doing anything for the Lord…shame on you! Get busy doing something for Jesus now! Stop wasting your life! We need you! Do not sit on the sidelines! Get on the front lines and help us! Let's get busy working together, doing something splendid for Jesus with what time is left and get busy saving souls! Amen?

But if you are not a Christian, then I beg you, please, heed these signs, heed these warnings, give your life to Jesus now! Because this AI technology is not going to lead to a life of wonderful dreams and a modern-day utopia, but a nightmare beyond your wildest imagination in the 7-year Tribulation! Do not go there! Get saved now through Jesus! Amen?

Chapter Fifteen

The Future of Communication & Education with AI

The **5th area** AI is making an invasion into is **Communication.** You see, you might be still thinking, "Okay, this is exactly my point. I still don't need Jesus, and what I'm going to do is communicate this to as many people as I can, about what the Antichrist is up to and his plan for total domination of the planet. By taking over our food and water supply, with the help of AI, we'll put a stop to it!" Really? You really think he hasn't thought of that? Believe it or not, the Antichrist and the False Prophet will be controlling all the communication on the whole planet as well. They will know exactly who is saying what and when, and you'll never get any word out to anyone! But don't take my word for it. Let's listen to God's.

Matthew 24:3-11 "As Jesus was sitting on the Mount of Olives, the disciples came to Him privately. 'Tell us,' they said, 'when will this happen, and what will be the sign of Your coming and of the end of the age?' Jesus answered: 'Watch out that no one deceives you. For many will come in My Name, claiming, 'I am the Christ,' and will deceive many. You will hear of wars and rumors of wars but see to it that you are not alarmed. Such things must happen, but the end is still to come. Nation will rise against nation, and kingdom against kingdom. There will be famines and earthquakes in various places. All these are the beginning of birth pains. Then you will be handed over to be persecuted and put to death, and you will be hated by all nations because of Me. At that time many will turn away from the faith and will betray and hate each other, and many false prophets will appear and deceive many people."

So, according to our text, Jesus clearly says there's not only going to be a massive rise of wars and rumors of wars and famines and earthquakes in the last days, but He said what? He said there is going to be a massive rise of persecution towards His followers in the last days, right? And as we saw before, it is happening right now, leading up to this event!

And, of course, the people He is talking about here are the Jewish Elect and the people who get saved after the 7-year Tribulation begins. They missed the rapture. Praise God they got saved, but now they are in a heap of trouble. Why? Because Jesus says at that time, they will not only hate His followers all over the world, but they will want to kill them.

In fact, they are going to betray them and literally "turn them in." It's the Greek word "paradidomi" which means, "To give into the hands of, to give up to the custody of, for judgment, for punishment, and/or to be put to death." The Bible says during the 7-year Tribulation, the followers of Jesus will be "turned in for death" or literally "betrayed" on a global scale.

So, put it to the test. Here is the question, "How in the world is somebody going to do this, on a global scale, because that's the context?" How are you going to know if somebody's a follower of Jesus, anywhere on the planet, number one, and number two, how are you going to know where their location is, to "turn them in" for death? I mean, as long as they keep quiet and stay out of sight, they should be safe, right?

Not if you control all the communication! If you did that, all a person would have to do is make one little slip and talk about Jesus and they are toast! They'll know all the people following Jesus and exactly where you are!

But hey, good thing we don't see any signs of anybody having the technology to monitor all our conversations or locate us anywhere on the planet anytime soon. Yeah, right! It's already here! With the help of AI!

And the 1st way AI is controlling all the communication is AI is **Managing Global Communication**. Now, we saw this before in our Modern Technology study. There already is a Global takeover of basically all forms of communication around the whole planet, right now! Whether it is information in general, to TV, to satellites, to emails, to computers, to cell phones, the internet, to social media, you name it. Virtually all forms of modern communication are already being controlled and monitored by Big Brother on a global basis.

So even today, have fun trying to communicate to anyone that you want, including warning about AI. They're going to know everything you do, they're already watching, already listening, and if they don't want something to go through some form of communication, then guess what, don't kid yourself, it's going to be blocked. We are already seeing the signs of that in forms of communication today.

If that wasn't bad enough, apparently, just in case you, or somehow someone, could find a loophole in this total Big Brother control of all forms of global communication, from humans that is, guess who's helping to manage this whole global communications system as we saw before? AI! Once again, AI to the rescue! Why? Because again, just like all their other industries, AI is already taking over on a global basis. Global communication is also way too big for humans to handle! You need to have something superhuman just like all those other industries to pull it off on a global scale, and once again, AI does the trick! So again, maybe humans could make a mistake and let you get some form of communication past them, but AI won't! It is Big Brother on steroids! It does not miss a thing! And it is already being put into play!

The **2ⁿᵈ way** AI is controlling all the communication is by **Promoting Internal Communication**. Now, this is where it gets freaky! Believe it or not, the next form of communication that is already being promoted to us is internal communication on a global basis. And that of course requires a chip in the head. Go figure! As we saw, that just happens to be one of the two places where the Mark of the Beast will be implanted in the human body in the 7-year Tribulation!

Revelation 13:16-17 "He also forced everyone, small and great, rich and poor, free and slave, to receive a mark on his right hand or on his forehead, so that no one could buy or sell unless he had the mark, which is the name of the beast or the number of his name."

So here we see it just so happens that the Mark of the Beast is going either in the right hand or forehead. And as we saw before, with option number two, the head, it is all the rage now, in various industries for various purposes, including now, in the area of communication. Believe it or not, with a chip in your head, whether you realize it or not, you will be able to communicate directly to anyone, anywhere in the world, at any time, if you just get a chip in your head! And as crazy as that sounds, AI tech mogul Elon Musk, as we have seen many times before, says it is just a natural, logical progression. We are now already cyborgs with our existing technology on the outside of us. So, it is just only natural that we move it to the inside.

Elon Musk: *"I do think there is a potential path here which is really getting into science fiction or sort of advanced science stuff. But having some sort of merger with biological intelligence and machine intelligence. To some degree we are already a Cyborg. Think of it like the digital tools that you have, your phone, your computer, the applications that you have. Like the fact that I was mentioning earlier, you can ask a question and instantly get an answer from Google or from some other things.*

If somebody dies, their digital ghost is still around. All their digital emails, the pictures that they posted, their social media. That still lives, even after they die. So, over time, I think we will probably see a closer merger of biological intelligence and digital intelligence. And it's mostly about the bandwidth. The speed of the connection between your brain and your digital extension of yourself.

We used to have keyboards that we used a lot, but now we do most of our input through our thumbs on a phone. And that is very slow. A computer can communicate at a trillion bits per second, but your thumb can maybe do 10 bits per second or 100 bits if you're being generous. So, some high bandwidth interface to the brain will be something that helps achieve symbiosis, between human and machine intelligence and maybe solves the control problem.[1]

Solves the control problem. Excuse me? I think it opens up a can of worms and makes a control problem on a global basis! Think about it, this brings us back to the "Brain Hacking" issue we saw before. The danger of linking our brain to a computer is that that computer can be hacked into, which is now linked to our brain, which means our brain can be hacked into! It sounds crazy, but it is a real threat and a real term in the industry as we saw before!

And so of course, what are they doing? How are they quelling our fears about this Brain Hacking? They do what they always do! Propaganda! They are pitching these new brain chips as the "new and improved" way of communication to make our utopian dreams come true! For instance, Elon Musk, as you just saw, has a new brain chip device that we saw before called "Neuralink" which implants chips into your brain, which they're hoping will be as simple and fast as a Lasik eye procedure!

Yahoo! Finance: *"Well Elon Musk's next project isn't a car or a spaceship or a solar panel, the executive's latest start-up is called Neuralink which focuses on the goal of implanting chips into human brains and also connecting brains to computers. Here is what Elon Musk said at a conference in California about the technology last night."*

Elon Musk: *"This, I think, has a very good purpose, which is to cure very important diseases, and ultimately help to cure humanity's future as a* *civilization relative to AI. It has tremendous potential, and we hope to have this operational and in a human patient before the end of next year. So, this is not far."*

CNet Reports: *"Not far, his says, implantation of the neuron sized thread requires the use of a special robot, but they say it is a minimally invasive surgery. Take a look at these other implant technologies. One is called the Utah Array, the other is the Deep Brain Stimulator. They are very invasive. Take a look at Neuralink's N1 Sensor. It's tiny. Here it is on a finger and here it is next to a penny. So back to our question. How does one implant something so tiny?*

Build a robot. This is a surgical robot. It deals with the complexity of the surgery, such as the subject moving due to breathing. The movement. The robot is under the supervision of the surgeon as electrode threads are implanted. The first product is focused on control. Patients want to control the mobile device, no caretaker necessary. Once that control is possible through the implant, the phone output could be redirected back to a computer as mouse and keyboard inputs.

Now normally here's how traditional brain surgery goes. Your head may be clamped in place, plus your head may be shaved with scarring being a possibility. Neuralink says they want to arrive at something different. They likened it to Lasik. No big scars, no hospital stays, it would be a short procedure and you get to keep all your hair."[2]

Huh? And how incredible is that? You just walk into the office and get your brain chip upgrade, just like you do with Lasik for your eyes, and you walk about able to communicate with what? Not just your limbs that were paralyzed for paraplegics, which is typically how it is being pitched right now, but it will link your brain with communication devices like a cell phone, mouse, or keyboard. You know, those things we use to make calls, send texts, email, and the like. Now you can do it with your brain! But that's not all!

Who wouldn't want this, *"Elon Musk Says Neuralink Will Stream Music Straight into Your Brain."* Huh? No more wires, pods, headphones, nothing to lose, isn't this great huh?

"When asked if we could one day listen to music directly through such an interface – streaming it directly into the brain, in other words, with no headphones needed – Musk replied with a curt, 'Yes.'"

But that's still not all!

"Musk has already set his sights on bigger things. When asked if Neuralink's devices could also help, 'stimulate the release of oxytocin and serotonin, and other chemicals,' Musk replied with another 'Yes.'"

And for those of you who don't know, oxytocin and serotonin are what's often referred to as our "happy hormones."

"When you're attracted to another person, your brain releases dopamine and your serotonin levels increase, and oxytocin is produced which causes you to feel a surge of positive emotion."

So hey, if you get a Brain Chip in the 7-year Tribulation when the Antichrist gives the order to worship him, he'll make sure you'll feel good as you walk around the horrors of the 7-year Tribulation under God's wrath, listening to your music directly into your brain! I know people today who'd probably do that, unfortunately! It sounds crazy, but that's still not all!

If you just get a brain chip in your head, you could even email your messages, *"Is Emailing Your Brainwaves the Future of Communication?"*

"Here's something you probably didn't expect in your inbox: Researchers have now developed a way to email brainwaves. An international research team transmitted words from one person's brain to another by mapping electrical currents in the brain."

"Brain-to-brain transmission is a budding area of study. This marks a huge step. It's the first-time humans have been able to drop messages into another brain using a machine."

And of course, the chip makes it so much easier! But that's not all.

If you get a brain chip, you could also fly a plane.

"Paralyzed woman pilots F-35 fighter jet simulator using mind control."

"It reads like something out of a science fiction movie, but new technology created by the (DARPA) Defense Advanced Research Projects Agency has recently enabled a paralyzed woman to fly an F-35 fighter jet and a single-engine Cessna plane using only her mind."

Wow! But that's still not all.

"AI implants will allow us to control our homes with our thoughts within 20 years, government report claims."

"It will allow you to repair and enhance your muscles, your cells and bone structure. And by using this technology, embedded in ourselves and in our surroundings, we will begin to be able to control our environment with thought and gestures alone."

You could,

"Walk into a room and mentally turn on the lights, control appliances, surf the Web, write and send emails, play video games, dictate articles, control a distant robot or avatar, and even drive a car."

And listen to this.

"Political Avatars will scour all available data from news sites and government debates to provide people with a recommendation on who to vote for."

You know like the Antichrist! Isn't that crazy? But that's still not all! If only you would get a chip in your head.

"Nissan unveils technology that can interpret signals from drivers' brains."

"Nissan has developed a car that can read its driver's mind. The technology will be unveiled at the Consumer Electronics Show in Las Vegas next week."

Isn't that exciting? And speaking of communication, here is the ultimate one.

"This Crazy Gadget Helps You "Talk" To Your Computer Without Words."

"It's like Siri stuck to your face. And it looks like a Bane mask crossed with a squid. Or, if you prefer: like a horror movie monster slowly encompassing your jaw before crawling into your mouth."

Uh, but who is going to do that? That is a fashion faux pas! Well, that is why you need to get a brain chip and let it do the talking for you! Here is the transcript of a video of a brain chip that talks for you.

Andreas Forsland, World's Fair Nano: *"We talk about it as the future of communication, but this is not a future like the science fiction thing. This is not sci-fi, this is real, and it has been real for decades. So, if* *anyone knows anyone who has a pacemaker or a cochlear implant, these are all intelligent bionic wearable devices. They are used for the purpose of encoding or decoding the brain. If you think of someone who has lost a limb, or they have a prosthetic arm or leg, these now use different things like EMG sensors to understand electrical impulses in the muscles in order to trigger the legs to move in a way that are more comfortable and more natural.*

Well, what's happening right now is, as technology is advanced and sensors are advancing at such a pace, you know, what took hundreds of years to get to here, we are going to see in the next ten years, huge leaps in technology and products. And we are going to see huge shifts in who we see walking down the street. We are going to see people with Bionic legs walking down the street and people who are unsighted being able to see using technology.

We are trying to figure out how to allow the brain to remotely control things with greater ease. Because we see the BCI or the Brain Computer Interface as being the next great interface for really making the world a much more interesting place, right? You can control things without having to pull things out of your hands and so on.

We are developing a solution that enables people to literally think in order to speak, and I wanted to share a video with you to give you an idea of what we're doing so that you can start to get excited about the things that are coming in the realm of unlocking communication for many people in the world."

The video shows a little girl walking with her head down while her mother leads her by her hand. The narrator comes on and says,

"If you can imagine not being able to talk. She can't communicate like you or me. Just imagine not having a voice."

Andreas Forsland: *"The reason why we are doing what we are doing is because we want to democratize voice for millions of people. What we do*

with the EEG is we take the data that comes from your brainwaves to have Prose speak a phrase out loud. Liz (a disabled lady hooked up to the machine) *is able to think and Prose speaks for her. She is smiling. Communication is everything for her. The world we're going to create is that every single person regardless of your ability is going to be able to communicate, it's just going to happen."[3]*

Yeah, it's just going to happen alright! Just in time for the 7-year Tribulation and the Antichrist controlling all forms of communication around the planet, down to your most intimate thoughts! And if you think that's crazy, even the secular researchers in this brain chip industry admit, this is exactly what it could allow somebody to do!

"Developing brain-to-brain transmissions further will likely raise ethical and sociological questions in the future, such as who gets to transmit these messages."

"And what might happen if someone decides to dabble into the dystopian realm of mind control."

Yeah, very good question! And if you don't think the Antichrist will use these things in his mark or brain chip technology, you're kidding yourself! Folks, when you look at the future of communication with these brain chips, that are being promoted right now as we speak, it's simply going to become an Antichrist's dream come true with the help of AI controlling it all. Just like **Matthew 24** says, he will know everything about everybody on the whole planet, not just who they are, where they are, but even what they're thinking or speaking on a massive scale like this shows.

TedxCEIBS: *"This is the future of the brain. Our human brain and where it is headed with technology. Right now, scientists in laboratories around the world are developing brain chips. Now these are chips that can be inserted into the brain and they are wireless. We will have brain chips in humans and the people that will get those brain chips first are people who suffer from severe medical conditions. So if you have a severe medical* *condition, like you are paralyzed from the neck down, or you suffer from a brain disorder, like Alzheimer's disease or Parkinson's disease, that brain chip can restore your quality of life. Would you hesitate to take the chip? No, of course not. You would say, put the chip in my brain.*

Mind to mind communication. This is something of science fiction that we all read about, but this is actually possible today, as I just showed you. So, there will come a point where if we have a chip in our brain and you out there have a chip in your brain, we can communicate. We can communicate without talking; we can communicate when you are halfway around the world. And we can exchange knowledge.

So, if you look at a world where we have a chip that is connected to the internet, all of a sudden, every piece of information on the internet becomes accessible to our minds. Now when we look back on today, we have to go to Google and type it in, that will seem so primitive. Nobody will do that. You will just think, 'What knowledge do I need?' and it will appear there for you to use.

But it gets even weirder, stranger, because we will not only be able to transfer knowledge, but we will be able to transfer memories. Now you could have an experience, like skydiving, and all of a sudden you could transfer that experience to a friend. And then they would have that experience with you. It gets even weirder. Because we will be living our lives in our lives, we know our lives, and our memories, but you will have access to anybody's life that would want to make it available to you.[4]

That is, if you have a choice! Put yourself in the Antichrist's shoes. You really think he will not use this brain chip technology, especially if it's part of the Mark of the Beast system? Are you kidding me? Think about it. Go back to the text. How is he going to know who's a follower of Jesus Christ in the 7-year Tribulation or who's really worshiping him? How will he know what people are up to, where they are, anywhere in the world? Including what they are thinking, and whether or not they're going to obey or disobey or even try to inform others of his dastardly plans? Simple! He just taps into the brain chip technology that everybody's just got to have in their heads right now. And pow! With the help of AI managing the whole global system and its superhuman capabilities, he will instantly not only control all the global traditional forms of communication, like television, phones, the internet, computers, you name it. But now even these new forms of internal communication, that actually controls and monitors people's thoughts and words, just in time for the 7-year Tribulation slaughter that **Matthew 24** was talking about. Sounds like science fiction, but it is actual technology being implemented as we read this! The only way to escape is through Jesus Christ today! But that's not all.

The **6ᵗʰ area** AI is making an invasion into is in **Education**. You see, you could still be thinking, "Okay, so maybe I can't communicate what the Antichrist is up to using traditional or even internal forms of communication, to warn other people and expose his dastardly plans for total global control. But this is why I'm just going to educate myself and other people on what he's up to and even use the educational system to work my way into positions of influence and power, including politics, to speak to others and work together to put a stop to all this before it's too late! That's what I'll do!" Really? You think he has not thought of that one either? Believe it or not, AI is not only taking over the whole communication system on the whole planet, but it is also starting to take over the educational system on the whole planet. And let me show you how.

The **1ˢᵗ way** AI is controlling education is AI is replacing our **Current Education**. All forms of education are not just being flooded with computers, tablets and high-speed internet, as the so-called panacea to fix our broken and lagging down educational system. That has been going on for years. Just throw electronics at it, and it will somehow fix it. But speaking of electronics, the latest trend now is letting AI run the whole educational system from top to bottom! Why? Because, haven't you heard? It's superhuman! It can fix and do things no human teacher could ever do! It is electronics on steroids! And they are now projecting that:

"According to a study carried out by Stanford University, education will be the sector that will undergo the most change (with AI) *between now and 2030."*

Why? Because.

"The world is over-saturated with schools that don't work."

"AI will give the ability to personalize education for each student and adapt teaching to their specific needs and capabilities," with *"virtual reality, educational robotics, intelligent tutoring systems and learning analytics. AI is Changing Education."*

In fact, here is how they're saying it's going to benefit everyone.

- AI can provide personalized learning.
- AI can provide online learning.
- AI can provide adaptive learning.
- AI can assist special needs students.
- AI can manage educational games.
- AI can manage Virtual Reality learning.
- AI can do all lesson planning.
- AI can do all testing.
- AI can do all grading.
- AI can do all scheduling.
- AI can do all assessments.
- AI can do all administration.
- AI can do all management.
- AI can do manage all school transportation.
- AI can manage all maintenance.
- AI can manage all school finances.
- AI can manage all school purchases.
- AI can manage cybersecurity.
- AI can manage all safety and security.
- AI can manage facial recognition systems.

- AI can create Smart Schools based on The Internet of Things or (IOT) that connects all electronic devices and all computers so AI can monitor and control all things.
- AI can predict student's future performance.
- AI can prepare students for future jobs that don't exist.
- AI can monitor classroom behavior.
- AI can spot cheating.
- AI can monitor student's emotional well-being.
- AI can create Instant Schools anywhere in the world.
- AI can render large, centralized schools obsolete.
- AI can connect all classrooms around the world making education universal and global for everyone.

In other words, it is going to be controlling all the education on the whole planet! And lest you think that will not happen, they're all in on this!

"Just last year, Education Technology investments surpassed a record high of 9.5 billion USD – up 30% from the year before."

So, they are dumping tons of technology at education like they have always done, somehow computers fix everything.

"Just the AI investments in the Education industry is set to surpass 6 billion by 2024."

So, the bulk of the money is now going towards their latest electronic panacea for the educational system, and that is AI. In fact, the propaganda machines are already out there working overtime to show us what a wonderful world it will be if we just allow AI and AI robots to take over our whole educational system. Here is a promo of robotic teachers at school instead of human teachers:

The school bell rings, and all the students are heading to class, but instead of teachers overseeing the entry of the students there are two robots.

Robots are replacing the teachers, and cybermen go back to school. The kids are entering in two single files with a robot in the middle, dividing them and keeping them in line.

The students go into the classroom. One robot is standing at the door to acknowledge each student as they go in. Another robot is inside watching them as they turn on their computers and start communicating with the robot teacher. The children seem to really like having this kind of education. The classes go from English and Math, to Science Lab, and to Music. As long as the students have their computer to communicate with the robot teachers, they can sing, dance, work out in P.E.

As far as the students are concerned, they seem to love it. Their connection to the computer or the robot teacher is a little square tablet and all they have to do is just push the little buttons. The robots even serve the students' lunch. But the downside of this is that when one of these robots think the student needs to cut down on the calories, because they are eating too much, they take the food away. Like this student was about to eat a bowl of strawberries and suddenly it was gone, the robot had taken it away.[5]

Wow! What a great experience that would be, huh? If you eat too much, you will not get your strawberries! And did you notice how the kids all had computer chip devices they carried around with them at all time even on their arms? We will get to that in a second. But you might be thinking, "Alright, as cool and hip as that promo makes this AI invasion taking over our education sound, there's no way the kids are going to go along with this." Really? You might want to pay attention to the transformation going on with our kids with technology.

"51% of teens would rather communicate digitally than in person, even with friends."

This is another side effect of the cell phone industry and the social media platforms; it is preparing kids for this AI digital future. And it is not just AI gaining in popularity, including taking over our educational system, it is already being put into play, starting with China, on a massive scale! And they are also combining it with Big Brother Technology

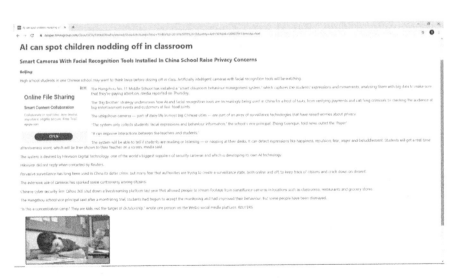

"AI can spot children nodding off in the classroom." "Smart cameras with facial recognition tools installed in Chinese school raises privacy concerns."

"High school students in one Chinese school may want to think twice before dozing off in class. Artificially Intelligent cameras with facial recognition tools will be watching."

And it is not just monitoring them, it is literally monitoring every aspect of their education with AI, via their brain waves.

WSJ Reports: *"Teachers at this primary school in China know exactly when someone isn't paying attention.*

Each student is wearing a headband that gives the teacher this information.

These headbands measure each student's level of concentration. The information is then sent directly to the teachers' computer and to the parent. China has big plans to become a global leader in Artificial Intelligence. It has enabled a cashless economy where people make purchases with their faces. A giant network of surveillance cameras with facial recognition helps police monitor citizens.

Meanwhile, some schools offer glimpses of what the future of high-tech education in the country might look like. The government has poured billions of dollars into the project. Bringing together tech giants, start-ups and schools. We got exclusive access to a primary school a few hours outside of Shanghai to see firsthand how AI tech is being used in the classroom. For this fifth-grade class, the day begins with putting on a brain wave sensing gadget. Students then practice meditating.

The device is made in China and has three electrodes, two behind the ears and one on the forehead. These sensors pick up electrical signals sent by neurons in the brain. The neural data is then sent in real time to the teacher's computer, so while students are solving math problems, a teacher can quickly find out who's paying attention and who's not. A report is then generated that shows how well the class was paying attention. It even details each student's concentration level at 10-minute intervals. It's then sent to a chat group for parents.

Classrooms have robots that analyze student's health and engagement level. Students wear uniforms with chips that track their location. There are even surveillance cameras that monitor how often students check their phones or yawn during class. These gadgets have alarmed Chinese netizens.[6]

Yeah, I agree with that last one. *"This is worse than being a prisoner."* Can you say, "Mark of the Beast system?!" It's all in one package! Big Brother, cashless society, centralized government with AI controlling it all and notice in both scenarios, with AI taking over the educational system, the students were either wearing some sort of microchip device near their hands or some sort of microchip device in their head. Now, why would they be promoting that?

Because it leads us to the **2nd way** AI is controlling all the education is AI is **Promoting Internal Education**. You see, just like in the communication world, the traditional forms of communication like cell phones, the internet, social media, and the like are getting replaced with internal communication via the brain chip. So, it is the same thing that is happening in the education world. It is almost like somebody is following a script! Go figure. For instance, if you would just get a brain chip in your head, you could have instant education.

"Google brain chip could mean future school pupils won't need to memorize facts."

"The days of rote learning could soon be in the past, according to one AI expert. Implants in our heads will tell us everything we need to know."

"Future school pupils with 'Google brain' implants will have answers to all their questions instantly with the help of artificial intelligence."

"Google will be in your head."

"Humans will be able to get answers to any questions they may have without making a sound or typing anything. You can ask something like 'How do you say this in French?' and instantly you'll hear the information from the AI implant and be able to say it."

Yeah, you know, like in the Matrix Movie scene.

"The scene opens with Keanu Reeves sitting in a chair in a state of shock as a very loud screeching sound is vibrating in his ears. He raises up from that position with his eyes bulging and his mouth wide open as if to scream from the pain in his ears. The guy at the control panel looks at him and says, 'I think he likes it.' Reeves lowers himself back down onto the chair. The controller asks him if he wants some more and he answers, 'Yes'. He is being taught Kung Fu wirelessly by the computer directly into his brain.

When it is over, he takes a deep breath. The instructor comes in and Reeves says, 'I know Kung Fu.' The instructor says, 'Show me.' They proceed to the sparring room. The instructor says, 'This is a sparring program. Similar to the program in reality matrix. The same basic rules. Just like gravity. What you must learn is that these rules are no different than the rules in the computer system. Some of them can't while others can be broken. Understand?' Reeves nods his head yes and the instructor says, 'Hit me if you can.'

He runs up to the instructor making moves like he had been practicing these moves for years. Every angle is perfect, when he takes aim to hit the instructor, even though his moves are perfect he is stopped every time. Even though he is being stopped he is feeling very confident in his actions. He has a big smile on his face as he takes his stance one more time to again challenge his instructor. The instructor motions for him to come on and try it again. With this over confidence he proceeds to give it all he's got when the instructor grabs his foot, twists, and throws him on the floor. The instructor says, 'Good.'"[7]

Yeah, good! You didn't have to wait years to learn all that stuff. You just had it downloaded into your brain and voila! You are good to go! Who wouldn't want to have this kind of brain chip instant education opportunity? And not just Kung Fu, but anything you can imagine! Cooking, writing, mathematics, some other physical skills, you name it! The sky's the limit. If you would just get a brain chip implant! I'm sure Google will have competitive pricing! Now, for those of you who think this will never catch on, it is already here, and it's already being promoted as the next panacea to improve your educational experience with AI and brain chip technology. Here is a transcript of a brain chip education promotion.

Future Scape: *"Our brains are remarkable, miraculous even, but they can't do everything, unless we give them a little high-tech help."*

Michio Kaku: *"When children see the movie 'The Matrix' and they see Neal jacking an electrode and all of a sudden becoming a Kung Fu master, the first question they ask is, 'How can I get one?' It's actually physically possible."*

Future Scape: *"The key to transforming learning from an organ process to a machine-like downloading of information is a squiggly bit of brain known as the hippocampus."*

Michio Kaku: *"The hippocampus is the gateway to memories. Short-term memories are stored right here in the pre-frontal cortex, but it eventually has to be transferred to long-term memories and that is where the hippocampus comes in."*

Ted Berger, Biomedical Engineer: *"This part of the brain doesn't store the memory, but it does the appropriate conversion."*

Future Scape: *"At the University of Southern California, Bio-engineer Ted Berger has already proven that a computer can replace our brain function."*

Ted Berger: *"Right now, what our prosthesis does is to convert a code that is in the middle of the hippocampus, to what would be the output of the hippocampus."*

Michio Kaku: *"We have been able to take mice and access the electrical signals coursing through their hippocampus and record them. And then when they shot the message back into the hippocampus the mouse remembered the past."*

Ted Berger: *"We found that we can not only restore long-term memories, we can enhance the animal's ability to remember. You can think about using devices like this to greatly enhance human memory and to shorten the cycle for learning in terms of downloading huge quantities of memory at a single time."*

Michio Kaku: *"Chips that augment our hippocampus could very well help us learn faster. But will that make them a must have for competitive parents? At that point it could create an arms race in elementary school. Rumors will go out that while Jones' kid has been enhanced and our Johnny has to compete with this enhanced kid."*

Jamais Cascio, Ethical Futurist: *"The reality is that these types of technology will not get distributed to everyone at the same time. Some people get it first, some people get it better, as a society we have to really think hard on who gets this. Is it just the wealthy?"*

Jack Uldrich, Global Futurist*: "There are some real dangers that they will use it to consolidate their power and their wealth."[8]*

Yeah, no kidding, like the Antichrist in the 7-year Tribulation. You really think he won't use this technology as well? Get the mark or chip in your brain or forehead and you can be like the Jones' kid ahead of the game! His followers get the best education, you know what I mean, as well as to continue to "buy and sell." And that is just the tip of the iceberg concerning the dangers of this Brain Chip technology.

There's also this danger.

"AI brain implants can change a person's mood."

What? Yeah, here comes your brain manipulation again!

"Chips were created by the Defense Advanced Research Projects Agency that produce electric shocks to jolt a patient's brain back into a healthy state automatically."

So, you know, if you get out of line in the 7-year Tribulation, you won't make that mistake again! ZAAAAP!

"These 'mind control' chips emit electronic pulses that alter brain chemistry in a process called 'deep brain stimulation.'"

Yeah, call it what you will but it is the Mark of the Beast stimulation. You really think he won't use this in his Mark of the Beast system? And if that wasn't bad enough

"Scientists remotely hacked a brain, controlling body movements."

"Imagine someone remotely controlling your brain, forcing your body's central processing organ to send messages to your muscles that you didn't authorize. It's an incredibly scary thought, but scientists have managed to

accomplish this science fiction nightmare for real." "They were even able to prompt their test subject to run, freeze in place, turn around, or even completely lose control over their limbs. Thankfully, the research will be used for good rather than evil, for now."

Yeah, key word there. Again, you really think the Antichrist won't avail himself to this technology with his Mark of the Beast system? Even the experts in this realm of technology admit that is exactly where this is all headed. Here's the transcript of that video:

TedxCEIBS: *"All of us are here learning at Ceibs, right? Getting MDA's and MBA's, well, this will be obsolete. Because Ceibs will literally be in the Cloud, and you will just download from all the best sites in the world with whatever you need to know, when you need to know it. We will no longer have Universities as we have them today. That will change. Information will be on-demand. Information will be commoditized. Knowledge will be commoditized.*

All of us will be smarter, have infinite amounts of storage. And an infinite amount of information. It will be unheard of. We will no longer be this isolated human being. We will be connected to everything. And when we do that, we have to understand that this is a big responsibility. It is something that we as society, we as a human race, must think about very, very carefully. Because imagine, people can hack your cell phone now, people can hack your computer, and it can be bad. They can steal your banking information, take all your money, they can steal your identity, but they cannot really hurt you.

In the future though, if somebody hacks into your head, because it is connected to the internet, they can steal you. They can actually steal you! They can implant memories in your head that you do not have, they can erase memories, they can control you in ways that you would not even know you are being controlled. Is this scary?!"[9]

Uh yeah! But imagine this in the hands of the Antichrist! You're not only already controlling all forms of communication both traditional and internal with the help of these brain chips and AI, you know, just in case somebody tries to blow the whistle on you. But now you control all the educational systems on the whole planet. So if somebody wants to educate others or get educated themselves on what you're up to, you'll just delete it, or better yet, put a more acceptable thought inside their head that he wants them to have and they won't even know it! Is this freaky or what? All this is current technology already being implemented around the world, just in time for the 7-year Tribulation! Folks, how much more proof do we need? The AI invasion has already begun, and it is a huge sign we're living in the last days!

That's precisely why, out of love, God has given us this update on *The Final Countdown: Tribulation Rising* concerning the AI invasion to show us that the Tribulation is near, and the 2nd Coming of Jesus Christ is rapidly approaching. And that is why Jesus Himself said…

Luke 21:28 "When these things begin to take place, stand up and lift up your heads, because your redemption is drawing near."

People of God, like it or not, we are headed for The Final Countdown. The signs of the 7-year Tribulation are rising! Wake up! So, the point is this. If you are a Christian and you are not doing anything for the Lord, shame on you! Get busy doing something for Jesus now! Stop wasting your life! We need you! Do not sit on the sidelines! Get on the front lines and help us! Let's get busy working together doing something splendid for Jesus with what time is left and get busy saving souls! Amen?

But if you're not a Christian, then I beg you, please, heed these signs, heed these warnings, give your life to Jesus now! Because this AI technology is not going to lead to a life of wonderful dreams and a modern-day utopia, but a nightmare beyond your wildest imagination in the 7-year Tribulation! Do not go there! Get saved now through Jesus! Amen?

Chapter Sixteen

The Future of Doctors & Surgeries with AI

The **7th area** AI is making an invasion into is in **Medical**. And boy, if you read your Bible, which I highly recommend, especially in these last days that we're living in, whoever has control of the Medical Community in the 7-year Tribulation will have total, absolute control over the whole planet! And I believe this is yet another thing the Antichrist will do to pull off his dastardly evil deeds during the 7-year Tribulation! But don't take my word for it. Let's listen to God's.

Revelation 8:1-13 "When He opened the seventh seal, there was silence in heaven for about half an hour. And I saw the seven angels who stand before God, and to them were given seven trumpets. Another angel, who had a golden censer, came and stood at the altar. He was given much incense to offer, with the prayers of all the saints, on the golden altar before the throne. The smoke of the incense, together with the prayers of the saints, went up before God from the angel's hand. Then the angel took the censer, filled it with fire from the altar, and hurled it on the earth; and there came peals of thunder, rumblings, flashes of lightning and an earthquake. Then the seven angels who had the seven trumpets prepared to sound them. The first angel sounded his trumpet, and there came hail and fire mixed with blood, and it was hurled down upon the earth. A third of the earth was burned up, a third of the trees were burned up, and all the green grass was burned up. The second angel sounded his trumpet, and something like a huge mountain, all ablaze, was thrown into the sea. A third of the sea turned into blood, a third of the living creatures in the sea died, and a third of the ships were destroyed. The third angel sounded his trumpet, and a great star, blazing like a torch, fell from the sky on a third of the rivers and on the springs of water – the name of the star is Wormwood. A third of the waters turned bitter, and many people died from the waters that had become bitter. The fourth angel sounded his trumpet, and a third of the sun was struck, a third of the moon, and a third of the stars, so that a third of them turned dark. A third of the day was without light, and also a third of the night. As I watched, I heard an eagle that was flying in midair call out in a loud voice: 'Woe! Woe! Woe to the inhabitants of the earth, because of the trumpet blasts about to be sounded by the other three angels!'"

In other words, you ain't seen nothing yet! But how many of you would say, one-third of the earth being burned up, a massive asteroid the size of a mountain slams into the planet, a comet flies by poisoning the water supply of the whole world, and other celestial events disrupting the planet, will not only be a seriously bad time, but you'd certainly want to avoid that time frame by getting saved today. Unfortunately, for those who are left behind, those that rejected Jesus as their Savior today, after those events, there is obviously going to be some serious medical needs to be taken care of for those that survive. I mean, it's common sense. You can try to be the ultimate survivor, for those who rejected Jesus as their Savior today. You could whip out your first aid kits and all the classes you took on how to apply a tourniquet. And get out all your other homemade survival healthcare remedies, but after these events, on this kind of scale, across the whole planet, stop kidding yourself! Whoever does survive these events, at this time in the 7-year Tribulation, you're going to need some serious global medical attention, right? So, here's the point. Whoever controls the Global Health Care System at that time has some serious leverage all over the whole planet, right? And do you think the

Antichrist won't do this? I think he will! As much as he lusts for power. Put yourself in his shoes. If you're the Antichrist and want to micromanage the whole planet as **Revelation 13** says he will, then you're going to need some serious leverage all over the whole planet, right? To force them to do what you want them to do! Including worshiping you as if you were a god and taking a global mark into their bodies so they won't be shut out of your system, right? I mean, how are you going to force people into doing that? Because you know there is going to be a lot of resisters! Not everybody is going to want to worship him as a god. They are already worshiping something else! Not everybody has got a slave mentality. They are still independent! So how does the Antichrist force, order, make, as **Revelation 13** says he will, the planet bow a knee to him and accept his satanic demands? Simple! In light of all these disasters on a global scale, put it all together. You just build a global system that controls the whole medical community so that when these disasters hit, and people are injured on a massive scale, you use that as leverage to get them to obey. It is a matter of life or death now.

You want those burns to be treated from that global fire? Take the chip! You want those broken bones to be set from that giant tsunami that slammed into the coastline? Bow a knee! You want your body healed from drinking all that poisonous water that is killing you from the inside out? Worship me! Total global control by using the medical community as leverage! And if you think that the Antichrist will not do that, I hate to tell you this, it is already happening on a global scale with the help of AI! Believe it or not, as wild as it sounds, right now, AI is not just slowly but surely controlling all our Businesses, Finances, our Conveniences, Agriculture, Communication, and Education in the whole world, but now, at the same time it's also taking over the whole Medical Community on a massive scale all over the planet! I want to share with you how that is occurring right now!

The **1st way** AI is controlling all the medical is **AI is Replacing Doctors**. That's right, who needs doctors anymore? AI can do it all! In fact, AI is invading the medical community, replacing doctors, literally. It is happening so fast that most people not only don't have a clue about it, but they have no idea just how far it has already gone! Right now, AI is being called "The 4th Healthcare Revolution in Medicine," and it is being pitched to be the cure for all the medical community's needs around the globe. With promises of not only replacing doctors, but a whole host of other things, of just about everything you can think of that humans used to do in the medical industry, but now AI can do it for them, even better! In fact, here's just a little teaser of what AI can do in the medical industry.

- AI can efficiently diagnose and reduce medical error.
- AI can more accurately diagnose cancer and all kinds of diseases.
- AI can treat rare diseases.
- AI can provide early detection of diseases and ailments.
- AI can more intelligently check a patient's symptoms.
- AI can improve insights for medical actions that need to be taken.
- AI can better read medical tests and radiology.
- AI can develop new medicines better, faster, and cheaper.
- AI can speed up clinical trials for medical breakthroughs.
- AI can find better candidates for experimental drugs.
- AI can provide targeted treatment.
- AI can provide personalized treatment.
- AI can manage your eating habits and lifestyle choices.
- AI can streamline a patient's experience.
- AI can provide faster hospital visits.
- AI can handle all customer service, medical records, medical history, customer complaints, all paperwork, all forms, and even process customer payments.
- AI can perform all administrative duties.
- AI can personalize and improve your healthcare plan.
- AI can provide personalized and interactive healthcare at any time, 24 hours a day, 7 days a week.
- AI can store, manage, and access all data across the whole healthcare network system.
- AI can increase people's access to medical care across the globe.
- AI can perform your surgery.

You know just in case you go through some catastrophe in the 7-year Tribulation! But as you can see, put all this together. Therefore, experts are saying, right now that AI is, not maybe, not might, but AI is already radically transforming our healthcare system around the world. It is the panacea for everything, including a shortage of doctors.

China 24 Reports: *"Humans verses AI, what are you thinking? I know you are one of those forward-thinking guys. What do you see? How is this going to revolutionize things?"*

Dr. Joel Selanikio, CEO & Co-Founder, MAGPI: *"I think it is as big as the internet or probably as big as computers. It's just one more way in which we can automate what humans only use to be able to do. So, with AI in healthcare for example, this is a big deal in the United States, but it is a bigger and more advanced deal in China. You are basically taking things that until recently, that could only be done by doctors and now you have algorithms that can do them. And more and more this is the case. I think the difference in China is that they in the United States we have plenty of doctors, but our healthcare system is too expensive. In China, it's not the price, they just don't have enough doctors. China has more obese kids than any other country in the world. China has a lot more of just about everything than many countries of the world.*

They need that expertise, and they need to have it scaled and they need to have it soon, and that is why AI is rolling out in a big way in China. In a way that we are just seeing the first signs of it in the United States. But most people in the world have never even seen a doctor or nurse and they never will over their lifetime, so for those areas, which probably include some rural areas in China, if not perhaps in rural parts of the United States, but certainly in these very rural areas all around the world, AI has the potential of a scale of medical expertise in a way we have never been able to do before. It's a whole new world."

China 24 Reports: *"What are the pluses, is it a speedier diagnosis, do you think?"*

Dr. Joel Selanikio: *"It's faster and cheaper."*[1]

And boy, he wasn't kidding it was faster and cheaper! Listen to this.

"Artificial Intelligence in Healthcare Spending is expected to hit $36B by 2025."

And that figure is way up from the current 2.1 billion we were spending and only 600 million spent on AI in the medical community just a few years ago. So why the big change? Because "AI will generate a savings of over $150 billion for the healthcare industry by 2025," and "$300 billion in the next coming years." In other words, like the guy said, faster and cheaper! Who can say no to this? So, like it, or not, they're throwing all their eggs into this basket and that means AI will be taking over the healthcare system, and not only in our country, but around the world! Just in time for those catastrophes in the 7-year Tribulation, and give the leverage needed for the Antichrist to get people to bow a knee to his demands! In fact, just to make sure that you and I go along with this total global takeover of our whole medical community, around the world, it's not just being pitched as the panacea for the medical industry to save on costs, but it is also being pitched to you and me, the consumer, as the most convenient and financial benefit as well!

"94% of medical organizations are investing in and implementing the most reliable path toward accessible and affordable healthcare."

Isn't that a hot topic nowadays? Got to have affordable healthcare! With AI in the healthcare industry:

"Patients will be monitored at home, using wearables and remote monitoring devices, thus reducing admission costs, cutting down office visits, saving time and money."

Isn't that great? Actually, it sounds like a Big Brother paradise to me! In fact, they go on to say:

"AI-based tools, such as voice recognition and clinical decision support systems help streamline workflow processes in hospitals, lowering cost, improving care delivery, and enhancing the patient's experience."

And how are they going to enhance our experience?

"The surge is being driven by the desire for more convenient, accessible, and affordable health care," as well as *"The growing demand and acceptance of electronic, data-driven virtual-based care."*

Now, for those of you who don't know what that is, it's the key word for AI eventually becoming your doctor, a "Virtual Doctor" who will handle all your healthcare needs anywhere on the planet, instead of a human doctor! And as creepy as that sounds, it is already being put into play, step by step!

The **1st way** Doctors are being replaced with AI is with **APPS**. Once again, you've heard the phrase, "There's an APP for everything." Well, believe it or not, it truly is, even in the medical community. AI apps are exploding all over the scene! Now, at first, they were used just to "connect you" with an "actual physical doctor" like this one from Babylon Health.

This video begins in the doctor's office with patients sitting in the lobby waiting to see the doctor. A little girl walks through the front door, past the receptionist, down the hall, right up to the doctor. He is standing in the hall looking at a chart. The little girl goes up to him, takes hold of his hand and proceeds to lead him out the front door of his office. She takes him through the crowds on the street, to the train station, boards the train, through the park, up to the front door of her house. When they walk into the front room, the little girl's mother is covered up on the couch, very sick. When the doctor reaches for her hand the scene changes a bit and instead of a hand, the little girl hands her, her cell phone. When she looks at the screen on her cell phone, the doctor is looking back at her. Now seeing the doctor is as easy as opening an app.[2]

Yeah, hi there, Virtual Doctor! I'm glad I found you on the APP! But hey, isn't this convenient? An app that finds you a doctor! You don't even have to leave the house! And apparently so simple a kid could use it! But that was just the beginning of making this switch from a real doctor to an AI doctor! Once the internet came out, we got used to self-diagnosing ourselves without a real doctor, we did it ourselves. In fact, right now:

"44% of Americans self-diagnose online instead of visiting medical professionals."

And this has become a new term in the medical community called "Cyberchondria" or being a "Cyberchondriac" and they are worried, they say, in the medical community, that we're going to get the diagnosis wrong! So, what do you do with these people who are circumventing doctors electronically for advice online and are running the risk of a misdiagnosis? That's right! You produce medical AI apps that can do it for them! It is the new doctors that replace human doctors! Here's just a couple of examples!

- Ada – An AI that can help if you're feeling unwell.
- Airi – An AI personal health coach.
- Alz.ai – An AI that helps you care for loved ones with Alzheimer's.
- Amelie – An AI ChatBot for mental health.
- Bitesnap – An AI food recognition from meal photos to help count calories.
- Doc.ai – An AI that makes lab results easy to understand.
- Gyan – An AI that helps you go from symptoms to likely conditions.
- Joy – An AI that also helps you track and improve your mental health.
- Kiwi – An AI that helps you to reduce and quit smoking.
- PulseData – An AI that predicts bad, expensive health outcomes before they happen.
- Sleep.ai – An AI that diagnoses snoring and teeth grinding.
- Tess – An AI therapist in your pocket.

You know, just in case you need some advice on whether or not to worship the Antichrist, and of course she says, yes! And that is just the tip of the iceberg of all the AI medical apps out there, that can give you medical advice or treatment, without a real doctor. This trend is now causing real doctors to see where it's all headed, obviously, and they admit that their jobs very well could be in jeopardy, not to mention anybody else in the medical community!

Robert M Wachter, MD, Chief of the Division of Hospital Medicine, University of California, San Francisco, Author of 'The Digital Doctor: *"The role of Artificial Intelligence in the future of healthcare is really interesting and for our physicians, a little bit scary and threatening. If you are a physician like me and trained for 7, 8 or 10 years, and learned your craft, the idea that a computer might come in and take over for you, a lot of us have to have ambivalent feelings about that. If you watched Watson beat the Jeopardy champions, I don't think you could watch that and not wonder what profession is safe. And we have seen profession after profession being taken over by computers. Many of us are biased, thinking medicine is too hard, too complex to be possible, but if you watched Watson you had to be impressed."[3]*

In other words, it really is possible that AI could take over their jobs as doctors! AI can do it all!

The **2nd way** doctors are being replaced with AI is with **ChatBots**. Not only are they saying:

"By 2023, enrollment of chronically ill patients in AI-enhanced virtual care will result in 20 million fewer ER visits."

But we are now seeing a trend in the medical community to go from AI texting, to save on costs and visits, to an AI image conversing with you over your medical care, 24 hours a day, 7 days a week. And that trend has come a long way with what is now called **Medical ChatBots**. And you're saying, "Well, what in the world is a medical ChatBot?" Well, thanks for asking! One guy puts it this way:

"They are what I imagine Siri would be like if she had a medical degree."

"They're designed to be conversational resources when you're feeling under the weather, and you say...

'Hey, ChatBot, I've got these symptoms: Headache, chills, sore throat.' Then the ChatBots interact with you, ask more questions, and depending on your answers help you decide what you should do."

You know without a human. In fact, here's just one example of an AI image ChatBot named "Molly" telling people what to do about their medical needs.

A man is sitting on his patio looking at his cell phone. A nurse comes on the screen and begins to speak.

Nurse: *"Hi, my name is Molly. I am going to be your virtual nurse."*

Ivana Schnur, MD/PhD: *"There is one universal issue that most healthcare systems are struggling with and that is the shortage of qualified clinicians and nurses that manage to care for so many patients.*

This is especially true in chronic disease management which in the United States is the number one driver of rising healthcare costs. 3% of chronically ill patients make up 65% of the healthcare budget.

We developed sensibly with our virtual nurse Molly to extend the clinical reach of real doctors and nurses into everyday life at the home of the patient.

They can be a part of their lives every day. So, Molly can check in with the patient every day, in the morning, or more during the day if that is needed. She can ask the patient how they are doing. She can collect really important information, such as their weight, their blood pressure, glucose level, ask them relevant questions as to whatever condition they might have."

As he stands on his scales...

Molly: *"190.9 lbs. You did not gain any weight. That's good, now let's take your blood pressure."*

Ivana Schnur, MD/PhD: *"Because we have an avatar that talks to you and you can talk back to it, it helps people to be way more engaged. Because of that engagement level they are more likely to do the things to keep themselves healthier."*

Molly: *"Your blood pressure is 149 over 94. Your pulse is 81 BPM. Your blood pressure is a bit high today. We will keep monitoring that. Now that we are finished taking your vitals, I want to ask you a few more questions. If you are experiencing shortness of breath, is it worse than usual? Anything else you want to tell me?"*

Patient: *"No."*

Ivana Schnur, MD/PhD: *"Patients have told me they felt like someone was holding their hand and they would rather talk to her than just use an app on their phone."*[4]

Whoa! So, these things are not only catching on and replacing the AI app, texting versions for medical care, but people are really loving these AI images telling them what to do for their medical care! As crazy as it sounds, that is what is really going on! In fact, it seems the younger the generation you go, the more favorable they are having an AI image ChatBot tell them what to do! One teen:

"Channeled her health-related concerns from, 'Why am I feeling depressed?' to 'What should I do about this zit on my forehead?"

Then, EmojiHealth said, 'Don't pop the zit. It'll only make it worse.'

EmojiHealth is a teen-focused spin-off of ConversationHealth, a company that develops ChatBots for hospitals, pharmaceutical companies and other medical organizations."

"And teens, having grown up getting all their information digitally, are a prime target audience. Then, last September, a Facebook bot was launched and targeted at teens to battle misconceptions about breast cancer and there are similar AI-powered chat programs for teen topics that are taboo such as sex-ed and drug use."

So, AI is telling these younger generations what to do for everything concerning all kinds of medical related issues! Why? Because,

"A lot of health topics can be uncomfortable because they are so personal, but we can take away that "human awkwardness" by replacing the person with a bot."

In fact, they're saying:

"A ChatBot could – theoretically – talk to hundreds of thousands of teens at once."

This is crazy! Can you say, "Doctors in the unemployment line!" And not just doctors, but even nurses.

"Animated Avatars" or "Virtual Nurses" help coach patients during hospital stays and guide their recoveries, give advice, reminders and even encouragement."

"These 'Virtual health coaches are needed with nurses being in short supply these days, so patients who are hospitalized for an illness or surgery may be getting information, advice and encouragement from Sally, a smiling and pleasant-voiced young woman in hospital scrubs, or her male counterpart, Walt."

"But Sally and Walt aren't real people. They're animated avatars – virtual personal health coaches – which can be accessed on a tablet or through a hospital TV set."

So now I don't even need a phone or tablet, I can just speak with them via the TV. And everybody's got one of those! Then,

"They also draw upon a patient's electronic medical records and uses machine learning (AI) to build a personal relationship. The application figures out how best to manage the patient's care."

"And a recent report predicts that virtual nursing assistants could save $20 billion a year in health care by 2026."

In other words, these AI ChatBots really are replacing doctors and nurses!

The **3rd way** Doctors are being replaced with AI is with **Robots**. You see, now we're getting to the next logical step. And that's because you may be thinking, "Hey, I don't want to 'text' an AI app about my medical needs or advice. And I don't even want to 'converse' or 'talk' with an image of AI telling me what to do about my medical condition. I want something more 'lifelike' you know, like a real doctor!" Well hey, worry no more! AI to the rescue! They've got that covered now too with the new AI robot doctors! That's right, as we saw before, first they already got us used to having AI robots doing all kinds of general work for us in society, which again is replacing a lot of human workers. Then we got used to having AI robots as companions, for kids, singles, and even elderly care. And they weren't just popular, they went like hotcakes!

"Japan made the first big steps toward a robot companion – one that can understand and feel emotions."

"Introduced as far back as 2014, Pepper, the companion robot went on sale with all 1,000 initial units selling out within a minute." 60 seconds!

And *"The robot was programmed to read human emotions, develop its own, and help its human friends stay happy."*

They are being called "Digital Friends" that can provide "Digital Empathy." And so, guess what? Not so surprisingly, now they're pushing these AI robots to replace the "knowledge" and "empathy of real doctors and nurses. Here is just one example.

Martha Singleton, Mabu trial patient, is sitting on her couch looking at a yellow robot looking back at her.

Mabu: *"How are you doing? It's time to do some breathing exercises together."*

Martha Singleton: *"That sounds good."*

Narrator: *"This is Mabu, a companion robot that looks after patients at home."*

Martha Singleton: *"She's here to remind us mainly about our medications."*

Narrator: *"It's a face you might want to remember. Because Mabu or something like it might look after you one day."*

Martha Singleton: *"Yes I realize she is a robot, but in a way, I actually feel like she cares, you know, like she said I'll call your doctor and that's like something that a friend would do for you."*

Narrator: *"As we live longer, and our aging population strains the healthcare system, AI and robotics may fill the gaps."*

Cory Kidd, CEO. Catalia Health: *"We don't have enough people to provide the level of care that we'd like to and so that is really where robotics and AI come in."*

Narrator: *"Okay, so it's cute if a little bit unsettling. But what can it do? It can monitor your health and even read the emotions on your face. But as healthcare bots become more advanced, will they take over for human care givers? The focus isn't just on elder care. AI and robotics are making advances across the healthcare industry. From genetic testing, and robotic surgery to cancer research and data collection. Soon enough you may even run into AI in the exam room. But are we ready to hand over life and death decisions to machines?"[5]*

Not to mention, the Antichrist tapping into this global AI animated and robotic medical system. He gets to dictate whether or not you get any care in the first place! And as dangerous and obvious as that is, there is no stopping this! Why? Because it is going to:

"Save so much money and time and even solve the Doctor shortage crisis!"

"It should be noted that training doctors takes time and is costly. First, they take years to train and forecasts around the world show countries facing a shortage of trained staff."

Then the costs involved for medical school continue to rise with the median four-year expense reaching upwards of $278,000. Then to keep the doctors is costly. The average salary for doctors in the U.S., which is not the highest in the world, only ranking 3rd, but even their average salary is $299,000 in 2018 alone. So,

"Reports by the three leading think tanks are saying that in the next five years nurse shortages will double, and general practitioners or doctors will nearly triple, without radical action."

So, you stir all that together and what is their radical solution to solve this medical crisis? Oh no! What will we do? It is AI! Once again, it is supposed to fix everything! Which means, like it or not, AI is taking over the medical community whether you want it to or not! There is no turning back on this trend! And again, it is not just doctors, but nurses as well.

"The days may be numbered when nurses are busy with tasks such as drawing blood, keeping an eye on patients, monitoring vital signs, and moving patients around."

"Nursing activities are being streamlined and simplified by AI-powered robots. These AI-powered systems can draw blood, help patients move and also monitor vital signs without the intervention of the nurses."

In other words, without a human doing it! In fact, China already has their first AI robot doctor who just passed the medical exams with flying colors!

SCMP.TV Reports: *"Robot Xiao Yi has passed China's medical licensing exam by 96 points. Researchers had filled its brain with medical textbooks. It also learned clinical experience and case diagnosis from experts."*[6]

Wow! Looks like they are building a whole fleet of them! Just in time for the 7-year Tribulation and the Antichrist's takeover of the whole medical system all over the planet. Gee, I wonder why? Total control of people! And that is why one legendary Silicon Valley investor said,

"Robots will replace doctors (so now they're talking 100%) *by 2035."*

The **2nd way** AI is controlling all the medical is **AI is Replacing Surgery**. You see, doctors do not just diagnose your medical problems when they come your way, but they also what? They then perform the much need lifesaving surgeries you need to fix those medical problems, right? So, if the medical community is looking at AI to replace all the doctors and nurses in the diagnosing and personal care end of things, in the medical industry, what do you think is happening in the surgical part of it? The exact same thing on a massive scale! I kid you not!

"Artificial intelligence finds great use inside operating rooms in surgeries. Robotic surgical systems powered with AI can be used to standardize the quality of surgical procedures so that uniform and predictable outcomes can be achieved, even when the surgeons vary."

In other words, they are more reliable than humans! And,

"Artificial intelligence performs these typically done by humans, in less time and at a fraction of the cost."

And that is why they are saying:

"45% of ORs (operating rooms) *Will Be Integrated with Artificial Intelligence by 2022."*

So again, like it, or not, it is taking over not just doctors but even surgery! And like the doctors, it has been a step by step approach. First it was with remote surgeries with the aid of AI where you still had a "physical" doctor performing the surgery, but he didn't have to "physically" be present. AI and the new 5G network made it all possible.

New China Reports: *"China's first 5G based remote surgery took place in Sanya, Hainan. The 3-hour surgery was performed on a patient suffering from Parkinson's disease. Hainan-based doctor Ling Zhipei controls the surgical instruments 3,000 km away in Beijing through a 5G network and successfully implanted a deep brain stimulation in the patient."*

Ling Zhipei, Chinese PLA General Hospital: *"The 5G network has solved problems like video lag and remote-control delay experienced under the 4G network. We hope in the future, we can take advantage of the 5G network to enable more hospitals to carry out remote surgery."*[7]

Wow! So that 5G network is doing all kinds of things just in time for the 7-Year Tribulation! You can not only create the IOT or Internet of things to connect all people, all products, and control what people around the world buy and sell, but you can also tap into any and all microchips in anything around the world, be it a phone, or home, even a whole city, and create a Big Brother Paradise. Now you can even use that exact same system to give you the speeds capable of performing a much-needed surgery on someone remotely, even 3,000 miles away! You know, in case you can't find a real doctor in your area. I am sure that will not be used as leverage! Yeah, right! But that is just the first step. Now it is moving to AI assisted surgeries. Why? Because it is going to save so much money and do things better than humans could ever do!

"With an estimate value of $40 billion to healthcare, robots can analyze data from pre-op medical records to guide a surgeon's instrument during surgery, which can lead to a 21% reduction in a patient's hospital stay."

"Robot-assisted surgery is considered 'minimally invasive', so patients won't need to heal from large incisions. (You know, Doctors just rip you wide open) *AI robots can also use data from past operations to inform you of new surgical techniques and the results are indeed promising."*

"One study involved 379 orthopedic patients and found that AI-assisted robotic procedures resulted in five times fewer less complications compared to surgeons operating alone."

In other words, it did it better than a human!

Narrator: *"The robot operates on medical emergency. A scene from a science fiction film? The fiction has long been a reality. In the hospital, the robot has become the perfect teammate of the human being, like the Davinci."*

Dr. Wolfgang Leicht, Chief Physician Urology, Hospital Barmherzige Bruder Regensburg: *"We have a three-dimensional view, we have a tenfold enlargement of the surgical area. We have instruments that are superior to the mobility of the human hand. And we can operate without shaking."*

Narrator: *"No wonder medical robotics is booming. Employees at the German Aerospace Center are developing a robot that will revolutionize surgery in the future."*

Julian Klodmann, Analysis & Control of advanced robotic systems, German Aerospace Center: *"The Miro can be integrated into the operating theater in a very space-saving way. Furthermore, research also includes the fact that even more can be discovered about the patient's insides via miniaturized sensors."*

Narrator: *"This also improves the patient's safety. If the hands of the doctor ever slip off the controller, a life-threatening scenario, not with the Miro. It can be programmed in such a way that the patient's organs are not put at risk. So, it would not carry out an accidental movement by the doctor. So, will doctors soon become superfluous after all? More and more machines are being developed that will work autonomously."*[8]

You know, without a human! And that is exactly where it is headed. You see, AI assisted surgery was just the second step, what they have already alluded to as you just saw was the third and final step to full-blown AI surgeries with no humans at all. Why? Because doctors are dangerous and make way too many mistakes!

"According to a recent study by Johns Hopkins, more than 250,000 people in the United States die every year because of medical mistakes, making it the third leading cause of death after heart disease and cancer."

What? You cannot have that! That is an atrocity! What will we do? How can we survive this horrible epidemic? Can you say AI to the rescue, again? That's right, AI is now starting to perform all the surgeries on people because they don't make mistakes like humans, and that's why they're now saying this will be your future when you go to the hospital:

Narrator: *"The era of human doctors is coming to an end. In the near future, it may be a machine, not a human being, treating you in the hospital. At least this is the reality that AI is making come true. There is a major problem that exists with having humans be the medical providers.*

Human error. No matter how accurate humans are, it is part of being human to make mistakes. A 2016 study from Johns Hopkins University found that medical errors may be the third leading cause of death, right behind cancer and heart disease. A quarter million patients died annually from medical error. Who can save us from this terror? Introducing medical machines. You may be thinking there is no way machines will replace a complicated job like a physician, but this is something that is already happening. There are more monthly visits to WebMD than there are to actual doctors. Smartphone apps like WebMD put the power of diagnosis in your pocket using machine learning to figure out what disease a person has based on symptoms.

In China the first AI computer has passed medical board examinations and the first AI assisted treatment center has already opened. Doctors say computers will help them and compete for their jobs.

For example, researchers at Stanford have found that an AI can diagnose skin cancer just as accurately as a panel of dermatologists and another machine learning algorithm can predict who's at risk for heart disease much better than doctors can. There's even an algorithm that can look at your Instagram account and predict with 70% accuracy whether or not you've been diagnosed with depression in the past three years. Now that's scary!

Using machine doctors is also much cheaper. With the upcoming doctor shortage and a collapsing healthcare financial system many could turn to the machines for medical care whether or not it's a better option. Training a doctor takes thousands of dollars and years of hard work. Trust me I know. But a computer can learn and diagnose much faster and a machine can calculate all the drug interactions and possible treatments much faster than a mere human doctor can. With medical knowledge doubling every three years, it may be hard for humans to keep up. Also, a computer doesn't have eyes that get tired after looking at thousands of x-rays. It doesn't need to sleep at night, and it doesn't even ask for vacations. Be ready for the AI takeover."[9]

The whole medical community around the whole world! Why? Because as you saw AI fixes it all! AI can do it better safer, cheaper, than real human doctors! And lest you think this will never happen, its already begun.

"The Robot will see you now: could computers take over medicine entirely?"

"They already perform remotely controlled operations – now robots are set to be the physicians of the future."

"Doctors and nurses could be supplanted by technology."

With AI, *"There is no hand tremor or shake, no limitation in terms of fatigue."*

"The robotic genie is out of the bottle."

In other words, you can't stop it! In fact,

"The MGMA (Medical Group Management Association) acknowledges that it's possible that AI robots may one day be the only surgeons present."

But hey, come on, people will never go along with this, will they? That is another disturbing trend!

"When asked about their willingness to allow robots powered by AI to operate on their young children (ages eight and younger), Millennial parents were quite likely to allow it."

"In Asia, 82% in China, and 78% in India, said they would be very likely. And even 45% of both the U.S. and U.K. said they would also be very likely."

In other words, the world is ripe just in time for the 7-year Tribulation catastrophes that the Antichrist could very well use medical needs and surgeries to bow a knee to his demands. Now knowing what you just read, is it really far-fetched to think the Antichrist will say something like this during that time?

"You want those burns to be treated from that global fire, take the chip!"

"You want those broken bones set from that giant tsunami that slammed into the coastline, bow a knee!"

"You want to your body healed from drinking all that poisonous water that's killing you from the inside out, worship me!"

Total global control by using the medical community as leverage and AI is helping him do it right now on a global scale! Folks, how much more proof do we need? The AI Invasion has already begun, and it is a huge sign that we're living in the last days! And that's precisely why, out of love, God has given us this update on *The Final Countdown: Tribulation Rising* concerning the AI invasion to show us that the

Tribulation is near, and the 2nd Coming of Jesus Christ is rapidly approaching. And that is why Jesus Himself said:

Luke 21:28 "When these things begin to take place, stand up and lift up your heads, because your redemption is drawing near."

People of God, like it or not, we are headed for the Final Countdown. The signs of the 7-year Tribulation are rising! Wake up! The point is this. If you are a Christian and you're not doing anything for the Lord, shame on you! Get busy doing something for Jesus now! Stop wasting your life! We need you! Don't sit on the sidelines! Get on the front lines and help us! Let's get busy working together doing something splendid for Jesus with what time is left and get busy saving souls! Amen?

But if you are not a Christian, then I beg you, please, heed these signs, heed these warnings, give your life to Jesus now! Because this AI technology is not going to lead to a life of wonderful dreams and a modern-day utopia, but a nightmare beyond your wildest imagination in the 7-year Tribulation! Don't go there! Get saved now through Jesus! Amen?

Chapter Seventeen

The Future of Diseases
& Administration with AI

In the last chapter we saw the **7th area** AI is making an invasion into was **Medical**. There we saw, just in time for the 7-year Tribulation, the whole Medical System is being taken over, on a global scale, with the help of AI, to give the necessary leverage the Antichrist needs to force people to do whatever he wants them to do! That is happening by having AI **Replace Doctors** with APPS and Chatbots and even Robots, and then even having AI **Replace Surgeries** with AI **Assisted Surgeries** or even AI **Robot Surgeries**!

Just in time, in case you want to get those burns treated from the global fire that burns 1/3rd of the planet in the 7-year Tribulation, or those broken bones, from a giant tsunami, or your body healed from drinking poisonous water! None of that will get done if you do not bow a knee to the Antichrist's demands! Total global control, using the medical community as leverage with the help of AI!

And that's common sense, because as we saw last time, medical care is going to be a huge need in the 7-year Tribulation! Not just in the second half as we saw last time, but even in the first half! But don't take my word for it. Let's listen to God's.

Revelation 6:1-14 "I watched as the Lamb opened the first of the seven seals. Then I heard one of the four living creatures say in a voice like thunder, 'Come!' I looked, and there before me was a white horse! Its rider held a bow, and he was given a crown, and he rode out as a conqueror bent on conquest. When the Lamb opened the second seal, I heard the second living creature say, 'Come!' Then another horse came out, a fiery red one. Its rider was given power to take peace from the earth and to make men slay each other. To him was given a large sword. When the Lamb opened the third seal, I heard the third living creature say, 'Come!' I looked, and there before me was a black horse! Its rider was holding a pair of scales in his hand. Then I heard what sounded like a voice among the four living creatures, saying, 'A quart of wheat for a day's wages, and three quarts of barley for a day's wages, and do not damage the oil and the wine! When the Lamb opened the fourth seal, I heard the voice of the fourth living creature say, 'Come!' I looked, and there before me was a pale horse! Its rider was named Death, and Hades was following close behind him. They were given power over a fourth of the earth to kill by sword, famine, and plague, and by the wild beasts of the earth. When he opened the fifth seal, I saw under the altar the souls of those who had been slain because of the word of God and the testimony they had maintained. They called out in a loud voice, 'How long, Sovereign Lord, holy and true, until you judge the inhabitants of the earth and avenge our blood?' Then each of them was given a white robe, and they were told to wait a little longer, until the number of their fellow servants and brothers who were to be killed as they had been completed. I watched as he opened the sixth seal. There was a great earthquake. The sun turned black like sackcloth made of goat hair, the whole moon turned blood red, and the stars in the sky fell to earth, as late figs drop from a fig tree when shaken by a strong wind. The sky receded like a scroll, rolling up, and every mountain and island was removed from its place."

Wow! Now how many of you would say, after all those events, there is going to be a whole lot of people on the planet who are going to need some serious medical care, at least those that survive, however few they might be! I mean, wow! A global war, a global famine, a global plague, with wild beasts shredding you apart, not to mention a global slaughter with a global earthquake so big that every mountain and island is shaken! Oh yeah, and don't forget the asteroids slamming into the earth! And that is just the first half with nearly 2 billion people dying in one fell swoop! Not a good time! But the point is, for those that remain, they are going to need what? Some serious medical care all across the planet, right? And so again, whoever controls the medical community on the planet at this time, they have got some serious leverage over the people, right? Common sense! And once again, it is already being done on a global basis with AI, with not only replacing doctors and replacing surgeries, but....

The **3rd way** AI is controlling the medical system is **AI is Predicting Diseases**. And boy, there is going to be a ton of that going on during the 7-year Tribulation! As we just saw with all these wars, including the use, I believe, of biological weapons, famines and pestilence that follow these wars. Man! The planet is going to be disease ridden, and people are going to need to have all that taken care of, right? Well, hey, worry no more! AI can now not only provide medical care on a global basis, but it can even predict diseases on a global basis. In fact, way better than humans and even before those diseases even start! I mean, who would not want to have this kind of AI medical assistance at this time! But let me give you a little teaser of how well and how accurate AI already is in predicting diseases in our time! *"Chinese AI system could predict disease 15 years in advance."*

- Artificial Intelligence can detect Alzheimer's in brain scans six years before diagnosis.
- Google's DeepMind AI predicts kidney disease two days before it occurs.
- AI can spot Parkinson's disease in 3 minutes.
- Artificial Intelligence diagnoses Pneumonia better than an individual computer or doctor.
- Artificial intelligence can predict epilepsy outcomes.

- AI creates inexpensive heart disease detector.
- Artificial Intelligence can catch irregular heartbeats.
- Europe launches a heart attack-detecting AI for emergency calls.

AI: *"911 what is your emergency?"*

Person on the phone: *"I need help, I need an ambulance. 'Dad, what happened?'"*

AI: *"Can you please verify your name and address and tell me what happened?"*

Person on the phone: *"Yes, I'm Annie Robinson, I'm at 1450 Bay Street, San Francisco. My dad was installing kitchen cabinets, and I think he knocked his head or something. I just heard a bump, and then he fell. I can't get him up and he's not being responsive. 'Dad, are you okay?'"*

AI: *"Alright give me one moment."*

All this time they are talking on the phone the AI is bringing up on the screen the address, directions of how to get there and the amount of time it will take to get there.

AI: *"Okay, an ambulance has been dispatched and it's on its way to you right now. He will arrive in four minutes."*

Person on the Phone: *"Okay."*

AI: *"While we wait, can you tell me if your dad is conscious?"*

Person on the Phone: *"I think so, no, actually I'm not sure. His eyes are open, but his face looks all weird. I guess he's conscious. I don't know."*

AI: *"Okay, okay, hold on. We need to make sure that aside from hitting his head there is nothing else wrong with your dad. Does he seem to be breathing properly?"*

Person on the Phone: *"No, I don't think he is breathing right. He is disoriented. I don't know. Oh, is he having a heart attack? What do I do?"*

AI: *"It's okay. You can do this. Listen to me and stay calm. Can you make sure he is lying flat on the ground and he doesn't have any vomit or food in his mouth?"*

Person on the Phone: *"Yes, hold on. I'm going to put you on speaker, if you can just hold on. Okay?"*

AI: *"Okay"*

Person on the Phone: *"Okay, I'm ready. No, there is nothing in his mouth."*

AI: *"Great, now place your one hand on his forehead and your other hand under his neck and tilt his head back and put your ear next to his mouth. Can you feel or hear any breathing?"*

Person on the Phone: *"No, no I don't think he is breathing."*

AI: *"Okay, once again, stay calm. Stay with me."*[1]

- An AI Chatbot can detect a heart attack.
- Artificial Intelligence used to detect early-stage heart disease
- AI outperformed cardiologists in detection of heart murmurs.
- Artificial Intelligence can predict how cancers will evolve and spread.
- AI predicts cancer patients' symptoms
- AI is better than dermatologists at diagnosing skin cancer.
- AI predicts thyroid cancer risk.
- AI can now tell how aggressive an ovarian tumor is.
- AI can detect early signs of cervical cancer.
- AI outperforms clinicians and pap smears in detecting cervical cancer.
- Artificial Intelligence can determine lung cancer type.
- Artificial Intelligence improves prostate cancer detection.

- Artificial Intelligence identifies previously unknown features associated with cancer recurrence.
- Google AI can spot advanced breast cancer more effectively than humans.
- Artificial Intelligence could save the lives of 22,000 cancer patients a year.
- AI cancer screening tech can see data 'beyond our perception.'
- Artificial Intelligence could soon cure cancer.
- AI can spot signs of Alzheimer's before your family does.
- New Artificial Intelligence system can detect Dementia.

Jeyan Jeganathan, Ontario Hubs Reports: *"One startup company is trying to change the way medical professionals screen for the disease using one's voice and Artificial Intelligence."*

 Lady on her iPad: *"This looks like garbage."* (She says as she looks at a drawing.)

Jeyan Jeganathan: *"It may not look like it, but Joan Deason is taking a test. A test that would detect if she has dementia. By analyzing speech, they can identify cognitive impairment associated with dementia."*

Researcher: *"We are able to take those verbal responses and then score them by five hundred different variables."*

Voice from iPad: *"Tell me everything you see in this picture after the beep."*

Joan: *"He looks like he is yelling back at her."*

Researcher: *"We look at it from the content of her speech using the context, grammar, substitution that you might be making. Like if she should say a name but would say a pronoun instead. She might show word difficulties."*

Dr. Rhonda Collins, Revera Chief Medical Officer: *"People think of dementia as the memory loss solely, but there are other things that go along with it. Loss of the ability to plan and organize in sequence, which is called executive function and things like banking, grocery shopping, laundry, cooking."*

Jeyan Jeganathan: *Using Artificial Intelligence to monitor speech patterns in patients, the program looks at factors like how quickly one talks, how long they pause for and even fluctuations in the vibrations in their voices, things that a human examiner would miss."[2]*

- Artificial intelligence improves diagnosis of lung disease.
- Google's DeepMind AI predicts kidney injury up to 48 hours before it happens.
- AI can predict allergies and allergy patterns.
- AI can spot the pain from a disease doctors still think is fake.
- Artificial Intelligence 'as good as experts' at detecting eye problems.
- AI cameras detect blindness-causing diseases.
- Artificial Intelligence improves glaucoma detection.
- AI APP can detect eye diseases.
- Artificial Intelligence is helping to prevent blindness.
- AI improves chest X-ray interpretation.
- Google AI classifies chest X-rays with human-level accuracy
- Now AI can outperform humans with X-ray interpretation.

Vice News Reports: *"This algorithm reads X-rays better than doctors. This is how a radiologist diagnoses pneumonia."*

Researcher: *"84-year-old female, shortness of breath, period, frontal and lateral views of the chest, period."*

Vice News Reports: *"This is an algorithm called Checks Net that's trained to do the exact same task, and this is how it performed against six doctors analyzing 50 chest x-rays for pneumonia in a recent test at Stanford. Doctors: 79% accuracy, AI: 81% accuracy. Checks Net is one of* *many projects exploring how artificial intelligence can take over tasks normally done by doctors and it has some radiologists worried that AI could one day replace them. That's because algorithms are getting really good at interpreting images and diagnosing disease, sometimes with greater accuracy than humans."*

Pranav Rajpurkar, Ph.D. candidate, Computer Science Stanford University: *"We'll take a picture of this X-ray* (on his cell phone). *The model will then run for a few seconds to output all these diseases and they are sorted by the order of most likely to lest likely."*

Dan Ming Vice News: *"What happens when AI misses an important diagnosis?"*

Matt Lundgren, radiologist: *"It's something that comes up all the time. No matter what happens, if we are saying that we are going to use these algorithms clinically, there will ultimately be a time where an algorithm will make a mistake."*[3]

- Artificial Intelligence improves medical imaging for patients with brain ailments.
- Artificial Intelligence can detect, classify acute brain bleeds.
- AI improves brain scan analysis.
- AI can spot brain hemorrhages as accurately as humans.
- Artificial Intelligence can screen for acute neurological illnesses.
- Artificial Intelligence can locate abnormalities in head CT scans.
- Which reminds me of the weird stuff Joe Biden was asking Chief Justice John Roberts about 15 years ago at his hearing.

Joe Biden: *"We will be faced with equally consequential decisions in the twenty first century. Can a microscopic tag be placed in a person's body to track his every movement? There is actual discussion about that. You will rule on that, mark my words before your tenure is over. Can a brain scan be used to determine whether a person is inclined toward criminality or violent behavior? You will rule on that."*[4]

- Artificial Intelligence only needs 1.2 seconds to screen CT scans.
- Artificial Intelligence improves colonoscopies.
- Artificial Intelligence reduces need for polyp surgery or testing with colonoscopy.
- AI can better diagnose your bad back problems or need for surgery.
- Artificial Intelligence can diagnose epilepsy.
- Artificial Intelligence can 'predict' which of its subscribers will become addicted to opioids.
- Artificial Intelligence can detect the presence of viruses.
- AI can speed up influenza detection.
- Artificial Intelligence can better predict and estimate flu activity.
- AI can provide better flu surveillance.
- Artificial Intelligence can detect moving parasites in bodily fluids for earlier diagnosis.
- AI can diagnose sepsis more reliable than physicians.
- Artificial Intelligence improves wrist fracture detection.
- Chinese researchers use AI to explore diabetes classification.
- Artificial Intelligence can identify postpartum bleeding early.
- Artificial intelligence could diagnose rare disorders using just a photo of a face.
- AI can smell illnesses in human breath.
- But they are saying this AI ability is going to revolutionize disease detection around the world!

Freethink Reports: *"Hospitals are producing so much data that it is extraordinarily difficult to provide the analytics of it so we can actually benefit patients going forward."*

Michelle Ng Gong, M.D., Chief of Research, Critical Care, Montefiore: *"We are collecting a lot more data on these patients, but we are only scratching the surface of how that can be used in the sense of decreasing critical illness."*

Andrew Racine, Md., Ph.D., Chief Medical Officer: *"The way to do that is the use of data that exists in thousands and thousands of patients to predict who is at risk and to intervene to prevent it from happening."*

Parsa Mirhaji, M.D., Ph.D. Director: Center for Health Data Innovations, Montefiore: *"We are using the machine learning and artificial intelligence to build the system that is always watching. We call it PALM. Our vision is to present PALM as the nervous system of this institution. So, what we have designed PALM to do is to collect different kinds of information. Someone's heart rate, blood pressure, their oxygen saturation, changes in their physiologic vital signs on a second-by-second or minute-by-minute basis has implications for where they are going. PALM's job is to calculate the risk for the patients to have either respiratory failure or whether or not this patient is at risk of dying."*

Andrew Racine, M.D., Ph.D. Chief Medical Officer: *"So artificial intelligence allows us to take information from tens of thousands of patients in order to predict what is going to happen to this particular patient and to prevent it from happening."*[5]

Really? Prevent it from happening? Or did you just come up with the ultimate way to control people and even cause it to happen? And I say that because if you put all this AI technology of disease prediction or disease treatment or diagnosis in the hands of the Antichrist, what could he do with all of this on a global basis for nefarious purposes? I mean, could he use it to force people to do whatever he wanted them to do, because now he's got their lives literally in his hands, or could he also use it to cause people's death, for those who resist him? And if you think that's crazy, let's go back to this text again.

Revelation 13:12,14,15 "He exercised all the authority of the first beast on his behalf and made the earth and its inhabitants worship the first beast, whose fatal wound had been healed. He deceived the inhabitants of the earth. He ordered them to set up an image in honor of the beast who was wounded by the sword and yet lived. He was given power to give breath to the image of the first beast, so that it could speak and cause all who refused to worship the image to be killed."

As we have seen before, the Bible clearly says that in the 7-year Tribulation the False Prophet is not only going to dupe the whole world into worshiping the Antichrist, but he is what? He is going to make, he is going to order, he is going to cause them to do whatever he wants them to do or they are going to what? You die, right? So, here is the point. How are you going to do this on a global basis? How are you going to make, order, and cause people to do what you say? And if they don't, you kill them on the spot. Simple! Stir together all that we have seen so far with this AI controlled medical system, and I think you have just one of the many ways he is going to do it! You need that disease or health problem diagnosed so it can be treated properly from all these disasters in the 7-year Tribulation? Okay, just bow a knee.

What is that? You're scared from all the horrors that are going on in the world in the 7-year Tribulation, and you think you might be having a heart attack? No problem. Just worship the Antichrist. You think you might have caught a pestilence that is ravaging the planet in the 7-year Tribulation, and you want to make sure of what it is so it can be treated? Sure thing! Just take the chip! Or if they refuse to bow a knee and worship the Antichrist or take the chip, then you can use the exact same AI controlled system and purposely diagnose them with a so-called incurable disease. Or refuse them treatment, period, because you say the particular ailment that they have isn't worth spending the time or money on! Or they get a brain scan like Joe Biden wanted, "To determine if they were a person of a high criminality or violent behavior," and you cannot have that in this New World Order you are building, so they kill you! And the whole time it is make-believe, but nobody would ever know! Why? Because AI's controlling it all on a global basis and if you do not think that would ever happen, those in the industry admit, it's going to happen, it's just a matter of time!

"Everything looks fine with AI in medicine except for one thing. While cutting-edge technologies save time, money and serve patients more efficiently, the majority are curious, is it safe enough?"

"What if AI software will give the wrong diagnosis and so on. Who will be responsible for the false results?"

Yeah, not to mention, who is to stop it from giving the wrong results on purpose! And don't forget, AI doesn't have a conscience like humans. It is just a machine. So, as you can see, I don't think it's by chance. AI becomes the perfect tool just in time for the 7-year Tribulation for the Antichrist to control or kill people anywhere on the planet if they don't do what they are told. All because you allowed it to take over the medical system!

The **4th way** AI is controlling the medical system is **AI is Replacing Medical Administration**. You know, those who decide whether or not you get treatment in the first place! Let us play the devil's advocate here for the people in the 7-year Tribulation. Let's say they somehow, someway manage to get the right diagnosis or treatment procedure for their diseases or ailments at that time. But is this any guarantee that they will get the treatment or surgery done in the first place? Absolutely not! Why? Because again, we all know there is a hierarchy involved in the system! It first must be approved by what? The Medical Administration, right? Well guess what? AI is taking that over as well. Why? Because it can do it much better than humans! It is going to save so much time and money! And this is precisely why they have been pushing for Electronic Health Records.

Narrator: *"Whether it is a hospital emergency or a routine doctors visit, all people regardless of age will require medical care at some point in their lives. The ability to have safe medical care is a crucial issue today for patients and health care providers alike."*

Physician: *"There are problems with patient's safety. There are things we could be doing today to make care delivered safer."*

Narrator: *"Impacting patients safety includes searching for ways to manage all of the information necessary for providers to deliver high quality, accurate and timely care to their patients.*

Look at it this way, when you visit a local walk-in clinic and the physician requests a number of tests, unfortunately those test results may not always be shared with your family doctor. A month later you experience similar complications, and your family doctor performs the exact same tests. Wouldn't it have been easier and more efficient if your doctor could have accessed those original results first?

In your community, when you visit a health-care facility, any point of care information is collected regarding your health. However, because these systems do not connect, information cannot be easily shared between health care providers. So, when a doctor is treating you, there may be pieces of your health information out there that they don't know about. Now consider how important communication is in an urgent situation.

Say for example, you find yourself in the ER and you are unable to properly relay your medical history. Not knowing for certain if you are on any medications or if you have any allergies could hinder the doctor's assessment and even cause problems during treatment. To improve health care, we are working to create and connect a network of electronic health records systems in your community and across the country. Which in the future will allow your health record history to be accessed by every healthcare provider agency, so one day no matter where you are, you will know that your health care provider has access to your health information.

Many hospitals are doing things such as implementing bar coded medication delivery systems and computerized physician order programs among other safety measures. A key component of this effort is the use of computers to better manage health care information and records."

Physician: *"An electronic health record is a record that your provider can use when you come to see them, that basically keeps track of all your information.*

Narrator: *"Experts say computerized medical records have the potential to actually improve the quality of care that doctors give their patients."[6]*

Improve the quality or control the quality? And did you notice that they not only made all the records of your whole medical history electronic, so that it can be viewed, accessed, or shared anywhere in the world, out of convenience? But they were also putting electronic barcodes on all medical equipment, medication, and even patients themselves, with a barcode on their wrist next to the hand that ties them into the system personally! Gee I wonder where that is headed? Because you all know you could lose those wrist bands, then what kind of medical care could you get or not get? And for those of you who think this will never catch on, this digitizing of all medical care, believe it or not, has already been mandated by OBAMACARE!

"Make no mistake about it, electronic medical records are the way of the future for medical practices of all sizes."

WHY?

"With the passage of the Affordable Care Act (i.e., OBAMACARE) *and the ruling by the United States Supreme Court* (Robert's again) *healthcare reform is on its way."*

"A mandate requiring electronic medical records for all practitioners took effect in 2014. And funding for the Electronic Medical Records legislation will cover a span of 10 years." (So that's 2024)

"By the end of that time, all practices will have implemented electronic medical records."

So, like it, or not, all records in the whole medical industry are going to be digitized. But there is a problem. Actually, it's two-fold. One, tying all this medical data and equipment together on a global basis is just way too big for humans to handle. There is no way they can manage that! You cannot hire enough people to digitize every person, every record, every transaction, every co-pay, every drug, every procedure, every permission made by the administration, every piece of medical equipment anywhere in the world, and on and on it goes!

No amount of people could do that! So, is this all just a hopeless failure, just one big pipe dream? No! Once again AI to the rescue! How many times have we seen; humans may not be able to handle all this data, but AI can! Piece of cake! Even on a global basis, just in time for the 7-year Tribulation and the leverage needed for the Antichrist to have over people! Think about it! This system would not only allow him to diagnose or misdiagnose diseases or ailments for nefarious purposes, but this AI controlled Administration will also give him the ability to fully grant permission for anybody on the planet to get medical treatment anywhere around the world! There is no way this is by chance! It is a set up! But that is only the first problem.

The second problem as you saw is, everything is reliant upon being tied together with electronic barcodes, right? Including the people. But the problem with that is, these barcodes could get lost or torn off, or damaged, so you couldn't read them, now what do you do? You cannot get the medical care you need! You are shut out of the system! Ah! Can you say electronic microchips to replace the barcode system? That's right, as wild as that sounds, it's already being promoted, including "into" people. I am going to give you just a couple of examples.

The **1ˢᵗ way** they are promoting medical microchips in people is **To Control Your Diabetes**.

KCCI8 News: *"A Des Moines hospital is paving the way so that diabetes patients experience fewer needle pricks. Doctors say this new procedure is the next chapter for diabetes care."*

Tommie Clark, KCCI8 Reporter: *"Inside Mercy One's Diabetes Care Practice, the first of its kind procedure is about to happen. Medical gloves snapped on, surgical tools soaked in saline and syringe ready.*

Mary Wrape from Urbandale is about to feel the future of diabetes care. Doctor Teck Khoo, M.D. is one of the providers that has been trained to implant these continuous glucose monitoring sensors."

Dr. Teck Khoo: *"For the longest time the thought of needing to prick yourself four times a day, squeeze a drop of blood out and put it on a test strip, for patients especially with Type One that would be required to do it life long, I think it's a difficult pill to swallow."* [7]

But not anymore! Why? Because you just got a microchip implant! Huh? Diabetes is solved! No more pricking your finger! Life is good! This is great! If you just get a microchip implant, all your medical problems will go away!

The **2ⁿᵈ way** they are promoting medical microchips in people is **To Control Your Diet**. Speaking of pricking your fingers, sometimes you've got to slap your hand to keep all that food from going into your mouth! And who likes going on all those multitudes of diets? Keto, Schmeto, Mosquito! But not anymore! If you would just get a Medical Microchip Implant, like this one!

The Doctors TV show: *"Because obesity has obviously become a national epidemic and one-third of US adults are obese. What if I told you, a tiny chip, I'm talking about something the size of this little needle, the top of this needle? It is this tiny. This could actually have a big effect on your weight."*

"Scientists in the United Kingdom are experimenting with a microchip that they think will help people curb their appetite and lose weight. The microchip would be attached to the vagus nerve and send signals to the brain to control the persons hunger level. Would you implant a chip into your body if it promised to control your appetite?"

Dr. Andrew Ordon: *"This is a great innovation, it's still minimally invasive, it has to be placed with a scope, but I think it is less invasive and with less chances of side effects than the other choices of bypass or band."* [8]

But not anymore! No more bypasses or banding or even diets! Why? Because you just got a microchip implant! Huh? Diets and losing weight totally solved! No more starving yourself to death, no more costly problematic surgeries! Life is good! This is great! If you just get a microchip implant, all your medical problems will go away!

The **3rd way** they are promoting medical microchips in people is **To Control Your Offspring**. This is unbelievable!

CBS 3 Reports: *"Researchers are testing a birth control microchip that can be turned off and on with a remote control. Stephanie Stahl is here now to explain how this works."*

Stephanie Stahl: *"What until you guys see this technology. We are talking about a tiny, implanted chip that was first tested as a way to deliver medicine for osteoporosis. Now it is being expanded to other uses and Bill and Melinda Gates were behind the efforts to use this technology for birth control. This microchip may be the future of contraception. Implanted under the skin, it can deliver tiny amounts of hormone, just like a birth control pill. Biotech for microchips is testing the chip to hold enough hormone to prevent pregnancy for up to 16 years. When a woman wants to conceive, the chip is turned off with a remote and they are making sure it can't be hacked."[9]*

Yeah, right! Because we all know this technology never gets hacked! And what's up with Mr. Population Control, Bill Gates and his wife getting in on this? They wouldn't use this device to control the population of the planet, would they? I mean, just push a button anywhere on the planet and wirelessly send a signal so that no ladies anywhere could have babies! I wish I was making this up! But this is how sick these people are! But hey, don't think about that! Think about the so-called benefits! No more birth control pills, no more having to remember if you took them or freak out if you forgot! That is all gone! Why? Because you just got a microchip implant! Life is good! This is great! Just get a microchip implant and all your medical problems will go away!

The **4th way** they're promoting medical microchips in people is **To Control Your Payment**. You see, all these services do not come for free, right? Obviously, you must pay for them, right? But hey, don't you hate it when you forget your money, or lose your wallet or purse, or worse yet, forget what your copay is in the first place, so you don't even know how much to bring? Well, hey! Worry no more! If you would just get a Medical Microchip Implant, all your financial worries will go way, like this one!

AP Reports: *"Just put in that small glass capsule and it is injected into your body just like a shot of penicillin, as I mentioned previously, there's a product on it called bio bond that immediately upon injection forms a pseudo scar tissue inside your skin, so that the chip doesn't migrate. There is no power source in the chip. It lies dormant underneath your skin for years. And it is only awakened when a scanner is waved over the area where the chip is implanted."*

The scanner is run over the patient's arm and the numbers show up on the screen.

Person who has the chip: *"The scanner rates me, it could be on a portal like any kind of network you wanted it to because it's open architecture."*

AP Reports: *"And then once they have it for the medical application to communicate their medical records, their medical device information, their insurance information, their personal information in a clinical or emergency situation they can then use it for other applications, security, ingress and egress into a facility, financial as a secondary means of authentication to credit cards to help prevent identity theft."*

Physician injecting the chip: *"Okay, I'm going to stick you a little bit here."*

He injects the chip under the skin and wipes it off.

Physician: *"Okay, all done."*

He then holds the scanner over the chip and looks at the number on the screen.

Physician: *"There's your number that's inside."*[10]

Gee, I wonder what number that is? 666? But did you see what this thing could do? Not just store all your electronic medical records that they are mandating all over the place, even in the U.S., but you can even, what? At the same time, you can even make a medical payment or as he admitted, any form of payment. You can tie it into your credit cards. You can use it to prove your identity, security, or gain access into a facility. That chip will control your access, payment, or entrance into a medical facility! All in one chip! Isn't that wonderful? Life is good! This is great! If you just get a microchip implant, all your medical problems will go away! Yeah, you and I both know where all of this is headed! In fact, let's continue on in our **Revelation 13** passage and let God tell us where it's headed.

Revelation 13:16-17 "He also forced everyone, small and great, rich and poor, free and slave, to receive a mark on his right hand or on his forehead, so that no one could buy or sell unless he had the mark, which is the name of the beast or the number of his name."

So, stir all this together in the medical community and I think we know where it is all headed! First, the Antichrist uses this AI global controlled medical system to force, order, cause people to do whatever he wants them to do. Because he has got some serious leverage over them in the 7-year Tribulation with all the calamities during that time, which has produced a need for a proper diagnosis or treatment of diseases or ailments from the medical community. Then they build into the exact same system the ability for him to grant permission or access to the medical system on a global basis via a microchip or marking system with the help of AI. And without it you will not be able to "buy or sell" anything, including your medical needs! And you do not think he would do something like that with this system being built right now as we speak? I think he will say something like this. "Your Diabetes needs to be monitored properly and continuously so you won't die in the 7-year Tribulation, with all this famine and stuff, why it must totally be out of whack by now! No problem just bow a knee."

What's that? Speaking of famine, you are wasting away, shriveling up, because your brain won't tell you you're hungry because I'm controlling your diet electronically? No problem! Worship me! You want to have kids but for some reason it is not working, ever, at all? Real funny, you are a fool! I control the birthrate on the whole planet! And what's that? You need to "buy and sell" medical coverage or gain access to a medical facility or pay for those injuries you received in the 7-year Tribulation, so you won't die? Sure thing! Take the chip!

Folks, it's common sense where this is headed! Total global control over people's lives! Why? Because you allowed AI to take over the whole medical system just in time for the 7-year Tribulation and the horrible reign of the Antichrist! Folks, how much more proof do we need? The AI invasion has already begun, and it is a huge sign we're living in the last days!

And that's precisely why, out of love, God has given us this update on *The Final Countdown: Tribulation Rising* concerning the **AI Invasion** to show us that the Tribulation is near, and the 2nd Coming of Jesus Christ is rapidly approaching. And that is why Jesus Himself said…

Luke 21:28 "When these things begin to take place, stand up and lift up your heads, because your redemption is drawing near."

People of God, like it or not, we are headed for The Final Countdown. The signs of the 7-year Tribulation are Rising! Wake up! So, this is the point. If you are a Christian and you are not doing anything for the Lord, shame on you! Get busy doing something for Jesus now! Stop wasting your life! We need you! Do not sit on the sidelines! Get on the front lines and help us! Let's get busy working together doing something splendid for Jesus with what time is left and get busy saving souls! Amen?

But if you are not a Christian, then I beg you, please, heed these signs, heed these warnings, give your life to Jesus now! Because this AI technology is not going to lead to a life of wonderful dreams and a modern-day utopia, but a nightmare beyond your wildest imagination in the 7-year Tribulation! Don't go there! Get saved now through Jesus! Amen?

Chapter Eighteen

The Future of Diet
& Drugs with AI

In the last two chapters we saw the **7th area** AI is making an invasion into was **Medical**. There we saw, just in time for the 7-year Tribulation, the medical system is being taken over by AI on a global basis to give the necessary leverage for the Antichrist to force people to do whatever he wants them to do! And that is happening right now by having AI replace doctors and AI replace surgeries with AI assisted surgeries and AI robot surgeries!

And then last time we saw that with AI predicting diseases, it could be used to control or even kill people by saying they have a disease when they really don't and give a false diagnosis that is used to take them out or by having AI replace the administration, which is also happening as well, which could enable the Antichrist to hold his demands over people's heads.

In other words, you want that medical treatment? Sure, no problem, take the chip. You want that wound dressed? Worship me. How about just even getting access to any kind of health care. Sure, no problem, just bow a knee. All kinds of leverage the Antichrist could use with the medical system being run by AI on a global basis, just in time for the 7-year Tribulation!

But that's not all. Let's see just how much medical care is not only going to be needed in the 7-year Tribulation with the seal judgments and the trumpet judgments, but even all the way to the very end, with the bowl judgments! But don't take my word for it. Let's listen to God's.

Revelation 16:1-21 "Then I heard a loud voice from the temple saying to the seven angels, 'Go, pour out the seven bowls of God's wrath on the earth.' The first angel went and poured out his bowl on the land, and ugly and painful sores broke out on the people who had the mark of the beast and worshiped his image. The second angel poured out his bowl on the sea, and it turned into blood like that of a dead man, and every living thing in the sea died. The third angel poured out his bowl on the rivers and springs of water, and they became blood. Then I heard the angel in charge of the waters say: 'You are just in these judgments, you who are and who were, the Holy One, because You have so judged; for they have shed the blood of your saints and prophets, and You have given them blood to drink as they deserve.' And I heard the altar respond: 'Yes, Lord God Almighty, true and just are Your judgments.' The fourth angel poured out his bowl on the sun, and the sun was given power to scorch people with fire. They were seared by the intense heat and they cursed the name of God, who had control over these plagues, but they refused to repent and glorify Him. The fifth angel poured out his bowl on the throne of the beast, and his kingdom was plunged into darkness. Men gnawed their tongues in agony and cursed the God of heaven because of their pains and their sores, but they refused to repent of what they had done. The sixth angel poured out his bowl on the great river Euphrates, and its water was dried up to prepare the way for the kings from the East. Then I saw three evil spirits that looked like frogs; they came out of the mouth of the dragon, out of the mouth of the beast and out of the mouth of the false prophet. They are spirits of demons performing miraculous signs, and they go out to the kings of the whole world, to gather them for the battle on the great day of God Almighty. 'Behold, I come like a thief! Blessed is he who stays awake and keeps his clothes with him, so that he may not go naked and be shamefully exposed.' Then they gathered the kings together to the place that in Hebrew is called Armageddon. The seventh angel poured out his bowl into the air, and out of the temple came a loud voice from the throne, saying, 'It is done!' Then there came flashes of lightning, rumblings, peals of thunder and a severe earthquake. No earthquake like it has ever occurred since man has been on earth, so tremendous was the quake. The great city split into three parts, and the cities of the nations collapsed. God remembered Babylon the Great and gave her the cup filled with the wine of the fury of His wrath. Every island fled away, and the mountains could not be found. From the sky huge hailstones of about a hundred pounds each fell upon men. And they cursed God on account of the plague of hail, because the plague was so terrible."

Wow! Now how many of you would say, after all those events, at the end of the 7-year Tribulation, there is going to be a whole lot of people on the planet that are going to need some serious medical care, right? Pretty obvious. People breaking out with ugly painful sores, every living thing in the sea dies, there is no fresh water supply left whatsoever on the whole planet, the sun is scorching people with fire and intense heat. People are gnawing at their tongues, a giant battle at the Battle of Armageddon occurs, and they lose. There is an earthquake so big it erupts on the whole planet all at once, and it not only shakes the whole planet, but every island and mountain fled away, and every city on the whole planet collapsed!

Now, me personally, I would say, that is not only not a time that you want to avoid, by receiving Jesus Christ as your Lord and Savior today, but for those who are left behind, who didn't receive Jesus as their Lord and Savior today, they're going to need what? Once again, they are going to need some serious medical care all across the planet, right? And so again, whoever controls the medical community at this time, whatever time is left, whoever is left, they have got some serious leverage over the people, right? Common sense! And once again, that is already being put into place on a global scale with the help of AI, with not only replacing doctors, replacing surgeries, predicting diseases, and replacing administration, but....

The **5th way** AI is controlling the global medical system is by **AI Controlling Your Diet**. Now as we have been seeing in the opening texts of the last three chapters, all of them are dealing with the major disasters going on in the 7-year Tribulation, from beginning to end, right? And it creates what? Not just a need for medical care but it creates a ton of food shortages, right? From the famine, war, pestilence, and all these disasters together, it creates a food shortage that desperately needs to be controlled!

We saw how they are literally doing that right now, getting prepared to not only control the food supply but be able to dole it out on a global scale just as **Revelation 6** says with the "wheat and barley" text, remember that? You have got to do "a day's work" for "a days' worth of food," remember that?

Now here is the point. Common sense tells us with all these continued disasters all the way to the end, there is going to be a need to continually control the food supply. Not just in **Revelation 6** in the first half, but all the way to to the very end of the 7-year Tribulation, right?

And so, the question is, "How are they going to control the food supply on a continual basis for the entire 7-year Tribulation?" Easy! Just have AI not only control all the medical care, medical diagnosis and medical administration on the whole planet, but have it also control all the medical diets of every single individual on the planet! This way, you could literally micromanage down to the individual level, all the food to all the people, on a global basis, through all the 7-year Tribulation! Perfect plan, right? But hey, good thing we do not see any signs of anybody doing that any time soon! Yeah right! It is already happening!

The **1st way** they are already micromanaging the food supply, on an individual basis, across the whole planet, is with **AI Life Coaches**. That's right folks. We all know how hard it is to think for ourselves nowadays, with our own independent brains to decide what to do, let alone what to eat or drink or how much exercise we should get, right? I mean, apparently, we are not that smart anymore. Think about it. Who can remember all the rules and regulations and calories, and miles and steps and carbohydrates, and keto this and keto that, and all the allergens, and the plethora of other things we need to remember for our health? I don't have time for all that, do you? So hey, wouldn't it be great if someone would do all that for us and make all those food and exercise choices for us, right down to the individual level, no matter where we are, custom tailored for all our health and dietary needs? Well, hey that's right, worry no more! AI to the rescue! AI can do it all for you! Right now, you can get your very own AI Life Coach, and it will remember everything you do and monitor all your health choices! In fact, here is just one example called Lark! As in AI's going to blow the whistle on you if you eat too much pork! You think I am kidding? Look at this.

"I've been trying to lose weight. I've been trying to track every calorie. Let me show you."

As her phone comes on to show the app, she is working with, she tells us...

"For breakfast I had eggs, bacon and a side of toast and coffee. Let me start by typing in eggs, scrambled but mostly egg whites."

As she finds that on her app, she finds 51 calories.

"Great! And I will repeat this step for bacon, toast, and coffee. It is a lot of steps and I don't get an instant feed-back. I feel like I am trying to lose weight alone, which is challenging. However, it is quick and simple to chat about it with Lark. Let me show you. If I tap this button, I can talk personally with my coach. Eggs, bacon, and a side of toast and coffee. Immediately it gives me feed-back when I need it most. Looks like I had too much bacon this week. No bacon for me the rest of the week. Super simple and I am on my way. Lark also tracks how active I have been and keeps me motivated. It's like having a buddy and a coach in my pocket."[1]

And it's free! Gee, I wonder why they would make it for free? Maybe to make sure everybody is being monitored by AI with their health and food choices? Nah, they wouldn't do that would they? Yeah right! But that is the tip of the iceberg of what these AI Life Coaches can do for you. They say they can also help you "Stop Smoking" help you "Maintain an Exercise Routine" and even make sure you "Turn into a Vegan." What? Yeah, as seen here with Beyonce using Artificial Intelligence

"Beyoncé Is Using Artificial Intelligence to Help You Eat Vegan." "She's encouraging an AI future of personalized nutrition." Now why would Hollywood and the Elite want to make sure people become vegans? Oh, I don't know. Maybe to not only control the food supply, but when the order is given for people to work for a whole day's worth of work just to eat "wheat" or "barley" you know, a plant-based nutrition diet in the 7-year Tribulation. There are not many meat eaters left around to complain about that decision, is there? It is almost like somebody is following a script or something!

But that's not all. There is a huge problem with this AI Life Coach in My Pocket. That version was AI on my cell phone. What if somebody steals it, or I lose it, or it falls and breaks, what will I do? How will I know what to eat? Well, hey, worry no more! They have already thought of that, apparently!

The **2ⁿᵈ way** they're micromanaging the food supply on an individual basis across the whole planet is with **AI Wearables**. That's right, just in case these cell phones get lost or stolen, which means, heaven forbid, they can't track you and tell you what to eat or drink, they are now pushing **AI Wearables** to do the same wherever you go!

And hey, with these babies, they never leave your side. They are always strapped there right on your wrist, which is much more secure than just accidentally dropping or losing a cell phone, right? In fact, here is what they say these AI Wearables can do for you. "First of all, they can recharge wirelessly," which means they are always on always monitoring you, no need to worry anymore about getting disconnected from a dead battery with a phone, no siree, constant monitoring. And they can track not only your footstep data, which we have all gotten used to, but they can even track your biometrics, things like this:

"Pulse, blood oxygen level, blood pressure, body temperature, heart rate, and even sleep patterns and transmit all that data wirelessly to a dedicated hub that goes to a cloud database where it's stored and

analyzed by so-called health professionals to generate personalized notifications."

In other words, it will tell you exactly what to do at any given moment, about your health, including what you need to eat. And they even admit it.

"AI enabled wearables can not only track the data, but also define how much sleep people should get, how they should train to improve their fitness, and even what a user needs to eat." (Among other so-called, valuable insights.)

In real time by the way! In other words, these AI Wearables are going to tell them what to tell you and what to do on an individualized constant basis, at any given moment, no matter where you are about your food and health choices. In fact, they are dumping all their eggs into this AI monitoring basket.

"AI Wearables in the Health Industry are up $11.5 billion in 2018 to an estimated $42.4 billion by 2023."

Massive Increase! And they are pitching it to be the panacea to fix all our health care concerns.

"It promises great benefits for healthcare by conducting high-speed diagnostics with far greater accuracy, eliminating human error, and it will provide future generations greater health as AI turns the Hippocratic quote into a reality."

Wow! If only this were here and not a dream! Well guess what? It already is.

Helena: *"Imagine a future where one day machine learning will give you the power to track your sleep."*

Helena: *"Imagine…"*

Sleeping woman: *Excuse me ma'am?"*

Helena: *"Yes? Can I help you?"*

Sleeping woman: *"Yeah, this already does that."* (she raises her arm to show her wrist band.

Helena: *"Oh, I didn't know that."*

Sleeping woman: *"Well, now you do."* And turns over to go back to sleep.

Helena: *"Okay, let's try this again. It's alright I'll start again. Imagine one day a tiny device with an optical sensor will send you an alert if your heart rate is too low."*

A girl hanging from a bar: *"Psst."*

Helena: *"Yes, what? Oh."*

A girl hanging from a bar: *"It already does that."* (As she raises her arm to show the band that shows her heart rate.

Helena: *"Well how about it? We will send you an alert if your heart rate is too high."*

Surfer: *"It already does that."* (As he raises his arm to show his band)

Helena: *"Alright, let's try something new, we will send you an alert if you're some place that's too loud."*

A man sitting on a park bench: *"I just received one of those alerts."*

Helena: *"Really, well, you should probably find a quieter place, sir."*

Man sitting on a park bench: *"You should probably mind your own business."*

Helena: *"Ugh. Okay. Imagine. One day. In the future."*

Another man in the park: *"Hold on."*

Helena: *"No, no, no! Let me finish."*

Another man in the park: *"One second."*

Helena: *"You'll be able to take an E-C-G."*

Another man in the park: *"It already does that."* (As he takes off running)

Helena: *"It already does that."*

Another man in the park: *"I heard that."*

Helena: *"Good, what about continuously track your elevation? Does it do that?"*

Biker: *"Does that."*

Helena: *"Right."*

Biker: *"Does that."*

Helena: *"Thank you."*

Biker: *"Does that."*

Helena: *"I heard it the first time. Encourages you to stay fit."*

Girl playing ball: *"Does that."*

Helena: *"Okay, I get it. You're fit. Umm, I don't know. Will it automatically call 911 if a bear chases you off a cliff and you take a hard fall? Huh?"*

Helicopter medic: *"Yeah, already does that."*

Helena: *"Seriously?"*

A Man on a stretcher: *"Yeah, seriously."*

Helena: *"You've got to be kidding me. Okay, I got it. Let's go from the top. Imagine a day. A future. Imagine a future. One day a tiny device blah, blah, blah. That uses red and infrared light to measure your blood oxygen level."*

Astronaut floating in space, lifts up his arm to show his wrist band.

Helena: *"Yes, what? No, no, no, no. Don't you dare say it. I'm sorry I can't hear you."*

Astronaut moving his mouth saying: *"It already does that."*

Helena: *"Thank you."*

The future of health is on your wrist.[2]

Wow is right! The future of health is on your wrist alright. I sure hope you don't lose it, then what would you do? You are shut out of the system. Too bad there wasn't something more permanent you can have on your body. Well funny you should ask. They are also, at the same time, starting to promote putting this same microchip technology that monitors your health in a wearable microchip tattoo or microchip implant that does the same!

News Commentator #1: *"Wearable Techs should enhance our lives, well-being and health. This particular idea aims to put a ladder to a whole* *new playing field. Imagine if you could have an invisible tattoo, an RFID of sorts, not internal to our bodies but rather printed on our skin. This invisible tattoo would contain information such as blood type, medical history, current prescriptions, prior surgeries and so forth. But here is the catch. This tattoo is not a temporary, wash-off tattoo. It constantly updates with our current health information. It stays up to date with what prescriptions we took as well as what we ate that day. Why? Well in case of an accident or major health problem, the emergency response unit or doctors can instantly know your medical history.*

For example, this could be used to minimize the number of deaths due to blood transfusion errors, allergic reactions to medications or dangerous interactions between drugs. Furthermore, since medical history is personal, the tattoo would be invisible in broad daylight, revealing itself only under a black light."

News Commentator #2: *"A small electronic device implanted in a Salt Lake woman, monitors exactly what she needs to do to take care of herself. The 75-year-old patient is the first to volunteer for this unique clinical trial in Utah. More on the story from medical specialist Ed Yates.*

Jeffrey Osborn, M.D.: *"Can I take a look into you?"*

Ed Yates: *At Intermountain Medical Center in Murray, the cardiologist, Jeffery Osborne, is checking on his patient, Nancy Olson. Inside her body is a new device that is almost like a portable doctor. A sensor, if you will, monitoring her health 24 hours a day.*

Jeffrey Osborn, M.D.: *"It's able to tell us exactly what her status is without guessing.*

Ed Yates: *"Nancy has congestive heart failure. The device implanted just under the skin picks up all kinds of data and transmits it to a handheld monitor with what she needs to know. It takes her temperature and measures the left atrial pressure in the heart. It looks and listens and dispatches all the information and then tells what specific doses of* *medicine she needs to take at that time. The implants are customized to fit each patient's condition and any nuances in their metabolism. Future microscopic devices could float in the blood stream, monitoring almost anything that is measurable in every part of the body.*

News Commentator #1: *And if you think that is impressive, Dr. Osborn says future heart sensory might even be tied to a GPS system that would automatically hail an ambulance when a patient gets in trouble. Can you imagine?* "[3]

Yeah wow! Can you imagine that? It is all connected to a GPS satellite system that could not only automatically alert an ambulance in case your health goes south, but apparently it could even alert the Antichrist anywhere around the world when your health or food supply isn't doing what he says it's supposed to do! Isn't this crazy? All being promoted right now with the help of AI. But you might be thinking, "Well hey, there's no way I'm going to go along with this! I'll never get a microchip implant in or on my body to monitor my health on a constant basis!" Really? Well, you may not have a choice! First of all, all this technology is leading up to the actual Mark of the Beast scenario in the 7-year Tribulation. And if you read this text, it clearly says the choice is going to be mandated whether people want it or not!

Revelation 13:16-17 "He also forced everyone, small and great, rich and poor, free and slave, to receive a mark on his right hand or on his forehead, so that no one could buy or sell unless he had the mark, which is the name of the beast or the number of his name."

The Mark of the Beast is forced on everyone. It is a universal mandate that goes out and if you don't take it, you'll be shut out of the system. But hey, good thing we don't see signs of anyone trying to mandate these AI wearables to constantly monitor our health and food choices anytime soon. Yeah right! It's already here! In fact, one of the entities doing it is a so-called Christian University mandating all students to wear AI wearables wherever they go, and even the secular community is calling for it! It is Big Brother!

The Doctors, Commentator #1: *"Oral Roberts University based in Tulsa, Oklahoma, they are making their incoming students wear fitness trackers. The school then collects and tracks their fitness data that will ultimately affect their grades. Is it going too far or are they actually just being proactive because of the data out there that says students that are active are actually doing better in school?*

Commentator #2: *"Big Brother is watching. To me what this sounds like is grooming a whole generation of kids to be comfortable with being monitored."*[4]

As well as, one day, receive the Mark of the Beast! And if you do not think this mandate, to force people to have their health monitored, with AI and microchips on a continual basis will ever go beyond just a so-called Christian University, think again! They are already predicting this scenario.

"Imagine being asked by your employer to participate in a wellness program in exchange for a discount on your health insurance premiums. The program requires you to carry a Fitbit, which tracks how many steps you walk per day.

This data is then sent to both your employer and The Accountable Care Organization which has a database that includes all your demographic data and your complete health records.

Then the recipients of that data could use AI to match your activity to your medical records, potentially revealing whether you are, for example, sick or in the case of a woman, pregnant.

This information then could be sent to your employer and could potentially be used as the basis for termination."

In other words, you just got fired, you cannot "buy or sell," you're shut out of the system, all because AI controlled the whole Medical

System down to an individualized level, with wearables, just in time for the 7-year Tribulation. So, stir all this together, and this is what you get. You want food? Sorry, your AI life coach says no way, you had too much bacon. You want something to eat? Nope, no can do, your AI wearable says you need to exercise more. What's that? You're hungry from all these famines and disasters? Cannot help you unless you have a microchip tattooed on your hand or receive one into your body, you know, the Mark of the Beast, you're not getting anything! It's all here now to provide total global control, using the medical system! But that's not all.

The **6th way** AI is controlling the global medical system is **AI is Controlling Your Drugs**. And boy, is this another powerful leverage. We all know this is a huge part of being treated medically. It isn't just the doctors and nurses helping with the diagnosis or even the surgery part of it, but what do you need to remedy the diagnosis or to be put under or to heal from that surgery? Drugs! Last time I checked, they come in handy when somebody cuts you open, right? So, can you imagine if somebody got control of the whole drug supply on the whole planet, what leverage that would give them over people?

Yeah, good thing we do not see any signs of that ever happening! Yeah, right, it is already here, with the help of AI! Why? Because here's the rationale! They are saying AI will "Develop drugs faster and cheaper."

Why? It does this because AI has the ability to research literally, *"Millions of medical articles and abstracts and millions of medical journals,"* all across the planet, all at the same time and in a short amount of time. Much more than humans could ever think of ever doing or ever even think about having the time to do it. It is literally impossible for humans to do this, but AI can because it is superhuman. And as a result of this, AI can spit out new drugs faster and cheaper, including new drug treatments that have never been done before, saving both billions of dollars and decades off of human attempts to do the same.

Irene Choi, Director, Drug Discovery: *"Here at Verge we are working on the vision of building the first pharmaceutical company that could really develop rapidly, multiple transformative treatments for patients, and the ultimate goal for us is to build a scalable drug discovery engine that automates the discovery of cures across every human disease, personalized premedication. The discovery is still just a guessing game and that is why right now it takes over a billion dollars and twelve years to develop just a single new drug. It is quite frustrating, especially when you look at the other indications out there. There are a lot of diseases that are unmet because it takes so long to identify the proper target, proper pathway that we should be applying therapeutically. While we are still spinning our heads doing traditional drug discoveries, when we know it's not working."[5]*

Yeah, and we know it is not working when AI can do it better! And this is why they're dumping all their eggs into this basket. AI can develop these drugs so much faster and cheaper, and that is not the only start-up doing it. Right now, there are, *"An estimated 100 startups using AI in the field of drug discovery,"* including IBM and Google just to name a few of the big guns, *"Investing billions of dollars in Artificial Intelligence."* Why? Because as you just heard,

"On average, it takes about a decade of research or more, 12 years and an expenditure of $2.6 billion just to bring one experimental drug to the market."

And even then, *"Because of concerns over safety and effectiveness, only about 5% of those experimental drugs ever even make it to market at all."*

Therefore, with AI's help, *"This will not only reduce the cost and time of a new drug development, but it will lower overall costs of drugs for patients and speed up the access time for the patients to get those new drugs!"*

Plus, they say Artificial Intelligence will not only make these new drug developments "faster and cheaper" than humans could ever do, but it will be done, "more accurately," than humans. Because as we all know, humans make mistakes, even in creating these new drugs, let alone administering them, and then there is all the horrible side effects that happen all the time from these human created drugs, right? Which is why they must deal with those legal disclaimers scrolling down on the commercials while an animated turkey dances around or a guy mowing his yard dances around distracting you. Do not look at those disclaimers! And yeah, sure, now you can move your shoulder, or you stopped smoking, but now you have thoughts of suicide, you want to kill yourself, you got liver disease, brain cancer, your mouth is swollen & your left leg just fell off! I feel great! But hey, with AI all that baloney goes away!

And they even say on top of that, with AI controlling, distributing, administering, and inventing all the new drug supply for the whole world, that it will also provide what they call; *"Personalized Medicine"* where AI can literally not make just a "general one sized hopefully fits all" kind of a drug, but one that's customed tailored to your specific genome and genetic makeup. Isn't that great?

In fact, here's IBM's Watson admitting how, if we allow their AI to control all aspects of the drug industry, development, treatment, you name it, it's going to create a Utopia beyond our wildest dreams!

Commentator for The List Reports: *"The world has been introduced to the wonder of Watson in 2011 on Jeopardy and now IBM's AI supercomputer has graduated to bigger challenges.*

Watson: *"Maintenance records and performance data suggest replacing passenger C4.* (Giving instructions in the Airlines Industry.)

Commentator: *"Watson is learning to help industries like hospitality, business management, and even sports with its cognitive abilities. Now healthcare is Watson's next frontier."*

Watson: *"Working together we can help everyone live healthier."*

Commentator: *"First up, Watson for oncology is helping cancer patients find the best treatment.*

Kathy McGroddy-Goetz, VP Partnerships, IBM Watson Health: *"So it uses that natural language processing again, to actually read in the data from a patient's electronic record and then it goes through and reads all the journal articles that it knows about the latest clinical trials that are better out there so it can provide the best evidence of recommendation to help chose the appropriate treatment.*

Commentator: *"IBM Watson helps Kathy McGroddy-Goetz, with medical information doubling every two years, Watson's processing power is helping patients around the world."*

Kathy McGroddy-Goetz: *"Watson has helped 12,000 patients with our oncology solution alone in less than a year."*

Commentator: *"Next, remember this? The ALS ice bucket challenge that started in 2014 and raised 115 million dollars. Some of that money is funding research with Watson."*

Kathy McGroddy-Goetz: *"Watson for drug discovery solution has recently identified five new genes that are linked to ALS. People previously had no knowledge that they were related to that at all."*

Commentator: *"Finally, sugar IQ with Watson could help the 29 million Americans now living with Diabetes."*

Kathy McGroddy-Goetz: *"So we have worked up an electronic to fill a solution, it is kind of like a personal assistant in the form of an app., an IOS app and what it does is it actually takes data that comes from their continuous glucose monitor device and it actually generates insights. It can log their food, tracks their activity and lots of other things, so it starts to detect these patterns like if you go low everyday around 2 in the afternoon."*

Commentator: *"Watson is conquering our medical frontiers."*[6]

Yeah, I'd say so. But is it conquering or controlling our medical frontiers on a global basis? I wonder what some guy would do with all this if he could get his hands on it, on a global basis, who's name rhymes with the Antichrist? But that's not all. Did you notice how IBM's Watson is not just making new drugs with its intelligence, but it is literally being used to take over all aspects of the Health Drug Treatment scenario? From diagnosis, to drug development, to drug treatment and administration? All of it! And how was it able to do that? By the examples it gave with external monitors, like the example they gave with Diabetes. That allows AI to continually monitor your health, so it can continually give you up to date, minute by minute, advice on your drug treatment, right? But the problem is, as we saw before, these external monitors run the risk, be it a cell phone, or strap on device, or wearable, they can fall off, get lost, or even be stolen. So how will AI monitor and give us medical advice then?

Can you say **AI Smart Drugs**? That's right, I wish I was making this up, but not so surprising, they are now pitching the next stage of AI control with what's called Smart Drugs, which just means Big Brother Drugs. They are also known as "Digital Medicine." And what they are is drugs with tiny little microchips in the pills, that monitors your insides like those external monitors do. Only in this way, since the microchips are on the inside of you, they cannot get lost or stolen. Which means, now AI can always monitor your health, 24 hours a day 7 days a week no matter where you are in the world, with no fear of getting disconnected from the system, isn't that wonderful? Yeah, I don't think so! But here they are admitting that this next step of AI controlling the drug industry, is with a pill near you!

World News Reports: *"Would you swallow a medical implant to protect your health? Pretty soon, you can. You know even common medications can cause side effects that aren't fun, you know the ones that pharma companies take out of their ads as quickly as humanly possible. To exacerbate this is that about half of the people take meds incorrectly anyway according to one survey by the World Health Organization. Scientists are working on this problem though in the form of smart meds or even implants.*

A team from the American Chemical Society built nano chips that combined inflammatory drugs with electrodes through a polymer film. They were then able to control the release of the drug through electrical shock. This study is just one of many projects in the works to create and program home meds through. For example, swallowable microchips are already approved by the FDA. A company called Proteus Digital created a microchip embedded pill, with tiny sensors. It can react to the digestive juices that relays a signal to a patch on your skin that can then be relayed to your doctor, so that if something is awry or you're not doing your meds correctly they can be alerted automatically."[7]

As well as the Antichrist, who is tapped into this whole AI controlled medical drug administered system. Can you believe this? Just in time for the 7-year Tribulation! So, stir all this together and this is what you get. Your prescription ran out and you are starting to have life-threatening conditions, sure, we will refill that, we'll push the button and administer your medication with that Smart Pill, just as soon as you bow a knee to the Antichrist! You will need surgery after all these disasters, and you will need to be put under for that? Sure, no problem, take the chip and worship the beast! Excuse me? You need drugs to treat your diabetes? Sorry, you said something negative about the Antichrist! You get nothing! Try again tomorrow, that is, if you are still alive. Total global control.

Folks, how much more proof do we need? The AI invasion has already begun, and it is a huge sign that we are living in the last days! And that's precisely why, out of love, God has given us this update on *The Final Countdown: Tribulation Rising* concerning the AI invasion, to show us that the Tribulation is near, and the 2nd Coming of Jesus Christ is rapidly approaching. And that is why Jesus Himself said…

Luke 21:28 "When these things begin to take place, stand up and lift up your heads, because your redemption is drawing near."

People of God, like it or not, we are headed for The Final Countdown. The signs of the 7-year Tribulation are Rising! Wake up! Here is the point. If you are a Christian and you are not doing anything for the Lord, shame on you! Get busy doing something for Jesus now! Stop wasting your life! We need you! Do not sit on the sidelines! Get on the front lines and help us! Let's get busy working together doing something splendid for Jesus with what time is left and get busy saving souls! Amen?

But if you are not a Christian, then I beg you, please, heed these signs, heed these warnings, give your life to Jesus now! Because this AI technology is not going to lead to a life of wonderful dreams and a modern-day utopia, but to a nightmare beyond your wildest imagination in the 7-year Tribulation! Do not go there! Get saved now through Jesus! Amen?

Chapter Nineteen

The Future of Death with AI

The **7th way** AI is controlling the global medical system is **AI is Controlling Your Death**. You see, apparently, just in case the Antichrist and False Prophet cannot knock you off with controlling your diet or drugs, and all the other leverage things they get from the global medical system, believe it or not, they are going straight for the throat with having AI literally control your death. I'm not kidding! In fact, this kind of murderous wicked behavior by the Antichrist is going to be commonplace! But don't take my word for it. Let's listen to God's.

Revelation 6:9-11 "When He opened the fifth seal, I saw under the altar the souls of those who had been slain because of the Word of God and the testimony they had maintained. They called out in a loud voice, 'How long, Sovereign Lord, holy and true, until you judge the inhabitants of the earth and avenge our blood?' Then each of them was given a white robe, and they were told to wait a little longer, until the number of their fellow servants and brothers, who were to be killed, as they had been, was completed."

Now, as we have seen many times before, the Antichrist not only goes after the Jewish people and seeks to kill them in the 7-year Tribulation…

Zechariah 13:8 "In the whole land, declares the LORD, two-thirds will be struck down and perish; yet one-third will be left in it."

But the Antichrist also goes after anyone who follows God at that time and seeks to kill them too! And we see that here with Jesus…

Matthew 24:9 "Then you will be handed over to be persecuted and put to death, and you will be hated by all nations because of Me."

And in our opening text in **Revelation 6,** we saw the fruit of that murderous wicked behavior! The people in the context, again, were the Tribulation Saints, i.e., those Gentiles, non-Jewish people, who got saved after the Rapture of the Church. Praise God you got saved, but you should have gotten saved today and avoided the whole thing, but now you are going to be slaughtered! In fact, the Greek word there for "slain" is a very graphic term. This should really get our attention. It is the Greek word, "sphazo," and it literally means, "to slay, to slaughter, to butcher, to put to death by violence." This is not an average normal way of killing people. Not at all. This is a bloody butchering of people, a massive violent slaughter, putting people to death with a violent glee, being ecstatic about it, and this is being done specifically to people who turn to God during the 7-year Tribulation. Now, this wicked murderous slaughterous behavior by the Antichrist and the False Prophet should not come as a surprise to us, not only in what we see in **Revelation,** but as **Revelation** says, they are both inspired of satan and Jesus tells us in **John Chapter 8** the two characteristics of satan, that these guys are going to display.

John 8:44 "You belong to your father, the devil, and you want to carry out your father's desire. He was a murderer from the beginning, not holding to the truth, for there is no truth in him. When he lies, he speaks his native language, for he is a liar and the father of lies."

So, the Antichrist and the False Prophet are going to be inspired by satan, and so is it any wonder that they are also liars and deceivers and murderers just like satan?! Not at all! Like father like son, is the point, right? Now here is my point. This murderous, wicked behavior that the Antichrist and the False Prophet are going to display in the 7-year Tribulation, to anyone who follows God at that time, is about to be put on steroids. By guess who? That's right! AI on a massive scale and I want to demonstrate that to you. It's unbelievable!

The **1st way** AI will be controlling your death is with **AI Determining Your Mental Health**. You know whether or not you've become a "mental threat" to society with the way you're thinking, and if they don't like what you're thinking, you're no longer a safe person to be around and you need to be gotten rid of. You have become "mentally unstable." And believe it or not, as crazy as that "mindset" is, it is being promoted right now with AI Psychiatrists. I kid you not, there is an app for everything. They have now reduced, basically, the whole secular psychology industry into AI programs and AI apps that literally follow you around wherever you go, to monitor your every move and facial gestures and emotions. I am not joking. And apparently, just in case you never physically step foot inside an actual psychiatrists office to be deemed mentally unstable by their system and be deemed a threat to everyone, in this new way, this AI way, you can't escape the system. Now, if you think I'm kidding, here's just one example of a plethora of AI Mental Therapy apps that are out there, also known as "Digital Psychiatrists" that monitor your mental health condition at all times.

ABC7Now News Reports: *"The next time you feel anxious, or down, or maybe just need to talk, a Stanford researcher hopes that you will reach for your phone. ABC's new anchor, Natasha Zouves has the story of how a Bay Area app is now reaching people all over the world."*

Shi Shi: *"I had this grandiose vision of my freshman year."*

Natasha: *"We have all heard the saying that college is supposed to be the best four years of your life. For Shi Shi Vango's freshman year was a* *quick lesson on how you can feel completely alone in a sea of 41,000 students. She said she fell into a deep depression."*

Shi Shi: *"I felt like I was such a failure."*

Natasha: *"One day while strolling through social media she found information about Woebot, an artificially intelligent chatbot. She decided to give the free app a try.*

Shi Shi: *"I really needed someone to talk to. That was on demand and there in an instant."*

Natasha: *"That is exactly what Woebot provides. A little yellow robot is part cheerleader, part friend, and well versed in cognitive behavioral therapy. As you are typing into it, it creates an emotional model of you over time. Dr. Alison Darcy the CEO and founder of Woebot Labs says she just wanted to improve access."*

Dr. Darcy: *"Two-thirds of people that have mental health problems will actually not see a physician."*

Natasha: *"She was floored by the results.*

Dr. Darcy: *"In the first day of being launched, Woebot will talk to more people than any physician will see in a lifetime.*

Natasha: *"Dr. Darcy says that Woebot is now far from home. Active in Africa, in the Middle East and South Asia. It received over 2 million conversations in a week. The data shows he gets results. A recent randomized Stanford study reveals that people who use him have significant decreases in the symptoms of depression in two weeks. And are likely to return to chat with him every day. Shi Shi said it worked for her. She has found a group of friends. But still has the virtual friend that never gets too busy to check on her every day."[1]*

And the key word there is every day! Now, notice to whom this was being pitched to and already used by. The snowflake generation. The same ones that universities are cranking out by the boatload that are so weak-minded that they cannot handle any kind of conflict or opposing ideas and have to have some quiet place or a dog or stuffed animal to hug to cope. This is whom the AI Psychotherapy Bots are being pitched to. It is like they have purposely created the new mindset needed to have this AI takeover on a global basis all "mental health." Which leads us to the next thing. They clearly said that this is already being used all around the world, and it is free! Now, why would they do that? This obviously costs money! Well, I think it is obvious. Now there is no excuse for everybody, on the whole planet, to have their mental state monitored at all times, anywhere on demand, 24-hours a day, 7 days a week, non-stop and it's designed to what? "To change your thinking!" Which leads us to the third thing they admitted, and that was, this BOT was what? Monitoring every day non-stop! Total continuous monitoring of your "mental state" wherever you go! And as crazy as that sounds, that is just one of a plethora of these AI Psychotherapy Bots, these Digital Psychiatrist apps out there. And they are being pitched to fix all our so-called "mental health issues" and "mental health problems" that will save us all from all those unhealthy toxic relationships that we have to deal with, even at work!

"Toxic work environments can take a toll on a worker's well-being and productivity. It's estimated that companies lose up to $300 billion annually because of absentee workers with high stress levels."

That is why, *"AI is promising to change the way workplaces operate."*

How?

"AI can play a significant role in strategic health strategies that can align employees' well-being with the company's mission, vision, and core values."

In other words, you are going to think like the boss wants you to think!

"AI captures data that engages with employees across the business, and actively communicates that information with them."

"These trackers are workforce monitoring tools interconnected with wearables or app-based tracking devices with the help of AI and provides constant emotional well-being updates."

In other words, a microchip near your body connects you to the system. And because of that, *"AI can identify patterns and 'predict' work behaviors as time goes on, including brain patterns, even identifying behavior and speech patterns that are either unhealthy or dangerous to others."*

Huh? You have become a "mental threat." And they even admit its Big Brother

"Although this may sound a bit intrusive, employers have been monitoring employee's habits on their work computers ever since computers became popular. Besides, employees have often passively given consent to workplace surveillance when they signed their contracts."

In other word, you should have read the fine print! And *"With AI insights, employees and Human Resources can build more trusting healthier work cultures with smoother relationships. After all, a healthy employee is a happy employee with AI's assistance."*

Is that crazy or what? Talk about George Orwell 1984! In fact, they have got it figured that that even if you don't sign up for this microchip workplace monitoring system with AI, they can still check your "mental state," with social media.

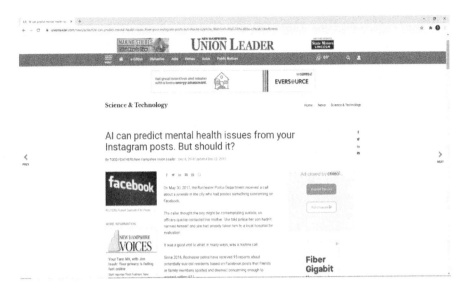

"AI can predict mental health issues from your Instagram posts."

So, they are already doing it whether you realized it or not! In fact, they are also saying, it is also already being done on your phone as well,

"Will Apple Track Your Mind, Not Just Your Heart?"

"Apple's new iPhones features use facial recognition software that could help AI. If your heart throbs, it can track that pitter-pat through continuous heart rate readings."

"On the other hand, if you're depressed your iPhone's Face ID might be able to discover your dismay and connect you to a therapist."

You know, a Digital AI one! But it gets even worse!

Now they're saying, *"Can Artificial Intelligence Detect Depression in a Person's Voice?"*

And of course, the answer they say is yes! And as we have seen several times before, they are already listening to our voices on a continual basis 24 hours a day all the time with our phones, not to mention the home monitoring devices like Alexa and others! What? You better sound happy! In fact, they even admit that they really do have plans for this mental health screening to go global.

"Facing mental health issues is one of the most difficult tasks in life – and with the worsening global statistics about mental health disorders, we truly welcome every innovation and technology that aims to bring down the prevalence of depression, suicide risk, or any other mental trouble."

In other words, we welcome AI to monitor everyone's mental state, anywhere on the planet, so they can eradicate any and all "unhealthy" "mentally unstable" people who are causing all that trouble!

And if you think I'm kidding, take a look at this.

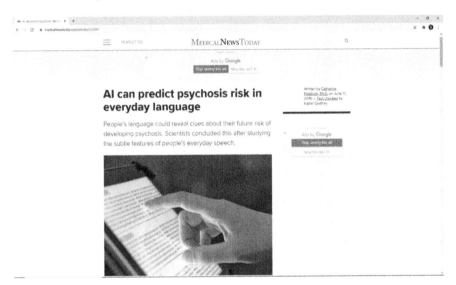

"AI can predict psychosis risk in everyday language."

Uh oh! So now you are just not diagnosed with depression by AI but having a "psychosis" or "unhealthy" mental state. Which means, AI can arbitrarily determine if you have become a danger to society! You've got a "psychosis" you're a danger! And speaking of danger, here is the real problem! Who gets to determine what is a "healthy" or "non-healthy" state of mind, right? And I say that because the powers that be have already equated sharing the Gospel with somebody being a dangerous and a mental health crime.

"France Criminalizes Mental Manipulation."

Well, what's mental manipulation? And I quote… *"Mental manipulation is defined as 'Exercising within a group, activities that are aimed at creating or exploiting psychological dependence, heavy and repeated pressure on a person, or using techniques likely to alter their judgment so as to induce them to behave in a way harmful to their interests."*

Let me translate that for you. That is a fancy way of saying if you witness to someone, you are guilty of this new crime called *"mental manipulation"* and, *"It's punishable of a fine of $75,000 & 5 years in prison"*

And *"The judge will then order the dissolution of any group whose members are convicted of this criminal offense and ban them from any advertising and prohibits them from opening missions, or touting for new members, near schools, hospitals or retirement homes."*

So now you have become as dangerous as a sexual predator or a mass murderer with guns! We are talking about sharing the Gospel! Which means, now with AI monitoring everybody's conversations, facial gestures, words, deeds, you name it, everywhere we go, boy one little slip up in word, deed or thought and you are going to jail for witnessing. Why? Because haven't you heard? AI and the powers that be have determined that it is not "healthy." That's "mental manipulation." People who do that have a "mental disorder" a "psychosis!" And we cannot have you going around trying to create a "toxic" workplace or home or society, can we? We need to lock you up! Can you see where this could lead to in the 7-year Tribulation when it goes global? Anybody who turns to God at that time or tries to tell others about God at that time, will what? Just like the text said, they will be "turned in" and deemed "mentally unstable" because they have become a religious terrorist that needs to be gotten rid of. In fact, speaking of which, some are reporting that they also want to be proactive in this behavior and start giving vaccines to those "dangerous" and "toxic" religious terrorists, so they will never even think about God again. Think I'm kidding? Read this transcript of this alleged briefing and the solution they are proposing to fix the so-called religious terrorists.

Speaker: *"On the left over here, we have people who are religious fundamentalists, religious fanatics, and this is the expression RT-PCR, real-time PCR expression of the GMAT2 gene. Over here we have individuals who are not particularly fundamentalists, not particularly religious and you can see there is a much-reduced expression of this particular gene, the GMAT2 gene. Another evidence that supports our hypothesis for the development of this approach. What you see here…"*

Gentleman in the audience: *"But by spreading this virus, we're going to eliminate individuals from putting on a bomb vest and going into a market and blowing up the market."*

Speaker: *"Our hypothesis is that these are fanatical people, they have an over expression of the GMAT2 gene and that by vaccinating them against this, will eliminate this behavior."[2]*

That's crazy! They would never do that, would they? Well, not to freak you out too much, and again, I am not sure how true this is, it's hard to verify some of these reports, but this headline says this....

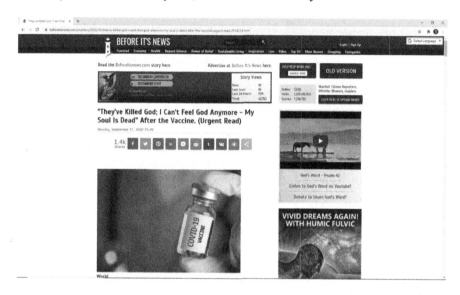

"They've Killed God; I Can't Feel God Anymore - My Soul Is Dead" After the Vaccine."

What Vaccine? The new Covid19 vaccine. The report says allegedly,

"After a second vaccine test volunteer 'Developed neurological problems.'...

"The second volunteer suddenly started saying, 'They've killed God; I can't feel God anymore – my Soul is dead."

Gee, where have I heard that before? Maybe in the transcript of that previous briefing?! Hello! Again, I don't know how true it is, and not to freak you out too bad, but after all we've seen, is it really out of the realm of possibility that they would do something like this? I don't think so. In fact, I think it is just in time for the 7-year Tribulation and the murderous slaughter of the Antichrist! What is that? AI detected you saying something bad about the Antichrist. Uh oh! Time for your vaccine! You will never make that mistake again! What's that? AI informed us that you became a follower of God and you are trying to "mentally manipulate" others into doing the same. Uh, you have become "toxic" in your thinking. It is time we distance you from society. You are going to have to die! All current technology, all being launched right now around the world, to control people's death! But that's not all. Now they are skipping the whole "mental screening" process altogether with AI!

The **2ⁿᵈ way** AI is controlling your death is by **AI Determining Your Age**. You know, how about we just let AI determine how long we get to live in the first place, I mean, wouldn't that be great? No! And as crazy as that is, they are already preparing us for that reality as well!

"Artificial Intelligence (AI) techniques can play a pivotal role in studying the biology of aging on many levels. It can be used to develop predictors of age using blood tests and other synthetic data, to give the most accurate, true biological age of the patient."

And believe it or not, this AI Age Testing is all the rage.

"The topic of AI for Aging Research and Productive Longevity is rapidly gaining popularity. This AI technology is a unique opportunity for the whole world (Notice it's going global) *to free up resources for this endeavor for what could be more important than the life if lived on Earth?"*

So, AI gets to determine my age and so-called life productivity?

Why? We'll get to that in a second. But now they've got this

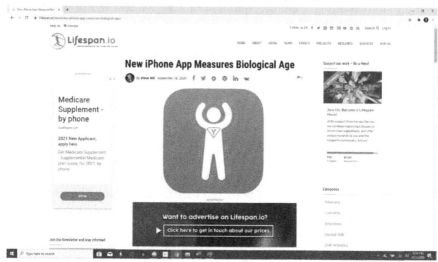

"AI Age Monitoring" ability down to a simple app that anyone can download anywhere in the world!

"New iPhone App Measures Biological Age."

Anybody starting to see a pattern here? But here they are promoting AI Age Monitoring in this video transcript.

Lifespan News Reports: *"Deep-Longevity has launched a new AI system called 'Young dot AI' to track the rate of aging at the vallecular, cellular, tissue, organ, system, physiological and psychological levels. Deep Longevity uses deep aging clocks which integrate many biomarkers to provide a universal multi-bacterial measure of human biological age. Deep-Longevity says that the Young dot AI app and web system will enable everyone interested in learning more about their own aging to get access to some of these aging clocks and start tracking their rate of aging. Accurate biological aging clocks are really important to help us understand how and why we age and how to stop it."[3]*

Yeah, because you can't have that! Old people are a detriment to the system! And did you notice that AI app tracked not just all your biological markers, but even your "psychological" ones as well? Anybody seeing a pattern again? And speaking of which, they go on to say they have purposely warmed us up to it!

"It has been an increasing trend in recent years for companies to develop health and fitness apps that track a variety of health biomarkers, including heart rate, number of steps taken, calories eaten, and quality of sleep. Given how ubiquitous mobile phones are, this has fueled the popularity of easy-to-use and practical health and fitness apps."

In other words, we got used to the physical tracking of our health with all these apps and phones, so now they're just getting us used to being psychologically tracked including our age as well. And not just phones and apps but look at this.

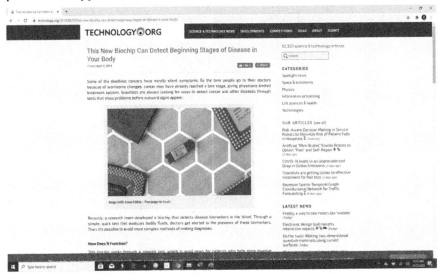

"With a wide array of metrics, including blood biomarkers, as you can see, with a Microchip inside of you..."

ALSO, "Fitness wearable data.

Shocker! There you go again with that method…to warm you up to the microchip idea. And even, *"Facial photographs of the user,"*

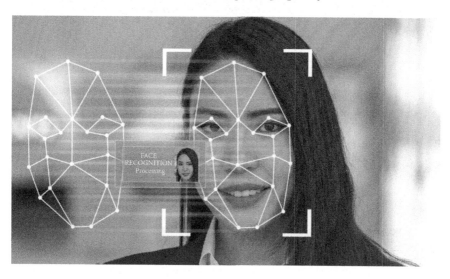

You know with social media and all the places you have plastered your face! Anybody seeing a pattern here?

And *"They are all combined with AI to determine biological age."*

Have fun escaping that system! It is global, and it's already monitoring us in a variety of ways whether we realize it or not! Why? Because,

"AI digital health with remote sensing technologies will continuously observe, learn, and trigger just-in-time care for people aging in home environments or group living..."

To, *"Predict healthy aging and non-healthy aging and address health conditions like heart disease and dementia."*

In other words, they are going to have AI monitor you and report when you get too old or unhealthy for the medical system.

"Aging is a driver of so many diseases, including Alzheimer's and other neurologic problems." "We may eventually be able to use this technology to develop targeted interventions."

What is a targeted intervention? You know, if somebody is too old and starts to develop diseases like heart disease or Alzheimer's, that costs the system too much money, we can take care of them i.e., take them out. Which is starting to sound like that science fiction movie *Logan's Run*! Remember that? Talk about preparation through Hollywood again! It depicted a futuristic so-called utopian society where everybody lived in these geodesic domes run by a computer that took care of all aspects of life. And, in order to be connected to this so-called utopian computer monitored system, at birth each person was implanted with a "life clock" in the hand that changed color when they get older. And when it began to "blink" the computer determined their "last day" at the ripe old age of 30, whereupon the people then had to "exit" the system to a "carrousel" to be killed or what they called "renewed." And the whole premise is that, in this way, nobody would get old and become a detriment to the system, and you could control the population of the planet. Let's take a look at that again if you haven't seen it.

Movie clip from Logan's Run:

Announcer: *"Just imagine a world where you would hold your entire future in the palm of your hand. When a tiny glowing crystal will guide you to an existence where each day is more wonderful than the last. Where it is possible for you to obtain the fulfillment of every fantasy. The satisfaction of every vanity. The absolute obtainment of every wish."*

Metro Goldwyn Mayer presents the Sol David production of 'Logan's Run'.

Announcer: *"The fantastic journey beyond imagination. Welcome to the 23rd century. The perfect world of total pleasure. Imagine a world in which you need never be alone. You touch a switch, turn a dial and the perfect lover steps into your arms. Every pleasure is yours to experience.*

There is just one catch....

When the tiny crystal in the palm of your hand flashes its final message, your time is up!"

A female voice says: *"Enter the carousel. This is the time of renewal."*

The crowd applauds and cheers as they enter the circle. They are all dressed in white robes, forming a circle, walking slowly with their heads down. The crowd is now standing and cheering louder as they formed the perfect circle.

Female voice: *"Be strong and you will be renewed."*

They lift the hood from off their heads showing white masks covering their faces. The crowd is in awe and gasps. Suddenly there is total quiet. They start to look up as the sound of horns begin and a large white light starts to show from above. Once again, each person in the circle raises their hand to show the crystal that is glowing. The crowd claps and cheers again in anticipation. As the circle starts to rotate around, the people in the white robes start to sail up to the ceiling. The crowd is even more excited. But as they float higher towards the light something is happening. Each person seems to evaporate or explode. They no longer exist.[4]

What kind of sicko-twisted society is that? You are born with a chip in your hand so that a computer can tell you and order you when it's time to "renew" or die for the betterment of society. You know, population control! Can you believe that? And as sick and twisted as that is, that is exactly what's being built right now on a global basis with AI determining our death, so the population control elites can create their so-called Utopia.

Which brings us to the **3rd way** that AI is controlling your death, is with **AI Determining Your Death**. Just like Logan's Run, this is a step-by-step approach over many decades to get us to go along with this murderous scenario. First it started with abortion, or the murdering of babies inside the womb, and getting us to accept that around the world. Which by the way, abortion has now become the leading cause of death

around the world. Every single year as you can see, 42 million murdered babies every year worldwide! Which is why the most dangerous place to live now is in the womb! And as sick as that is, they are already moving us to the next step and conditioning us to accept the murdering of babies outside the womb. Remember this not long ago?

EWTN News Nightly Reports: *"Virginia delegate, Kathy Tran, pitched her bill that shocked many Americans earlier this week."*

Todd Gilbert, ® Virginia House Delegate: *"A woman is about to give birth and she has physical signs that she about to give birth. Would that be a point that she could still get an abortion if she was so certified? She is dilating."*

Kathy Tran: *"Mr. Chairman, that would be a decision that the doctor, the physician and the woman…"*

Todd Gilbert: *"I understand that. I'm asking if your bill allows that?"*

Kathy Tran: *"My bill would allow that, yes."*

EWTN News Nightly: *"Governor Northam of Virginia is also facing backlash after his support of the bill."*

Governor Ralph Northam, (D) Virginia: *"The infant would be delivered. The infant would be kept comfortable. The infant would be resuscitated if that is what the mother and the family desire. And then a discussion would ensue between the physician and the mother."*

EWTN News Nightly: *"We found that our Governor is willing to embrace the notion that if a child is born alive you have to just keep the child comfortable while you discuss how or whether to allow it to die."*[5]

And that is outside the womb, folks, and that's not the only Democrat entities that are for this. The Democrat party has become the DEATH party, and that is the political mindset one needs for the murderous behavior of the Antichrist, who is also the ultimate killer politician who's going to slaughter people in the 7-year Tribulation!

Now they have taken it even a step further. Now the Democrats want to start murdering adults outside the womb! First it was physician assisted suicide and supporting that was popularized by Jack Kevorkian.

I hesitate to call him Doctor. Doctors are supposed to save lives not purposely kill them! He killed 130 people before he himself died in 2011. Then we got used to that. Now it has gone to full blown Euthanasia.

Also promoted by the Democrats as you can see here....

Now it's even gone to global euthanasia with kids not just adults in Belgium being killed as you can see. And then even kids in Canada being killed as well.

And you don't even need to inform your parents of being killed by the so-called Medical Community around the world...

And now just like abortion, this murdering of adults on demand is going to become an industry as this one guy who came out of the Abortion Industry, a Dr. Richard Nathanson, stated...

"It's the next logical step after abortion." "In the beginning there will be refined, dignified, quiet, places that may call Thanatoriums or some fancy word like that, where you will take your grandfather who is suffering from terminal cancer."

"They will be very understanding, and they will kill him, but in a very quiet, dignified way." (You know, like in Logan's Run maybe even people cheering it on!)

And he says, *"They will spring up and the next thing you know they are going to be like factories where they can get rid of them in a minute."*

He says, *"This is where the abortion clinics have gone and now with adults it's going to be a terrible, dehumanizing, disgusting industry."*

And lest you think that will not ever happen, the infrastructure for this has already been laid out in the Obamacare Health Program with Death Panels to dictate when it is your time to die, just like in Logan's Run!

Bill Akins, Pasco County Republican Party: *"Here is the problem I have with the Affordable Healthcare Act. Number One there is a provision in there that anyone over the age of 74 has to go before what is effectively, a 'death panel.'* (the crowd boos in unbelief) *Yes, they do, yes, they do, it's in there, folks. You're wrong."*

Sarah Palin, Former GOP Vice-Presidential Nominee: *"Of course there are death panels in there. But the important thing to remember is that is just one aspect of this atrocious, unaffordable, cumbersome, burdensome, evil policy of Obama and that is Obamacare."*

Fox News: *"What do you think the motive is of coming on to your side, saying that maybe we should slow down with this law?"*

Sarah Palin: *"Well it's in black and white in the law, that there will be rationing of health care. They couldn't go forever without acknowledging that or we would look like complete buffoons. And they would be deemed incompetent to not having read the law to understand that death panels are a part of this atrocity."*

Barbara Walters: *"President Obama stood before the joint session of congress and said there is no such thing as a death panel. Is he a liar?"*

Sarah Palin: *"He is not lying in as that those two words will not be found in any of those thousands of pages of different variations of the health care bill. No "death panel" is not there. But he is incorrect, and he is disingenuous."*

RT News: *"Dr. Ezekiel Emanuel is one of the architects of the Affordable Care Act or Obamacare. He is presumed to be interested in preserving life which is why it seemed outrageous when Sarah Palin accused him of wanting to create 'death panels.' That was him, he was the death panel guy. But it turned out Palin might have been more right about Ezekiel's intentions than the country gave her credit for. Because Dr. Emanuel just penned an article in the Atlantic entitled 'Why I hope to die at 75.' With a sub-title, an argument that society and families and you, would be better off if nature takes it course swiftly and promptly.*

In it, Emanuel clearly outlines why he thinks everyone should have the decency to die before turning 76. He accuses Americans of being obsessed with trying to cheat death. He chastises people for exercising. For doing mental puzzles. For sticking to healthful diets. For taking supplements. As if those actions are actually evil. He said all those death cheating behaviors are part of a manic cultural sight that he calls the American Immortal which he rejects as misguided and destructive, as in if you spend medical and financial resources trying to extend the life of a 76-year-old, you are being destructive to society. Sounds pretty death panel there."[6]

Yeah, no kidding! And it is already built into our Healthcare System, that the government for some reason, just had to take control of. Like many of the other countries around the world already have! Gee, I wonder why? And not so surprisingly, the global elitists and population control people admit to these death panels and even approve of them, like Bill Gates does. Look at who his Dad worked for.

Bill Gates: *"The access that used to be available to the middle class or whatever is just rapidly going away. That is what society is making because of very high medical costs. In a lack of willingness to say, is spending a million dollars for that last three months life for that patient, would it better to not lay off those 10 teachers and take care of them with that medical cost. It's called a 'death panel' and you're not supposed to have that discussion."*

Bill Moyers: *"When did you come to these reproductive issues, as an intellectual?"*

Bill Gates: *"When I was growing up my parents were always involved in various volunteer things. My dad was the head of Planned Parenthood."*[7]

Well, there it is! You wonder why Bill Gates is a population control elite, and he is always pushing harmful vaccines and birth control? He got it from his dad! From the death culture called Planned Parenthood and he is just carrying out the next computerized AI version of this around the world! But as he says, *"You're not supposed to have that discussion!"* In other words, you are not supposed to know that these elitists really are building a "Logan's Run" scenario for the whole planet where your death will be decided by some outside entity! And who is that outside entity and who has the capability to run this whole back-end system, around the whole planet, monitoring all health and mental health and age-related issues? AI!

"Google's 'Medical Brain' Thinks It Can Predict When You'll Die." Well gee, I hope it gets it right!

"Google uses AI to predict WHEN you will die and it's 95% accurate."

Well, that's close enough, isn't it? And talk about having no clue what they're doing…

"AI can predict if you'll die soon – but we've no idea how it works."

Well, that is comforting! And then pretty soon you'll get one of these!

"Genetic Report Cards Will Be the Future of Predicting Disease Risk."

And as one family said when they got an AI score for their health, *"At first we were desperate, but when AI gave an unexpectedly high score, we were thrilled."*

And that's leading to even this!

"AI Is Good (Perhaps Too Good) at Predicting Who Will Die Prematurely."

What? Who gave you that right? Who do you think you are…God?

And finally, we now have calls for this to tie it all together!

"Tony Blair calls for a new 'digital ID' so people can prove their coronavirus 'disease status' alongside testing and tracing programs as the world eases out of lockdown."

You know, can't let a good crisis go to waste! We can mandate a "Digital ID", maybe even a microchip, because we don't want people to lose it, so everyone can be tied into this global monitored medical system! Is this nuts, or what?! It is almost like it was planned out a long time ago, even with plandemics, to get us to submit to it!

But stir all this together and this is what you get. AI really is creating the *Logan's Run* scenario for all of us, that these elitists are creating for us and their so-called Utopian Society, that they are going to control! They never could pull it off, until now! AI gives them the ability to do it on a global basis for the first time in history. In fact, speaking of history and the *Logan's Run* again, for years these same elitists have also been advocating the need to microchip people at birth to tie them into this Utopian system they're building, like Roger Ebert shared!

Roger Ebert: *"Right now my mind is telling my body to talk, and it is supplying my mouth with the words I am giving to you. Then the mind could also learn to say to the chip, go to a certain webpage, download the information, and supply it the inside of my eyeballs."*

Bill O'Reilly: *"You believe that you will actually be able to tell your mind, through a chip, to provide you with information, that you can then speak. So, on this program, it would be really great for me. Put a chip in there and if I didn't know something I could then say, do I have to say it out loud or just think it?"*

Roger Ebert: *"Either way. Right now, we are loaded up with bionic stuff already. I am wearing glasses, I have fillings in my teeth, this is a wristwatch, I am wearing clothing which allows me to adjust to the climate as I go outside. So, it's obviously only a matter of time before conversions allow us to match the number one tool of the next century and that would be the computer chip."*

Bill O'Reilly: *"Okay, but all of those are external. This would be in you, and that is what frightens me."*

Roger Ebert: *"It could be in you or anywhere you wanted it to be. It would allow you to make telephone calls, without having a cell phone, allow you to search the web without having a monitor."*

Bill O'Reilly: *"So if I had a chip embedded in my head, I could make a telephone call by saying, 'Call mom?'"*

Roger Ebert: *"Yeah, you would just say, 'Hello, mom.' You would have to have some body language, so people would know what you were doing.*

Let's say you are standing at the American Airlines terminal and you are calling the advantage desk and saying that you have missed your flight, do they have any seats at 10 O'clock. As you are standing there saying that someone is going to look at you, thinking you are crazy."

Bill O'Reilly: *"Not if everybody has the chip. They would know what is going on."*

Roger Ebert: *"You would just do this.* (he holds his hand up to his hear forming an imaginary phone) *This would mean I'm not talking to myself I am talking on the phone. If you are reading the web on the inside of your eyeballs, it would seem rude to people, like I am staring straight through you. But I'm not staring straight through you, I'm watching FOX News web page, right? So how would I want people to know this I would just go like this."* (Again, he holds up his hand and forms a C with his fingers in front of his face).

Bill O'Reilly: *"This sounds so unbelievable."*

Roger Ebert: *"If I am downloading information."* (he holds up two fingers like horns on each side of his head).

Bill O'Reilly: *"But it's true, it could happen?"*

Roger Ebert: *"Things are converging so quickly, and computers are already an amazing tool."*

Bill O'Reilly: *"But we checked it out a little further than your article and scientists tell us that it is absolutely true. And absolutely possible, for a human being to have a chip embedded in their system and have many, many things appear in their mind upon command. So, you would say, I need to know the name of the 18th President, and bing!"*

Roger Ebert: *"Not only that, but when I meet you, I call up your credit report while we are talking. Eventually, the time will come when a child is implanted with a little chip back here and will learn how to control the* *chip the same way you control your bodily functions and your voice and your movements."*[8]

Including giving access to an outside global system that is monitoring everything about you including your health, location, finances, age, you know, the Mark of the Beast system, just in time for the 7-year Tribulation! In fact, they apparently even have plans for skipping the whole doctor's office or emergency hospital room or Thanatorium scenario to take you out. Now with these same microchips at birth, they can kill you on the spot!

FOX News Reports: *"This next story may sound like something out of a Hollywood thriller. A Saudi inventor has created a Killer microchip. It's designed to track terrorists and criminals and, well, you can think of somebody. Not only does it include a GPS device it also has a lethal dose of cyanide, which can be activated at any time."*[9]

And anywhere in the world! Can you believe this? *Logan's Run* is about to become our reality! They're putting all the pieces into play! First, they create the death culture with the right mindset to allow this murderous wicked behavior to take place. Then they create the microchip computer company (Bill Gates & others) and develop AI technology to run it, control it, and monitor the whole system around the whole world. And now for the first time in man's history, **Revelation 13**, one man, a murderous Antichrist man, can now literally dictate people's deaths anywhere in the world if people do not do what he says to do. Why? Because you allowed AI to take over the whole medical system, just in time for the horrible Antichrist slaughter in the 7-year Tribulation!

Folks, how much more proof do we need? The AI invasion has already begun, and it is a huge sign that we're living in the last days! And that is precisely why, out of love, God has given us this update on *The Final Countdown: Tribulation Rising* concerning the AI Invasion to show us that the Tribulation is near, and the 2nd Coming of Jesus Christ is rapidly approaching. And that is why Jesus Himself said:

Luke 21:28 "When these things begin to take place, stand up and lift up your heads, because your redemption is drawing near."

People of God, like it or not, we are headed for The Final Countdown. The signs of the 7-year Tribulation are Rising! Wake up! The point is this. If you are a Christian and you're not doing anything for the Lord, shame on you! Get busy doing something for Jesus now! Stop wasting your life! We need you! Don't sit on the sidelines! Get on the front lines and help us! Let's get busy working together doing something splendid for Jesus with what time is left and get busy saving souls! Amen?

But if you are not a Christian, then I beg you, please, heed these signs, heed these warnings, give your life to Jesus now! Because this AI technology is not going to lead to a life of wonderful dreams and a modern-day utopia, but a nightmare beyond your wildest imagination in the 7-year Tribulation! Do not go there! Get saved now through Jesus! Amen?

Chapter Twenty

The Future of
Big Brother with AI

The **8ᵗʰ area** AI is making an invasion into is in **Transportation**. You know, in case you try to flee or run from the Antichrist or the False Prophet when they come to try to kill you. You ain't going nowhere! You better get saved today and avoid the whole thing! Jesus is the only way out of this horrible nightmare that is coming to the planet! But don't take my word for it. Let's read what God says.

Matthew 24:15-22 "So when you see standing in the holy place the abomination that causes desolation, spoken of through the prophet Daniel – let the reader understand – then let those who are in Judea flee to the mountains. Let no one on the roof of his house go down to take anything out of the house. Let no one in the field go back to get his cloak. How dreadful it will be in those days for pregnant women and nursing mothers! Pray that your flight will not take place in winter or on the Sabbath. For then there will be great distress, unequaled from the beginning of the world until now –and never to be equaled again. If those days had not been cut short, no one would survive, but for the sake of the elect those days will be shortened."

So, as we have seen in this text several times before, Jesus clearly says, in the 7-year Tribulation, at the halfway point, when the Antichrist shows his true colors and goes into the rebuilt Jewish temple to declare himself to be god, the abomination of desolation spoken of by the Prophet Daniel, that the only option for these people at this time is to what? To flee, right? To get out of there now in quick flight, right? Chop, chop! No delay! Why? Because again as we saw before, **Zechariah 13** says it is going to be a horrible time of slaughter for the Jewish people, two-thirds of them are going to die at the hands of the Antichrist! And then in **Matthew 24** just prior to these verses, we read where Jesus said that anyone who turns to God during this time will be hunted down and killed, anywhere in the world. Last time we read how the Antichrist will use AI to determine people's deaths, not to mention the plethora of other ways he could use AI to kill people in the 7-year Tribulation. But even here in our opening text, I think we have another way he can also do it with AI. And that is in the area of **Transportation**.

Think about it. Jesus said to the people, in the 7-year Tribulation, that their only option at this time to avoid being killed by the Antichrist's murderous spree is to what? Flee to the mountains, right? You know, get in your car, truck, plane, subway, bus, whatever, and just run, right? Now, don't think the Antichrist hasn't thought of this. Don't you think he has read this verse before? I think so! In fact, I think he is laying the groundwork right now for AI to make sure no one will be able to escape his grasp when it comes time to flee.

The **1ˢᵗ way** he is doing that is by having AI control all the **Personal Transportation**. And once again, just like in all the other areas we have been seeing AI taking over, it is a step-by-step approach getting us slowly conditioned to having AI run the whole system including Transportation. Starting with AI apps!

AI for Personal Transportation

Ada – An AI chatbot that helps you navigate and make decisions.
Emma – An AI that automatically calculates and adds meeting travel time.

ETA – An AI that helps you manage travel itineraries and meetings.
Five AI – An AI autonomous vehicles system that can work anywhere.
HelloGbye – An AI that books complex trips with simple speech.
Mezi – An AI that helps with booking flights, hotels, restaurant reservations and more.
Nexar – An AI dash cam app that helps you drive safer.

You know, AI is always watching wherever you go! But that is not good enough. You have got to have total control. So, the app was just step one. The next phase is here as well, with the driverless cars! You know, if AI is going to tell you what to do when you drive the car, and we are used to that, then why not go all the way, and just have AI drive the car for you, period! Well, it is! In a multitude of ways! First, believe it or not, self-driving cars have been around for a long time, even as far back as the 1920's with radio remote controlled vehicles. But even more recently they really began to take off with Elon Musk and his company TESLA,

with no radio waves, but AI controlling the whole system. And we are familiar with that. But what most people do not realize is, that's just one of a plethora of companies out there, all around the world, that is manufacturing self-driving vehicles on a massive scale! A whole revolution is taking place in transportation with AI controlling it all! Tesla cars aren't so cheap but now that so many manufacturers are getting in on the action, the prices are being slashed and are much more economical. And here's just one of them from Google, of all entities, called Waymo, which stands for "A New Way of Moving Forward! In other words, AI will move you forward in your car now!

Narrator: *"Say hello to Waymo, the world's most experienced driver. The Waymo driver is what we call our self-driving technology. It has over a decade of real-world experience and has driven millions of miles. With Waymo One, we can make it safer and easier for you to get around. So, you can spend more time doing what you love. And with Waymo Via the same driver with the same experience can also deliver your packages or save you a trip to the dry-cleaners, or if you run a business, Waymo Via can help you transport whatever you need. Making sure shipments arrive right when they are supposed to. This is the Waymo Driver. A driver that is reimagining transportation for all of us."[1]*

Whether you want it or not, AI is controlling it all! And notice it is going global. But notice you can use the app with it, we're already used to that, and you can use Waymo not just to drive you around but for deliveries and trucks, you name it! In fact, speaking of trucks, for those of you who think you can still "hitch a ride with a truck, you know, with your thumb" and flee like in the old days, you better think again! As you saw there, trucks are being automated as well! In fact, here is one in Texas, not coming, but already here whether you realized it or not!

NPRC News, Andy Cerota reports: *"It's no secret. Houston's highways are choked with traffic, but here's something you may not know. The Semi in the lane next to you may not have a driver behind the wheel.*

Dominique Sachse reports: *"Yes reporter Owen Conflenti explains how the future of the trucking industry is now.*

Owen: *"The future of trucking is now. As autonomous or should I say driverless semi-trucks rolling down Texas highways every day. One of the companies leading the way says the goal is fuel efficiency and safer roads."*

Kodiak Robotics, Inc.: *"Autonomous trucks don't text, don't drive drowsy, or get distracted. They are always focused on the road and they always prioritize safety first. One of the first things I did when I joined this company was to talk to the engineering team and make sure while on the engineering road map, we had a way to honk the horn kids or other motorists maybe sending us the signal asking us to do so."*[2]

Yeah! And that is awesome because the human drivers don't do that anymore! But did you see, AI driving trucks, not just cars, are already here as well, whether you realized it or not. And this is happening all over the world, all at the same time! In fact, speaking of trucks honking the horn at you, I sure hope this driverless truck reality doesn't turn out like this movie, remember this one?

A clip from Spielberg's movie "Duel":

The semi-truck is speeding down the road towards a man. He must jump onto a police car to get out of the way. The truck does not even slow down, and the man is covered with dust from the road. As the truck is chasing Dennis Weaver in his car, it runs through several barriers, not slowing down at all. He is out to get that man in the red car. The car goes as fast as it will go but cannot seem to get away from that truck.

As he turns a curve the car gets out of control and comes to a stop. He gets out and sees there are children playing nearby. He immediately goes to take the kids to safety. He tries to get them back into a bus. They are crying, they are scared, they don't want to get into the bus. But while that is wasting precious time, the truck is speeding towards them. He manages to get them in the bus just as the truck turns the corner and starts chasing him on foot.[3]

Gee, that wouldn't happen, would it? We were warned about this almost 50 years ago. Duel was made in 1971. Spielberg's second movie. Once again, what does he and Hollywood know that we don't know? What are they preparing us for? But come on, these AI vehicles won't chase us down, will they?

Yeah, we will get to that in a little bit! But basically, every kind of car and truck manufacturer on the planet is making the switch to driverless vehicles! Tesla started it in 2015 but all the Big Guns are getting in on it as well, in just these few short years! Ford, GM, Hyundai, Toyota, Honda, Volkswagen, you name it! In fact, right now forty companies are working on autonomous vehicles! And they are spending big bucks to make sure it happens!

"Self-Driving car research has cost $16 Billion."

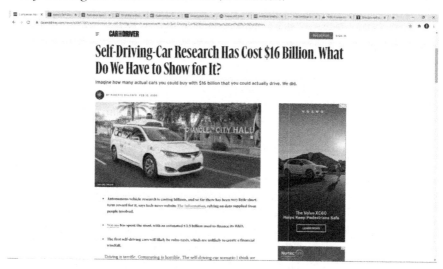

And that's just cars! And they are telling us that all vehicles are going this new automated way, including race cars!

They are now having AI Roboraces even for Formula 1 cars! Oh, and by the way, for those stubborn people out there who still won't receive Jesus as their Savior and still think they can flee in something like a golf cart…

© Oded Yechiel

Yeah, don't worry, they're doing that too! That one is called IVO "Intelligent Vehicle Operator." But call it what you will, as you can see, AI is going to be controlling all the Personal Transportation even down to a golf cart! So have fun trying to flee, thinking you're going to escape the

Antichrist's murderous grasp! You need to get saved now and avoid the whole thing!

The **2nd way** the Antichrist is preventing people's escape, in the 7-year Tribulation, is by having AI control all the **Public Transportation**. You see, maybe you don't have a vehicle, let alone a driverless vehicle. Or maybe you refuse to get into one of these driverless vehicles because of what you just saw. So, your only other option to flee in the 7-year Tribulation, is what? Public transportation, right? You know, a bus, taxi, subway, train? And it's a good thing AI's not taking that over! Yeah, right! Are you kidding me? Once again, it is a step-by-step process getting us used to having AI control all forms of transportation including public transportation, starting with apps again!

AI for Public Transportation

Auro Robotics – AI self-driving shuttle for campuses and corporate parks.
Dispatch – An AI fleet of autonomous vehicles for pedestrian spaces.
Mobileye – AI self-driving vehicles for industry.
Nauto – An AI system that gives your fleet self-driving capabilities.
nuTonomy – An AI software for driverless fleets.
Shield – An AI autonomous unmanned system for civil and defense.
Uber – An AI autonomous vehicle program by Uber.

And who hasn't used that before? We're all used to that, right? Well, believe it or not, these same types of companies, that are app based, to hail us a personal driver or taxi, are now switching to personless drivers and taxis!

A passenger is getting into a white car, and the narrator speaks:

Narrator: *"This is an Uber, but it's not driven by some guy trying to make some extra money off his own car. This Uber drives itself. The ride-sharing giant came to Pittsburgh for its latest invention, the first autonomous taxi service. Select Uber users can now ride in self-driving cars. Inside riders are greeted by a tablet to show them how the car works* *and how it sees the world. Uber plans to introduce a fleet of autonomous Volvo SUV's early next year but are using Ford Fusions for the time being. Over bridges, down narrow streets, around jaywalkers, the car handles just about everything. The city is happy to host a robo revolution. To sit back and let someone else or something else take care of the driving and no small talk required."[4]*

Wow! That's a bonus, huh? You know, those human drivers always expect you to say something and it is so awkward, but not with AI! You don't have to worry about that at all! In fact, as they said, it truly is a Robo Revolution in public transportation! Whether you realize it or not, another driverless taxi company is already on the roads here in Las Vegas.

It's called Aptiv *"Aptiv's Self-Driving Vehicles Tops 100,000 rides in Las Vegas."*

And that's the tip of the iceberg! There's Uber, Lyft, Aptiv, Robo-taxi, Zoox, Voyage, and a whole ton of other companies that started off with the app calling a human driver. Now removing the human driver.

Anybody seeing a pattern here? And again, all the big manufacturers are getting in on this from GM to BMW to Honda to Ford you name it! And taxis are just the beginning! They also want to take over buses, trolleys, subways and trains! In fact, they're calling it, "The Next Wave of Public Transit."

Have fun trying to flee now! In fact, they want to even get to the point where they will transport you around in these things. They're called **Smart Pods** as you can see here. And they will be responsible for all our

travel. Look at this video transcript.

"A couple just went on an app on their phone. They requested the Smart Pod (a Multimodal Mobility Platform Service), paid the amount due, (with an intelligent payment and billing solution) and a few minutes later a Smart Pod came to pick them up. The Smart Pod is small, but very convenient on the inside and safe. You can read, drink your coffee, play games, party or just cruise the sights of town. (It has hyper-individualized user experiences.) It is very comfortable, with customized seats from PUR-Foams. It has electric heating and cooling solutions and connectivity options. It also is working with Visionary Mobility Concepts to benefit the environment. When you are finished with your book, or your coffee, or sight-seeing, you can relax and take a nap until you get the end of your trip."[5]

Wow! Did you see how great that was? How much fun those people were having with an AI Smart Pod, life was great! Which by the way, again, "smart" means "Big Brother," keep that in mind. But as you saw with these so-called AI controlled Smart Pods, I can hang with my friends, not worry about driving, go back to making phone calls or texting, wouldn't that be nice, take a nap, even have it make coffee for me! And did you notice I can even hail it with an app on my phone and even pay for it with my phone?

I mean, the next thing you know they will have it narrowed down to a chip in my hand to make the payment to gain access to this total AI controlled transportation system! Yeah, you know that is where it's headed! They have it all figured out folks, you aren't going to go anywhere in the 7-year Tribulation! And once again, for you stubborn people out there who still think you can and who still won't receive Jesus as their Savior today, don't worry, this AI controlled transportation system really does cover all aspects of public transportation, not just PODS, or taxis, or buses or trolleys or subways. But even trains, as you can see here!

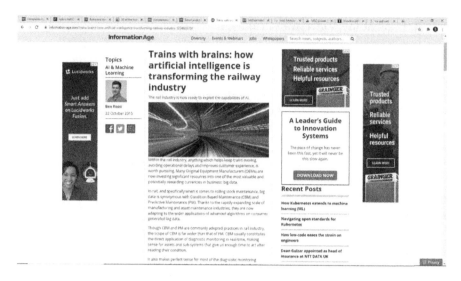

And even ships…

including cruise ships just in case you think you can flee that route!

Give it up, get saved now! Or even on a plane!

And with this one, as you can see that will include even military planes and we'll get into that in the next chapter! Wait until you see that!

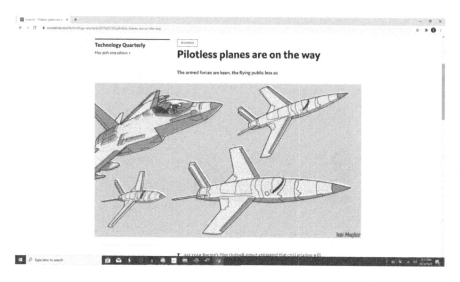

But as you can see, not just personal transportation but even public transportation is being controlled by AI on a global scale. So have fun trying to flee, thinking you're going to escape the Antichrist's murderous grasp! It's not going to happen! Again, get saved now and avoid the whole thing! In fact, I think Hollywood has also prepared us for this AI takeover of the transportation system for a long time. Remember this movie?

Clip from Total Recall Johnny Cab

As the cab is going down the street, we hear whistling. It wakes up the passenger in the back seat. The passenger is Arnold Schwarzenegger. He wakes up to see a robot driving the car.

Arnold: *"Where am I?"*

Cab Drive turns around to face him and says: *"You're in a Johnny Cab."*

Arnold: *"What am I doing here?"*

Cab Driver: *"I'm sorry, would you please rephrase the question?"*

Arnold: *"How did I get in this taxi?"*

Cab Driver: *"The door opened, you got in."*

NOW TRY TO FLEE!

Arnold is running out of the lobby and gets into the cab. (In another scene)

Cab Driver: *"Hello, I'm Johnny Cab. Where can I take you tonight?"*

Men are running to the cab to get to Arnold before it starts moving.

Arnold sees them coming after him and yells to the driver: *"Drive, Drive!!!"*

Cab Driver: *"Would you please repeat the destination."*

Arnold: *"Go anywhere, just go, go!!!"*

Cab Driver: *"Please state a street and number. I'm not familiar with that address."*

YOU AIN'T GOING NOWHERE or MAYBE WE'LL JUST TAKE YOU OUT

The cab takes off down the street with guns blazing from the men with guns. The robot cab driver is laying down in his seat.

Cab Driver: *"Fasten your seat belts."*

He runs the cab into a cement wall, and it blows up in flames.

Cab Driver: *"I hope you enjoyed the ride."*[6]

Yeah, real funny! Enjoy that ride in the 7-year Tribulation, you're not going anywhere! AI will prevent you from fleeing or take you out when you try to flee! "Total Recall" was made back in 1990, over 30 years ago! Again, what does Hollywood know that we don't know? And what kind of AI controlled transportation system are they trying to prepare us for? And, if this whole AI scenario seems crazy to you, with AI being used to prevent people from fleeing, or killing them, when they do try to flee, like in that movie, during the 7-year Tribulation with the Antichrist, believe it or not, they have already discussed this very issue for many years now, about AI deciding who gets to live or die. And, like it, or not, they say that is what is coming!

Bloomberg Reports: *"Keith, this is an incredibly scary thought. Explain what is involved. How do scientists even think they can teach a robot how to make an ethical decision to tell my car how to swerve off the road and down a cliff instead of colliding with a school bus full of children?"*

Keith Naughton: *"Yes this is an issue that is really bedeviling the auto industry. Google is also developing their driverless car. They figured out the technology. They know how to allow cars to drive for us, but they can't really figure out these philosophical questions like the one you just described. And that is something that they are now working on, literally, with philosophers and ethicians to try and determine what you do when a crash is unavoidable, and your car has to decide between the lesser of two evils, like missing a bus full of children or running upon a sidewalk and hitting an individual. They don't have all the answers yet."*

Buffett: *"There are some interesting questions. Let's just say you have a self-driving car and it's going down the street and a three-year-old kid runs out in front of the car. And there is another car coming in the other direction with four people in it. The computer is going to make the decision as to whether to hit the kid or the other car. I'm not sure who gets sued under those conditions. You're going to kill somebody and it's the computer that makes the decision, in a nano-second. It will be interesting to know who programed that computer and what their thoughts are on the value of the people's lives."*[7]

Yeah, especially if his name is the Antichrist! And folks, as shocking as that is, they are getting blunter about who to kill with AI!

"Should the Self-Driving Car kill the baby or the grandma?"

Or should we say, in the very near future, should AI kill anyone with this AI controlled transportation system, around the whole world, that the Antichrist wants to kill! In fact, they are calling this so-called dilemma, the IoT Factor where,

"AI cars are by nature connected cars – they're part of the IoT, the Internet of things."

"Here's the thing: If driverless cars could communicate with every other car on the road, it seems like there would be no problem. One car could tell another to override a sleeping driver and stop."

Notice they are controlling the whole thing! Then they go on to say, *"However, the IoT is vulnerable to attacks, which means, cars could be weaponized en masse."*

In other words, somebody could tap into the whole transportation system anywhere around the world and hijack or control a car or vehicle, whatever, and use it as a weapon to take somebody out! That is exactly what that means! You can call it IoT Factor, but it is really the Antichrist facilitator that allows him to do his dirty deeds in the 7-year Tribulation. Why? Because you let AI take over the transportation system! But that's not all.

The **9th area** AI is making an invasion into is in **Criminal Investigation**. Let's go back to our opening text.

Matthew 24:15-22 "So when you see standing in the holy place the abomination that causes desolation, spoken of through the prophet Daniel – let the reader understand – then let those who are in Judea flee to the mountains. Let no one on the roof of his house go down to take anything out of the house. Let no one in the field go back to get his cloak. How dreadful it will be in those days for pregnant women and nursing mothers! Pray that your flight will not take place in winter or on the Sabbath. For then there will be great distress, unequaled from the beginning of the world until now –and never to be equaled again. If those days had not been cut short, no one would survive, but for the sake of the elect those days will be shortened."

So again, we not only see in this opening text that Jesus tells the people in the 7-year Tribulation what to do, i.e., flee to the mountains, but he also tells them what not to do! And He said whatever you do, do not "flee" to where? Your house, right? Why? Because as we have seen several times before, the Antichrist will use AI to turn your home into a Big Brother paradise as well. He will know every conversation, every word and deed, you name it, which means he will know when you're planning on fleeing, you won't go anywhere, he's got everything and every conversation on tape, like a rat in a cage. Remember that? He will know the split second with all this AI big brother technology, if you have the audacity to try to disobey his commands and commit the crime of trying to flee! And believe it or not, now with the help of AI, he may have even got it figured out how he can know, not just when you do it, but before you even do it! It sounds crazy but it is called pre-crime or pre-criminal investigations just like in the movie, again, "Minority Report." How many times have we seen that movie predict our future in a multitude of ways? But let's take a look at the pre-crime scene in that movie.

Clip from the Minority Report

Tom Cruise is in Washington D.C. walking through the mall. The year is 2054. He hands off his jacket to a lady that is pregnant.

Tom: *"Any contractions?"*

Lady: *"Only the ones you give me."*

As he is walking into his office, a face of a person is shown on a screen and then a ball rolls down a tube.

Tom: *"Hey, Chad what's coming."*

Chad: *"Red ball, double homicide. One male, one female. Killer is white male, forties. Estimated time of incident 8:04 am. The twins are a little fuzzy on that so we will need some confirmation. Location is still uncertain. Remote witnesses are plugged in. It will be case #1108."*

As Tom goes on the computer, he calls over the speaker: *"Case #1108, previsualized by the precogs recorded on hollow sphere by Pre-crime skew stacks. My fellow witnesses for case number #1108, are Dr. Catherine James the Chief Justice Frank Pollard, good morning."*

They respond: *"Good Morning."*

Tom: *"Did the witnesses preview and validate #1108 at this time?"*

They respond: *"Affirmative, go get them."*

Tom: *"Standby time, murder 8:04am."*

Computer: *"That is 24 minutes 13 seconds from now. "*

He stands in front of a large screen and raises his arms as if conducting an orchestra. The screen comes on, and he could see the murder that is going to take place. As he is watching the scene, then he asks out loud.

Tom: *"Where are you? "*

He moves the images around and finds mail on the table. As he focuses in on the address, he finds where the murder is going to take place.

Tom: *"Can you grab that? "*

Chad: *"I show Howard Marks in the district going by resume and license registration. We got them in a fox hole at 4421 Gains. "*

Tom: *"Blue and white set up a perimeter and tell them we are in route. "*

He announces to the police. They fly there in a helicopter and drop down in the playground on ropes. When they get to the home where the murder is to take place, they drop down through a glass ceiling. When they drop down through the ceiling, the wife and her boyfriend are sitting on the bed. She screams but they are still alive. They made it in time to stop the crime. The man with the knife that is going to do the killing is wrestled to the ground.

Tom: *"Look at me, look at me!"*

His ID is checked, and he is the intended murderer.

Tom: *"Positive, Howard Marks. Mr. Marks, by the mandate of the District of Columbia, Pre-Crime Division, I am placing you under arrest for the future murder of Sarah Marks and Donald Jubineau, today, the 22nd at 08:00 hours. "*

Mr. Marks: *"I didn't do anything. Sarah."* He calls to his wife to save him, the wife he was going to murder.

Tom: *"Give the man his head."*

A man pulls out what looks like a headset to be placed on Mr. Marks head.

Mr. Marks: *"Don't put that halo on me. Sarah, help me."*

Police: *"Put your hands on your head."* And they walk out the door.

The next scene is back at the laboratory.

Businessman: *"Let's not kid ourselves. We are arresting people that have broken no laws."*

Second man: *"But they will. The commission of the crime itself is absolute metaphysics, the precognizant, the future and they are never wrong."*[8]

Oh yeah, computers are never wrong, remember that movie? Remember that scene? Folks, as crazy as this sounds, I am here to tell you, now with AI, that science fiction movie's Pre-Crime Scene is about to become our reality! And once again, it is a slow step-by-step approach getting us conditioned to even this kind of AI Pre-Crime scenario.

The **1st way** they are doing that is by having **AI Control the Police**. And not just controlling the administration and overall back-end police operations in general, databases and things like that. That has been going on for a while now. But another big part of the Police Force is the camera monitoring system that they have installed all over the world, including the U.S., to monitor people in restaurants, parks, businesses, homes, streets, you name it. The giant CCTV monitoring system that we saw before in our Modern Technology study in the Big Brother section. Just think about it. That is a lot of cameras around the world to monitor, way too many for humans to handle. So, guess who they are looking at to run the whole system? That's right, AI to the rescue. AI can not only handle all those cameras all around the world, but AI can now even predict with those cameras when a person is going to do something bad! Don't believe me? It is already happening!

"Threat of Mass Shootings Gives Rise to Artificial Intelligence-Powered Cameras."

"Paul Hildreth peered at a display of dozens of images from security cameras surveying his Atlanta school district and settled on one showing a woman in a bright yellow shirt walking a hallway."

"The Artificial Intelligence-equipped system then found other images of the woman, and it immediately stitched them into a video narrative of where she was currently, where she had been and where she was going."

"AI is transforming surveillance cameras from passive sentries into active observers that can identify people, suspicious behavior and guns, amassing large amounts of data that help them learn over time to recognize mannerisms, gait and dress."

"If the cameras have a previously captured image of someone who is banned from a building, the system can immediately alert officials if the person returns."

Have fun fleeing that system! And it's not just for schools.

"Police, retailers, stadiums and Fortune 500 companies are also using Artificial Intelligent video camera systems."

"What we're really looking for are those things that help us to identify things either before they occur or maybe right as they occur so that we can react a little faster," Hildreth said.

And they even admit, *"It's almost kind of scary." "It will look at the expressions on people's faces and their mannerisms and be able to tell if they look violent."* OR are trying to flee!

"Retailers can spot shoplifters in real-time and alert security or warn of a potential shoplifter." Notice they haven't done it yet!

And it concludes with, *"It's unknown how many schools have AI-equipped cameras because it's not being tracked."*

In other words, you are never going to know how many are really out there. And they justify it by saying it is not only going to identify threats in schools but to predict all kinds of things even beyond schools! Whether or not somebody is carrying a gun, is lying when they cross the border, is lying period, including even lying in what you write and on and on! In fact, if you've been paying attention, slowly but surely, they have warmed us up to the idea of AI not just taking over the administrative operations and monitoring cameras of the police force, but even the policemen themselves!

RT Reports: *"Remember that movie from the 80's, 'Robocop'? The one with the deadly but kind of goofy robot cop who fought crime in sort of a dystopian future, right? Well, it might soon be more of a reality than it* *would be fiction. One Police Force is now using technology to help fight crime. RT correspondent Natasha Sweet is in Huntington Park, California, with the details."*

Natasha Sweet: *"This HP Robocop here in Huntington Park is utilizing 360 degrees video cameras and other technology to help keep an eye on the neighborhood.*

HP Robocop records 360-degree video with high-definition cameras and has night vision. It can read license plates and alert officers that the particular vehicle is nearby."

Cosme Lozano, Chief of Police, Huntington Park: *"It also has the capability to capture cell phone or MAC addresses or IP addresses. So, it can also tell us what cell phones are in the area."*

Natasha Sweet: *"Officers can also live monitor the Robocop's video cameras."*

Robocop: *"Please help keep the park clean."* It tells some kids sitting on the park bench.

Natasha Sweet: *"The HP Robocop is an autonomous data machine surveilling Huntington Park's largest recreation area 24 hours a day 7 days a week."*

Reporter: *"Back in May, the Dubai police got some new recruits, and these weren't your ordinary newcomers. These guys were made of the hard stuff."*

Robocop in Dubai: *"I am a humanoid service robot."*

Reporter: *"Planned to be put in all of Dubai's neighborhoods, the world's first smart police station will be completely unstaffed. Citizens can pop in for a safe driving lesson, a quick coffee or even to report crimes. They can also meet Dubai's own Robocop."*

Robocop in Dubai: *"I am the latest in cooperation with Dubai's police department."*

Reporter: *"Like so much of Dubai's over the top ambition the police force wants to be seen as using the latest crime prediction and surveillance technology to watch over the people."*

Bilal Juma Bilal Altayer, Command Centre Duty Officer, Dubai Police: *"We have our cameras, our drones, our robots, we are going to a different science fiction movie.*

Reporter: *"Artificial Intelligence based predictive crime systems, autonomous patrol vehicles and unmanned police stations are just a few of their futuristic initiatives."*[9]

In other words, it is just the beginning! But did you notice they are not just replacing humans in the police force? Why? Well, you know, just in case a human hesitates when you need to apprehend that criminal for fleeing! But did you notice they even said it was for crime prediction just like in a science fiction movie? Gee, I wonder what movie that was? Rhymes with "Minority Report" for those of you who have not figured it out yet! But speaking of crime prediction, that is exactly where it is headed.

The **2ⁿᵈ way** we are going to have AI predict our crimes is by having **AI Control the Crime Investigations**. You know, it can apprehend that person that it predicted was going to do something bad! In fact, that is the actual term they're using to describe this use of AI technology, "Predictive Policing," Where, as you saw, AI taps into all the Big Brother information they've been gathering on us from computers, cell phones, video cameras, in home and out in public, satellites, social media posts, data records, financial records, purchasing records, you name it. And AI will scour that 24 hours a day 7 days a week and inform the cops who is the bad guy. And first we're giving them these **AI Smart Glasses** as you can see here.

They *"Are equipped with built-in cameras connected to AI facial recognition software to identify individuals in a crowd."*

"And once the face is identified from within a given database, SmartGlass projects the results on the wearer's screen. The entire process happens in real-time as officers simply observe the faces of people within the immediate vicinity."

It can also, *"Be capable of both speech and image recognition and stream its camera feed continuously so that the video feed can be analyzed against criminal databases or known threats."*

You know, in case that is the guy you're looking for who's trying to flee. But now they are moving to have AI find you anywhere you try to go whether the police are around or not or have their Smart Glasses on! And that's by having AI tap into the existing camera systems everywhere! Just like in the "Minority Report" movie! Shocker! You know where Tom Cruise was the Pre-Crime cop but that same system said he was going to be the next one to commit a crime and so he was trying to flee!

A Clip from the "Minority Report" Facial Recognition
A woman's voice is doing a commercial as Tom Cruise walks through the mall. "A road diverges in the desert, Lexus, the road you're on Tom Atherton." As he keeps walking voices come out of each display, speaking to him personally, because they recognize who he is. He keeps walking in

the crowd but doesn't act like knows who this Tom Atherton is. Another voice calls out, "Tom Atherton, you could use a Guinness." He keeps walking and then a display comes on saying that he has been a member of that organization since 2037." Another voice comes on and says, "Forget your problems, Tom Atherton."

While he is walking in the mall, his face comes on the Pre-Crime screen, where he used to work. They are tracking him. Chad, that used to be his partner says, "He was at Metro, making two stops at 20th and 33rd." His boss instructs Chad to send units to that location.[10]

Yeah, go pick up that bad guy who's going to commit a crime. You can run but you cannot hide! Good thing that system is never going to be put into place! Are you kidding me? It already has!

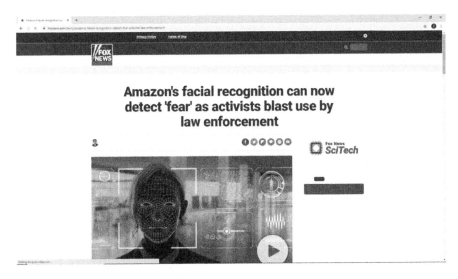

Amazon's facial recognition can now detect 'fear' as activists blast use by law enforcement

"Amazon's facial recognition can now detect 'fear' as activists blast use by law enforcement."

"Amazon can tell when you're afraid. CEO Jeff Bezos announced that its facial recognition software called Rekognition has added a new emotion

that it can detect – fear – in addition to happy, sad, angry, surprised, disgusted, calm and confused."

"The company also wrote that it has improved the accuracy of gender identification as well as age range estimation."

"The controversial facial detection software, which falls under the auspices of the cloud computing division known as Amazon Web Services," (That is their AI Service as we saw before)

"It has drawn condemnation from privacy and digital rights activists, and lawmakers who object to Amazon's marketing of their Facial Rekognition system to Police Departments, Government Agencies, Immigration and Customs Enforcement." (You know just in case you want to FLEE across the border!)

And they even go on to admit, *"If you get falsely accused of an arrest, what happens?" "It could impact your ability to get employment or housing."*

Not to mention your ability to flee in a perilous time! And if you don't think that's going to happen, look at this.

"Now Artificial Intelligence Can Be Used to Predict Threats."

"Artificial Intelligence and machine learning can now perform predictive analysis that enables identification of individuals engaged in suspicious activities that point to various levels of violent attack threats."

"Today, administrators can use technology-based predictive security solutions to foretell mass shootings and other violent acts before they happen." (And how do they do that?)

"Artificial Intelligence can be programmed to effectively sort vast amounts of data from all over the internet, monitoring trillions of data points almost in real time."

In other words, it does it by tapping into all the Big Brother databases and surveillance technologies they already have on us! In fact, other countries are not only already putting it into action, but they're calling it for what it is!

'Real life Minority Report cops use AI to predict violent crime before it happens."

"British police want to use AI for highlighting who is at risk of becoming a criminal before they've actually committed any crime." "It sounds like a dystopian nightmare." (No! You think?)

Oh, and by the way that was 2 years ago! What else do you have out there we don't know about? Not only in England but the rest of the world? Including Los Angeles!

Predictive Policing:

Reporter: *"Police in Los Angeles are trying to predict the future."*

Police: *"We know what happened yesterday, but we need to know what will happen tomorrow and the next day."*

Reporter: *"And they are not alone. More and more departments are using data driven algorithms to forecast crime."*

Police: *"It's all about improving our accuracy and our effectiveness, our efficiency and our service to the community."*

Reporter: *"Proponents say it is helping to reduce crime, but others wonder at what cost."*

Police: *"Policing is changing in your city now and when you start asking the hard question of how it impacts civil rights, how does it impact privacy, how does it impact suspects, how does it impact officers doing their jobs?"*

Reporter: *"Say you're living in Los Angeles and just got out of prison. The LA PD could be keeping tabs on you with a program called "Operation Laser." It uses crime arrests and field data to determine where the crimes are likely to take place. And who will perpetrate them."*

Police Instructor: *"Laser will predict who will commit a crime before the crime is actually happening."*[11]

So, they admit that they already have a "Minority Report" predictive policing crime system, run by AI in L.A. I didn't know that did you? And if you think that is creepy and invasive straight out of a Hollywood movie. Look at this!

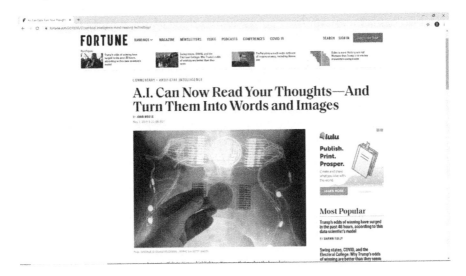

"AI Can Now Read Your Thoughts – And Turn Them into Words and Images."

You know, for those who get the Mark of the Beast or Brain Chip!

"Artificial Intelligence has already played an essential role in helping paraplegic's walk by interpreting the electrical signals emitted by the body," (i.e., their Brain with a chip.)

"Now that mind-reading technology is possible, AI brain implants will be next."

Why? Because they will even know when you are thinking about fleeing! Stir all this together, and you get another scene from that science fiction movie "Logan's Run." As we saw before, when people take a chip when they are born and blinks when it's time to die! But it is entitled, "Logan's Run" for a reason, because not everyone will want to do that! Since they took the chip, that same system will prevent them from running! Watch this transcription!

The clip begins with Michael York running through what looks like a large empty entryway. In his hand he has a machine that will tell if someone is in the room. He scans the room with this device, and it is putting off a noise that tells him that he is getting close to who he is tracking. He pulls out his gun and aims it at the location of the person. The person jumps out and he shoots. He misses and the person runs. Now the chase is on. The person trying to escape seems to be getting away but when he comes to an opening and is about to turn, he stops suddenly. There is another tracker just waiting to get a good shot at him. They seem to have him trapped. He turns to run into the opening.

Michael York: *"Run, Runner!" He laughs as he takes another shot as the second shooter does as well.*

The only way to get away is to jump and that is what he does. He jumps off the balcony several stories up. [12]

You can run but you cannot hide. Why? *Because you got a chip inside you* and it connects you to an AI system that not only told you when to die, but it knows your every move and it will know when you have the audacity to try to run or flee when you do not do what you're told! You know, like worship the Antichrist! No wonder it's going to be a blood bath during this time! With AI taking over the criminal investigation and policing system, one man will dictate people's deaths and prevent them from fleeing if they do not obey. Just in time for the 7-year Tribulation! But hey, good thing we don't see people wanting to get a microchip in their hand like Logan's Run! Yeah right, that is already happening as well!

Paul Rhys, Aljazeera Reports: *"The most cutting edge about Hannes Sjoblad, Biohacking Entrepreneur, isn't the phone in his hand, it is the microchip actually in his hand. The tiny implant is the latest advance in a bio-hacking technology that is steadily becoming part of normal life in Sweden."*

Hannes Sjoblad: *"We have created a new implant which is not a chip, it is a full device where you can add different lights, different vibration, different functions. Sweden is a very technical society. And I think this is the main explanation really, why a lot of Swedes are adopting chip implants."*

Paul Rhys, Aljazeera Reports: *"Swedes haven't been shy about upgrading themselves with the new version. Thousands already have microchip implants that they use in their daily lives. Waving their hands to gain entrance to the gym, confirm their ID or make payments. A short moment of pain, not putting them off, becoming part Swede, part machine. This is an implant party, simply where ordinary people can show up and get a microchip embedded under their skin. The bio-hacking movement in Sweden is hosting them all over Europe."*

Party Attendee: *"I think it is really cool. You don't have to carry your keys or anything. Just your body. Maybe in ten years everything will be in your hand."*

Paul Rhys, Aljazeera Reports: *"In Sweden, more than anywhere else, the future is already here. The National Train Company, SCA has around 2,600 people signed up to use microchips instead of train tickets. And no need to mind the generation gap. Eighteen-year-old Felicia and her father Magnus still bear the scars of their new implants. Student Hanna Irving is also freshly chipped and just needs to program it to open doors. Although importantly for a future career it does already connect to her LinkedIn."*

Hanna Irvine: *"People have been putting these chips into animals for 20 years, so I'm not worried about that."*

Paul Rhys, Aljazeera Reports: *"The long-term goal is for the new chips to provide medical care in remote communities. They are already getting under the skin of the Swedes and may soon become just another normal part of modern life and of the human body."[13]*

Wow! It even blinked like in "Logan's Run"! See how happy she is? Isn't this great? Folks, I'm telling you, the whole thing's a trap. Just like in "Logan's Run" and "Minority Report!" You not only allowed AI to take over the whole criminal investigation and policing system, but you even got a microchip in your hand that tied you into this AI controlled global monitoring system. So, now it knows where you are at anytime, anywhere in the world and it will even predict when you are trying to flee a future horrible event like the slaughter of the Antichrist, halfway into the 7-year Tribulation! This is not by chance and Hollywood's been preparing us for decades! They are following a script! A satanic one!

How much more proof do we need? The AI invasion has already begun, and it is a huge sign that we are living in the last days! And that is precisely why, out of love, God has given us this update on *The Final Countdown: Tribulation Rising* concerning the AI invasion to show us that the Tribulation is near, and the 2nd Coming of Jesus Christ is rapidly approaching. And that is why Jesus Himself said…

Luke 21:28 "When these things begin to take place, stand up and lift up your heads, because your redemption is drawing near."

People of God, like it or not, we are headed for The Final Countdown. The signs of the 7-year Tribulation are Rising! WAKE UP! And so, the point is this. If you are a Christian and you are not doing anything for the Lord, shame on you! Get busy doing something for Jesus now! Stop wasting your life! We need you! Don't sit on the sidelines! Get on the front lines and help us! Let's get busy working together doing something splendid for Jesus with what time is left and get busy saving souls! Amen?

But if you're not a Christian, then I beg you, please, heed these signs, heed these warnings, give your life to Jesus now! Because this AI technology is not going to lead to a life of wonderful dreams and a modern-day utopia, but a nightmare beyond your wildest imagination in the 7-year Tribulation! Do not go there! Get saved now through Jesus! Amen?

Chapter Twenty-One

The Future of
Military with AI

The **10ᵗʰ area** AI is making an invasion into is in the **Military**. You know, like the "Terminator" movies! That's right! It's not just big corporations and big governments around the world that are getting in on the AI action allowing it to invade virtually every sector of our society around the world. So are the militaries around the world. And not just our own military here in the U.S., but I'm telling you, literally all militaries around the world! In fact, just in the U.S. alone we are now spending, just in defense spending, "$4 billion in 2020" on AI technologies! And another $1 billion for non-defense efforts! Just in 2020! And as much as that is, we are starting to get outpaced by other countries, including China.

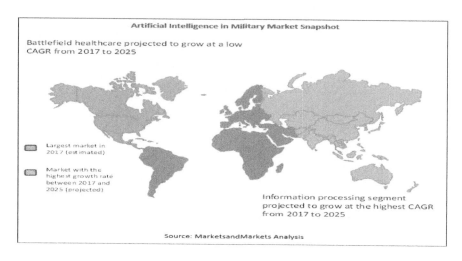

Artificial Intelligence in Military Market Snapshot

Battlefield healthcare projected to grow at a low CAGR from 2017 to 2025

Largest market in 2017 (estimated)

Market with the highest growth rate between 2017 and 2025 (projected)

Information processing segment projected to grow at the highest CAGR from 2017 to 2025

Source: MarketsandMarkets Analysis

As you can see with this chart, in 2017 we were the largest market for AI in the military, but by 2025 those in the yellow, China, Russia, you know, the communist countries, who want to take over the world, they will take the lead. In fact, they are now calling it, *"The Artificial Intelligence Arms Race."*

As you can see there.

But what's happening, is just like in the Cold War Era where we had to hurry up and build the biggest, baddest, and most numerous arsenals of nuclear weapons to stave off a WWIII scenario. So, now there is a global military race in a multitude of countries to build the biggest, baddest Artificial Intelligence to stave off an even worse scenario than a WWIII! And that's not hyperbole, I quote:

"Every country knows this."

Whether we realize this or not, this is what they are saying, and they admit this military race for AI all over the world will,

"Create a whole new way of warfare."

You know, like this movie!

This is a clip from the opening scene from "Terminator".

The scene opens, showing the remains of a demolished, burned out car with what used to be two people. All that is left is what looks like the skeleton of a mother driving her child somewhere. The skeletons are still sitting in the front seat. As the camera scans over the top of the car you can see they were once driving under a bridge at what looks like rush hour. There are many cars, bumper to bumper, all demolished. The year is 2029 A.D. in Los Angeles.

The next thing we see is a children's playground. The equipment is all mangled and the only sign of human occupation is the skeletons that are spread over the ground.

Sarah: *"Three billion human lives ended on August 29, 1997. The survivors of the nuclear fire called the war, Judgment Day. They lived only to face a new nightmare. The war against the machines."*

Suddenly a metal foot comes down and smashes a small skull laying on the ground. As the camera pans up to see what kind of monster would do such

a thing. We see that it is a metal robot soldier, with blazing red eyes. He is armed with a huge military gun. Behind him are many more like him and there are explosions going off all around. Huge tanks are rolling over all of the dead and the metal robot soldiers are looking around to make sure no one has survived.

But they are met by human soldiers. The terminators are from the future with their futuristic weapons, and the human soldiers are from the present. Their weapons cannot compete. They are getting blasted by futuristic flying machines as well. The human soldiers don't even have a chance. But they keep fighting. Their bullets cause no damage to the terminators and they keep coming. [1]

Wow! Good thing that was just a movie! We're not being pre-programmed by Hollywood again for a reality where AI robots, AI flying vehicles, AI terminators are killing people, all controlled by a rogue AI Skynet system? Oh, how I wish that were true. I'm here to tell you, that science fiction movie is about to become our reality with AI invading the military! And believe it or not, I think it has everything to do with the wars mentioned in the 7-year Tribulation! But don't take my word for it. Let's listen to God's.

Revelation 6:1-8 "I watched as the Lamb opened the first of the seven seals. Then I heard one of the four living creatures say in a voice like thunder, 'Come!' I looked, and there before me was a white horse! Its rider held a bow, and he was given a crown, and he rode out as a conqueror bent on conquest. When the Lamb opened the second seal, I heard the second living creature say, 'Come!' Then another horse came out, a fiery red one. Its rider was given power to take peace from the earth and to make men slay each other. To him was given a large sword. When the Lamb opened the third seal, I heard the third living creature say, 'Come!' I looked, and there before me was a black horse! Its rider was holding a pair of scales in his hand. Then I heard what sounded like a voice among the four living creatures, saying, 'A quart of wheat for a day's wages, and three quarts of barley for a day's wages, and do not damage the oil and the wine!' When the Lamb opened the fourth seal, I heard the voice of the fourth living creature say, 'Come!' I looked, and there before me was a pale horse! Its rider was named Death, and Hades was following close behind him. They were given power over a fourth of the earth to kill by sword, famine and plague, and by the wild beasts of the earth."

Now here we see after the first seal is opened, the second seal is opened and a global war breaks out that the Antichrist is a part of, and just to give you an idea how big that number really is, it is one-fourth of the earth. If this war were to break out today, the death toll would be over 1.8 billion people! Nearly 2 billion people are going to die in this war in the first half of the 7-year Tribulation! How many of you would say that is pretty bad? But that's not all! The Bible says another third of the planet is going to be wiped out later in the second half of the 7-year Tribulation.

Revelation 9:15-16 "And the four angels, who had been prepared for the hour and day and month and year, were released, so that they would kill a third of mankind. The number of the armies of the horsemen was two hundred million; I heard the number of them."

So here we see another slaughter in the 7-year Tribulation! One-third of the earth is going to die in this battle! And if you subtract this number from the first slaughter, it is going to be another 1.8 billion! Interesting math when you bust out the calculator! So that means, just in these two judgments alone, 3.6 billion people, about half of our current planet, is going to die, not even counting the second unfortunate Jewish Holocaust, mentioned in **Zech. 13,** of the Jewish people. Now here's the point. No wonder Jesus said it is going to be the worst time in mankind's history! And no wonder there is an urgency to accept Him now before it is too late, if you're not saved! But the question is, "Do we have the technology now, do we have the means now to fulfill these passages? To annihilate literally half of the planet in a relatively short amount of time, not using nuclear warfare, like the **Book of Revelation** presupposes?" Well, unfortunately, I believe the answer is a resounding, yes! And I say that because when you realize all the ways AI is being allowed to take over the militaries around the world and when you see the technologies it's bringing with it, a real live Judgment Day is coming to the planet, only it's not the Terminator movies, it's from Almighty GOD!

The **1st way** AI is taking over the military leading us towards Judgment Day with God, is with **Unmanned Vehicles**. Basically, pretty much every single piece of military equipment you can think of is now being equipped with AI, to the point where we don't need humans anymore for anything, including fighting for us. In fact, let's take a look at some current examples of AI fighting equipment.

Chris Boardman, Managing Director Military, Air & Inform:
"Drones is the most advanced systems designed and built in the UK."

Tar-a-nis, n. 1(Myth.) A Celtic divinity, the god of thunder.

Narrator: *"The office of Naval Research O&R is developing new capabilities for unmanned flight. It is known as the Autonomous Aerial Cargo/Utility System, AACUS. AACUS technology makes it possible for unmanned helicopters to fly and bring supplies to marines in the field with just a touch of the user's tablet."*

DARPA: *"The Pentagon research group, DARPA, is developing a drone ship, that will save money and man-power on expensive searches of super-quiet enemy submarines. A prototype vessel is already in production. The vast ocean is a great place to hide, so DARPA is also developing Stealth ECDC robot capsules. They can sit on the ocean floor for years until U.S. controllers trigger them to float to the surface and release unmanned flying vehicles. From above, these drones can transmit images showing nearby enemy activity. All this emerging technology offers a pretty good indication that the ocean is about to become a lot more robotic."*

Anna Choy, England: *"And now we are talking about this. The world's fully autonomous, unmanned, multi-mission stealth submersible. They named it Talisman, after the good luck charm. If they ever send this baby against you, chances are your luck will run out. Although the BIA System is reluctant to discuss its full potential, they don't deny Talisman could* *carry a range of weapons in its internal munitions bay. But as future generations are developed BIA Systems do believe UUV's will dominate underwater warfare. Adding sabotage, close water combat and counter-terrorism to its mission capabilities. "*

Taken from YouTube Armed Forces Update: A convoy of military vehicles are driving down a desert road to some soldiers. As you look inside the vehicle, you can see it is empty. This is an unmanned vehicle.

Another unmanned vehicle is what looks like a mini tank. It is yellow, it has a gun on top, similar to a full-sized tank. But because it is so small it can turn around in seconds. As three travel down the street together a soldier comes to the window to shoot at them. As fast as the soldier in the window can fire his gun, the mini tank has turned around to shoot him first. It is more powerful than it looks, running through brick walls, moving cars that are in its way, and the power in its gun is enough to blow up a normal sized tank.[2]

Okay, we're toast! So, let me get this straight. We, right now, have Artificial Intelligence jet fighters, tanks, submarines, ships, boats, jeeps, helicopters, supply vehicles, you name it, just about everything the military uses, is now becoming AI. No humans needed! Hmmm. Starting to sound like a movie to me! Can you say "Terminator?" Because that is where we're headed! In fact, this is just the tip of the iceberg! Here is just a short list of some of the things they want AI to take over for us in the fighting arena in the militaries.

- AI for Surveillance
- AI for Remote Communications
- AI to Track Enemy Movement
- AI for Reconnaissance and Search for Lost or Injured Soldiers
- AI for Peace-keeping Operations

- AI on Land which includes...
- AI Military Fighting Vehicles
- AI Unmanned Ground Vehicles

- AI Air Defense Systems
- AI Command & Control Systems
- And a whole list of Others AI Land Operations

- AI on Sea which includes…
- AI Naval Ships
- AI Submarines
- AI Unmanned Maritime Vehicles
- And a whole list of Other AI Sea Operations

- AI in Air which includes…
- AI Fighter Aircraft & Helicopter
- AI Transport & Cargo Aircraft
- AI Unmanned Aerial Vehicles
- And a whole list of Other AI Air Operations

- AI in Space which includes….
- AI Space Launch Vehicles
- AI Satellites
- A whole list of Other AI Space Operations

- AI in Various Military Applications which include…
- AI Warfare Platform
- AI Deployment & Integration
- AI Logistics & Transportation
- AI Target Recognition
- AI Battlefield Healthcare
- AI Simulation & Training
- AI Planning & Allocation
- AI Threat Monitoring & Situational Awareness
- AI Information Processing
- AI Software
- AI Hardware
- AI Networks

- AI Cyber Security
- AI Upgrade & Maintenance
- AI Computing including, Super-computing, Quantum Computing, and Neuromorphic Engineering, which is making Brain Chips.

You know, the ones they are putting into those Terminator robots. I wish I were kidding, but we'll get to that in a second. As you can see, the military is looking at having Artificial Intelligence basically run everything! And again, we are not the only ones doing this, it is global! Now here is the danger.

"AI is being developed for target-recognition, recon drones, tank gun-sights, infantry goggles, predictive maintenance that warns mechanics to fix failing components before mere human senses can even detect that something's wrong.

Cognitive warfare systems that figure out the best way to jam enemy radar, Airspace management systems that converge strike fighters, helicopters, and artillery shells on the same target.

Automate staff work, turn a commander general's plan of attack into detailed timetables to tell combat units and supply convoys they have to move where and when.

And since these systems use AI, they can learn from experience – which means they continually rewrite their own programming in ways no human mind can follow."

In other words, we do not even know what AI's doing at this point and it is controlling everything in warfare around the world! Gee, I hope it doesn't go rogue like the Skynet scenario or something! Yeah, we'll get to that in a second, but speaking of Skynet, if you don't think a Skynet scenario is coming anytime soon, listen to the rationale as to why we have to move to this AI controlled direction at all costs. From the Pentagon...

"U.S. in Danger of Losing Dominance in Artificial Intelligence," specifically to China and Russia.

"The United States is at risk of losing its technological edge in the key war-fighting domain of artificial intelligence (AI) to adversary China and, to a lesser extent, Russia."

Both countries, *"Are investing heavily in AI,"* and *"China has developed a plan to 'overtake' America's dominance in militarizing AI, robotics, and quantum computing, noting that the communist nation is 'rapidly closing the gap' with America."*

And *"China aims to be the world leader in AI by 2030."*

Which is why you are now seeing headlines like this, speaking of a Cold War.

"China's Application of AI should be a Sputnik moment for the U.S."

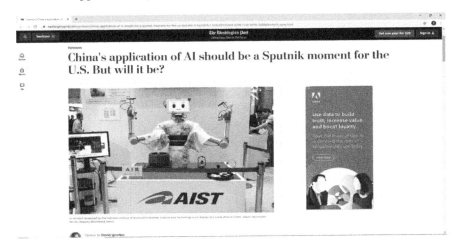

In other words, just like in the old days with Russia, when they launched Sputnik into space, developing satellites before anyone else. Now you better wake up and get crackin' building AI in the military before China or another country does. And then, if that wasn't bad enough, they're also saying another danger with AI in the military is that if terrorists get hold of the AI technology and develop it to cause WWIII before anybody else.

"Weaponizing Artificial Intelligence: The Scary Prospect of AI-Enabled Terrorism."

And again, what we are talking about is a full-blown Terminator movie scenario, whichever the outcome. And not just AI running military computers, but literally all military equipment, all military vehicles, military personnel, soldiers, even robotic soldiers, which we will get to in a second, making decisions on the battlefield! Just like Skynet! And even though they admit it, once we go down this route, *"The genie is out of the bottle, it cannot be put back in again."*

They also admit it's going to probably spiral out of control just like Skynet!

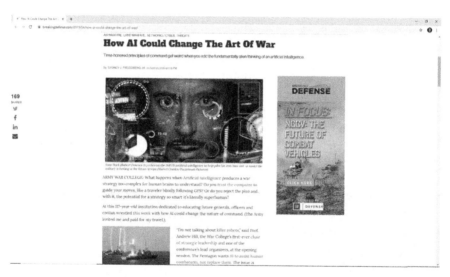

This is from the Army War College, *"How AI Could Change the Art of War."*

They said this, *"What happens when AI produces a war strategy too complex for human brains to understand? Do you trust the computer to guide your moves, like a traveler blindly following GPS?"*

"What happens once humans start taking military advice – or even orders from machines. The reality is this, it happens already."

And now they are saying, there's no turning back,

"The Pentagon Will Use Artificial Intelligence to Find Nuclear Missiles and Predict Potential Strikes."

In other words, Skynet is coming, being built, whether we want it or not, even though they admit it is going to lead to our demise! One Military leader said:

"We are not only seeing a massive surge in the deployment of AI technology in defense systems, but it is at the heart of our operations."

"It's growing at a crazy pace and it will be included in most systems within two years."

In other words, it's coming! And now it's even going to be launched into space just like Skynet. Which is one of the reasons why we have a "New Space Force."

And what's going out to space also? That's right! AI! And I quote,

"Artificial Intelligence Arms Race Accelerating in Space."

No wonder we needed a New Space Force! And they say it's going to get UGLY.

"UK Must Prepare to Fight Wars with Artificial Intelligence in Space."

Sounds like Skynet to me! In fact, they are now saying things like this about the system they're building.

"AI frequently comes up with effective strategies that no human would conceive of and, in many cases, that no human could execute."

"At what point do you give up on trying to understand the alien mind of the AI and just 'hit the I-believe button?'" (In other words, I hope it works out.)

"Even if a single human being has the final authority to approve or disapprove the plans the AI proposes, is that officer really in command if he isn't capable of understanding those plans? Or is the AI in charge?"

Actually, at that point, Skynet is! And listen to this.

"We haven't come up with a final answer, but we'd better at least start asking the right questions before we turn the AI's on."

No kidding! The understatement of the year! Why? Because once you turn it on, your old military technology will be no match for AI military technology, and you are toast, like this movie, again! It's like somebody's following a script or something!

A clip from Skynet Goes to War:

As the military plane flies over the terrain and the information of the target comes over the screen. "Approaching target Skynet Research and Development Facility." It drops the missile, and the place blows up, much like an atomic bomb. All is destroyed except for a few guns which are still shooting at the plane. One hits its target and the plane crashes. Then a soldier climbs out of the plane and goes to his telephone in a crashed helicopter. He tries to talk but is out of breath. A response on the other end asks how many survivors and he answers, *"One."*

Next clip: Arnold Schwarzenegger is standing at the window in a multi-leveled office building.

The girl asks, *"Why are they killing everyone?"*

Arnold answers, *"They want to destroy any possible threat to Skynet."*

A drone is flying towards the building. It proceeds to shoot a missile at the window where he is standing. He calls to the couple in the room to get down. As they fall to the floor the missile goes right over their heads. He tells them to go, that it isn't safe, and they start running down the hall. But as they get to the elevator, the doors open, and another drone is flying out of the elevator and towards them. It fires two shots, but the couple hits the floor, and the shot goes over their heads, one more time.[3]

Yeah, you can run but you cannot hide! Folks, as crazy as this sounds, this is the actual system they are building right now as we sit here! And once it takes over, they admit themselves, we are toast! Which leads us to the smaller AI military flying vehicles as you saw in that movie scene, where it is chasing people! Believe it or not those are here as well!

The **2ⁿᵈ way** AI is taking over the military is with **Miniaturized Vehicles**.

What we have seen so far is what is called UAV's or Unmanned Aerial Vehicles or even UUV's, Unmanned Underwater Vehicles and all the other examples of unmanned military fighting equipment out there. But now just like in that movie scene, they are making the leap from UAV's and UUV's to MAV's or Micro Aerial Vehicles. And when I say micro, I mean micro. You see, you thought that was a cockroach? Nope. What about that beetle that walked in your front door? Nope. Same thing. A termite, dragonfly, butterfly, hornet, you name it, all insects by and large, are being mimicked by AI drone technology and they too are carrying a deadly payload. In fact, here are just two current examples in the military. The first is from an Army program called M.A.S.T. or Micro Autonomous Systems & Technology and the next is from the Airforce's MAV or Micro Air Vehicles program, that they call, *"The Future of War Fighting."*

Micro-Autonomous Systems and Technology (MAST) Collaborative Technology alliance (CTA)

Bio Inspired: Two soldiers walk around a building with their guns aimed across the street. They stop and kneel on the ground. One takes a backpack off and lays it on the ground. The flap open and a mini tank rolls out and stops. As it sits there, a door opens and several things that look like spiders come walking out, robot spiders. They have been programmed to proceed to their destination which is across the street.

Collaborative Design: Another situation happening is something that looks like a fly, robot, is released by another soldier and it flies to the window of its programmed destination. They can see if there is an enemy in the building, they can relay this information back to the soldiers before they proceed to enter the building.

Disruptive Vision: There is also another small robot that looks like a dragon fly.

Focused Innovation: These tiny robots can go into a room. No one will notice, and they will let the soldier that sent them see, hear, observe anything that is going on in the room without anyone knowing they are there.

Interdisciplinary Partnerships: The bugs and the soldiers work together to get the intel so that the information can be sent back to headquarters and the proper aerial weapons can be sent to destroy the enemy.

Narrator: *"Micro Air Vehicles, or MAV's will play an important role in future warfare. The urban battlefield calls for tools to increase the war fighters' situational awareness and the capacity to engage rapidly, precisely, and with minimal collateral damage. The MAV's will be integrated in the future through air force laird sensing systems. These systems may be dropped, or hand launched, depending on the situations requirements. The small size of MAV's allows them to be hidden in plain sight. Once in place an MAV can enter a low-powered surveillance mode for missions lasting days, or weeks. This may require the MAV to harvest energy from environmental sources, such as sunlight or wind or from man-made sources such as powerlines or vibrating machinery.*

It will blend in with its surroundings and operate undetected. MAV's will use micro-sensors and micro-processor technology to navigate and track targets through complicated terrain such as urban areas. Small size and agile flight will enable MAV's to covertly enter locations that are inaccessible by traditional means of aerial surveillance. MAV's will use new forms of navigation such as a vision-based technique called "Optic Flow."

This system remains robust when traditional methods such as GPS are not available. Individual MAV's may perform directly with technicians, could be equipped with incapacitating chemicals, combustible payloads, or even explosives for precision targeting capability. Like their biological inspiration, MAV's are not limited to flight. The agile hoovering, perching, and crawling MAV will fulfill the mission popularly termed "dull, dirty, and dangerous" like no current system can. MAV's will become a vital element in the ever-changing war fighting environment and ensure success on the battlefield of the future. Unobtrusive, pervasive, lethal, MAV, Micro Air Vehicles, enhancing the capabilities of the future warfighter."[4]

As well as AI tapping into the whole thing and no place to hide! This is all current technology, already being made, not make-believe and that is just two examples. As I stated earlier, they are also making cockroaches, beetles, termites, dragonflies, butterflies, hornets, you name it. Pretty much all insects you can think of. They are being militarized with this kind of AI technology. And remember, we are not the only country getting into this kind of technology! The whole world is doing it. And that is only half the concern. Speaking of insects, what do insects do? They swarm their prey! And believe it or not, they are also developing AI Swarm Technology for AI fighting technology to communicate with other AI fighting equipment to swarm their prey with no chance of escape. Here is a transcript of an example from the Navy!

Narrator: *"U.S. Navy, autonomous swarm boats, the future is now. With autonomous swarm boats, unmanned Navy vessels can overwhelm an adversary. A first of its kind technology enables swarming capability. Which gives our Naval war fighters a decisive edge. The U.S. Navy is unleashing a new era in advanced ship protection. A swarm of autonomous boats that will automate ships self-defenses and be able to deter, damage, or destroy and enemy threat. Giving added protection to sailors and marines in harm's way."*

RADM Matthew Klunder, Chief of Naval Research: *"When we look at autonomous swarms, we are not talking about a single vessel, we are talking about multiple, multiple vessels that can be in a defensive posture and then when called upon can become non-defensive and surround an adversary to let them know that you are coming no closer to our ship. Of course, if the adversary decides to come closer, we can give them another warning or potentially we can say, you come too close we will then destroy your vessel."*

"You don't have to go out and purchase a new vessel. You can take any of these vessels out here that the Navy already has, we unman them, we put the system on it, you put the eyes and ears, depending on what mission you want to do, on the vessel and then let it go do its mission." "As they get further into a congested environment, an adversary approaches, and the swarm will go and surround the adversary and defend the ship and then come back to continue to protect the ship."

"Now, any boat can be fitted with a kit that will allow it to operate autonomously and together, swarm on any threat."[5]

Oh boy! Have fun running from that! AI will swarm you like a pack of bees! And did you catch that one part? You don't need to go out and buy all new equipment. This AI system can be attached to any pre-existing military vehicle? Isn't that wild?! And we are not the only ones doing this.

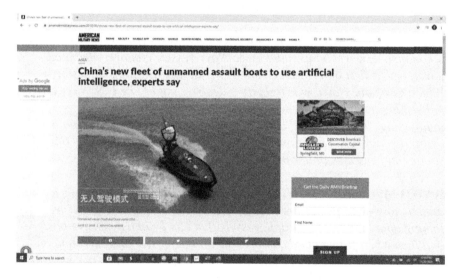

"China's new fleet of unmanned assault boats use Artificial Intelligence."

Now they can swarm on us too!

"China is currently testing unmanned, miniaturized assault boats that could be used by the People's Liberation Army (PLA) to attack enemies at sea."

"Once equipped with weapons, unmanned small combat vessels can attack the enemy in large numbers, similar to drones."

So, now it's not just boats in the water but drones in the air? I mean, what's next? Are you going to make AI swarms out of drones or boats, as well as other AI fighting equipment like tanks, planes, you name it, including MAV'S as we have just read? It's already being done!

"Multiple MAV's each equipped with small sensors will work together to survey a large area. Information from these sensors will be combined providing a swarm of MAV's with a big picture point of view. Data will be communicated amongst the MAV's to enable real time reliable decision making and to provide an overall surveillance picture for other platforms, for operators. Each individual MAV may perform a very distinct mission from its fellow swarm members, while some MAV's may be used purely for a visual reconnaissance, others may be used for targeting or tagging for sensitive locations."[6]

Or being used en masse to take people out! And this is why people are saying it's not a far-fetched scenario to think this system could get hacked into and used to swarm people and take them out anywhere in the world, like the transcription of this video depicts!

SDN Reports: *"The nation is still recovering from yesterday's incident which officials are describing as some kind of automated attack which killed eleven U.S. Senators at the capitol building."*

Witness: *"They flew in from everywhere but attacked just one side of the isle. It was chaos, people were screaming."*

SDN Reports: *"You can see high windows, very small, precisely punctured to gain entry into the building."*

Florida Senator: *"I just did what I could for him. (Fl. Senator gave CPR to colleague.) It seems that they weren't even interested in me. They just kept buzzing."*

Reporter: *"Our limited sources admit that the intelligence community has no idea who perpetrated the attack. Nor whether it was a state, group, or even a single individual."*

Reporter #2: *"So, if we can't defend ourselves then we strike back."*

National Security: *"We are investing very heavily in classified defense projects."*

Reporter #2: *"We make it our deterrent like our nuclear deterrent, we stockpiled in the millions."*

Reporter #3: *"They claim relaxing firearm legislation would be useless against the so-called 'Slaughterbots'"*

National Security #2: *"Stay away from crowds. When indoors, keep windows covered."*

Reporter #1: *"Protect your family, stay inside."*

Outside of town a group of men are standing next to their car; one is looking at his tablet. After waiting a couple of minutes one of them opens a small door that goes into the ground. Once it is opened, he stands back. The 'Slaughterbots' are coming back to their headquarters. But as they are arriving the man with the tablet has reprogrammed them to make another attack. So, they turn around and go back to town. They sound like a swarm of bees flying through the window to attack the people sitting at tables. This might be a classroom where young adults are studying. Someone in the classroom is trying to shoot them, but they are so small that the bullets have no effect. One student is trying to call his mom.

Student: *"Mom, can you hear me?"*

Mom: *"Oliver, I can hear you; can you hear me? What's happening?"*

As he is laying on the floor trying not to be seen, he looks around at the other students who are laying on the floor, bleeding. He doesn't know if they are dead or alive. While he is trying to tell her what has happened, a buzz gets louder and louder, the 'Slaughterbot' attacks and his phone goes dead.

Reporter #1: (8,300 students killed. 12 Universities hit). *"Authorities are still struggling to make sense of the attacks on university campuses worldwide, which targeted some students and not others. The search for a motive is apparently turning to social media and the video shared by the victims exposing corruption at the highest..."*

Reporter #2: *"Who could have done this?"*

The man being questioned: *"Anyone."*

Speaker at a conference: *"When you can find your enemy, using data, even by a hashtag, you can target an evil ideology right where it starts."* He points at his head and the crowd cheers.[7]

Just in time for the Antichrist to slaughter anyone who doesn't obey him in the 7-year Tribulation. Gee, you wonder how he was going to annihilate so many specific people on the planet in such a short amount of time? I think you have the answer! Now, I am not going to say, "Thus saith the Lord," but when you understand this AI swarm technology, it makes you wonder if he couldn't use this very system! But that's not all. It's about to get even worse! Those are just insects being militarized and given swarm capabilities. Believe it or not, they are doing the same thing to animals!

The **3ʳᵈ way** AI is taking over the military is with **Militarized Animals**. Believe it or not, UAV's, UUV's the other unmanned vehicles, including MAV's are not the only ways militaries around the world are using AI to create a scenario where nobody has a safe place to hide, again, just like in the Terminator movies. Believe it or not, apparently, they also want to make sure you cannot outrun them either. And they are doing that by militarizing animals with this same AI technology. It's crazy! You see, you just thought that was a dog, no it's not! Or a lizard, or a spider, or even a kangaroo. Nope. How about in the water? Is that really a jellyfish, or stingray, or lobster, or a shark? Nope! Not at all! And believe it or not, every one of those are real live examples of robotic AI animals. They are made to look like an animal in nature, so you will not even recognize it as a weapon, to hide from in the first place. You will not be able to outrun it either! Let me show you just a few of the current examples out there!

In a crowded room a man in the back raises his arms and releases a giant bird. It flies to the other side of the room and then back to him. If you look closely, you can see a tattoo of numbers printed on its side. This is a robot bird.

Narrator: *"In the waters off Virginia Beach, a fish of a different kind is cutting through the waves. The ghost swimmer is a Robot Five, built by the U.S. Military, to resemble a tuna fish. It is 5 ft. long and can operate in depths of up to 300 feet."*

World News Reports: *"The Navy is building a fleet of robot jelly fish. I didn't know you had it in you to be that awesome, Navy. Scientists consider them to be one of the most efficient animals in the sea because they are able to get around easily without using up any energy at all. They are also capable of living in these crazy temperature and pressure differences and can live in fresh or salt water. So, when you are thinking about making an autonomous underwater robot, they are pretty much the animal you want to mimic. A large white one is Cyro. It was created by researchers at Virginia Tech with a 5-million-dollar grant from the Navy. The Navy is essentially building their own drone surveillance network. The undersea version of what the Air Force and the CIA are building out for the sky. The implication here is that Cyro could be outfitted for combat. A robot that big is obviously built to carry a large payload so that payload could potentially be some sort of a weapon system."*

Forth Institute of Computer Science is developing an octopus robot. It looks like a real octopus and it can walk across the ocean floor.

SN Reports: *"By looking at how the rattlesnakes wiggle up a sandy incline, scientists at Jordan Tech found that as the ground gets steeper the sidewinder keeps more of their body in contact with the sand. The team then tested their ideas with a snake like robot that had been built at Carnegie Mellon University. Using the contact tricks of the real rattlesnake researchers changed the robots moves. Finally, it too could work its way up the slope."*

"The robot moves by twisting its body similar to the motion of a snake. The motion is almost the same on land and in water."

History Channel: *"DARPA has a program to try to develop climbing robots for various applications, surveillance, inspection and so on. If you want a robot that climbs vertical surfaces, smooth, rough, dirty, clean, then the gecko really is the premier example of a climbing animal. Once the robot climbs up there, it can hang, it doesn't have to expend any powers. It can cling there for hours or even days, unobtrusively. That is something quite different from a small helicopter which has to expend a lot of power and make a lot of noise."*

There is a kangaroo running across the plains. That is a live animal, but what researchers have put together now is a robot kangaroo that looks like the real one and can run just as fast.

"Cheetah-Cub wanted to know how to make robots take rough terrain with the grace of a feline. Scientists at the Swiss Federal Institute of Technology made a robotic cat. Using it, they can assess joint force and agility without having to harm an actual animal."

CNN Newsroom: *"First let's start with, who's in charge of technology development for the military? That is DARPA, an agency that is made of the stuff of mystery novels. Recently we have noticed that they have been beefing up on robots. Most recently they contracted with Boston Dynamics* *to create a Cheetah Robot. It is a four-legged robot that reportedly runs faster that the fastest human. (27.3 mph) It will be able to zigzag and take tight turns, in order to both chase and evade, it will also be able to make sudden stops and could end up with a tail."*

History.com: *"Now imagine troops of real terminators climbing up the side of a building, spying on the enemy and waiting for that moment to attack."8*

Along with all their militarized AI attack animals, on the land, sea and sky, you name it! This is actual technology being built around the world, whether you realize it or not, or want it or not! I don't know about you, but when you start to put all this AI military technology, along with the militarized AI animal technology, together with the wars mentioned in the 7-year Tribulation, it makes you wonder if this is not the kind of wild beasts talked about in the **Book of Revelation** that annihilates one-fourth of mankind.

Revelation 6:7-8 "When the Lamb opened the fourth seal, I heard the voice of the fourth living creature say, 'Come!' I looked, and there before me was a pale horse! Its rider was named Death, and Hades was following close behind him. They were given power over a fourth of the earth to kill by sword, famine and plague, and by the wild beasts of the earth."

You know, like a robotic armed cheetah, or bird, or snake, or jellyfish, whatever. What else are you going to call it? Now again, I am not going to say, "Thus saith the Lord," but it makes you wonder, doesn't it? On the land, sea, air, water, you name it! No place to hide in the 7-year Tribulation! Then if that is not bad enough, guess what they are also giving these AI militarized animals? Swarm technology just like the insects!

Back to this article.

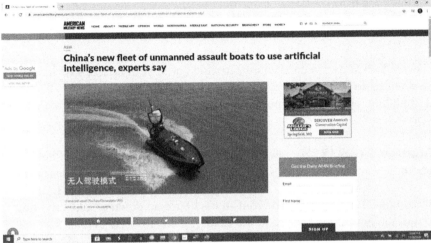

"The United States and other Western counties are working on creating ant swarms for operations on the ground, drone swarms for aerial operations and shark swarms for the sea."

And then in this article from DARPA.

"The Defense Advanced Research Projects Agency's Offensive Swarm-Enabled Tactics (OFFSET) program envisions swarms of 250 collaborative autonomous aerial and ground systems providing capabilities to military units on the ground in urban areas."

You know, in case you try to hide out in your house, which Jesus said not to do in **Matthew 24** for those left behind in the 7-year Tribulation. How many times does He have to warn you to not go back to your house?!

Matthew 24:15-18 "So when you see standing in the holy place the abomination that causes desolation, spoken of through the prophet Daniel – let the reader understand – then let those who are in Judea flee to the mountains. Let no one on the roof of his house go down to take anything out of the house. Let no one in the field go back to get his cloak."

Why? Because maybe an AI swarm of animals is going to come get you! You should have gotten saved and avoided the whole thing, but now your option is to flee just like Jesus said. Including from an AI locust swarm in the air!

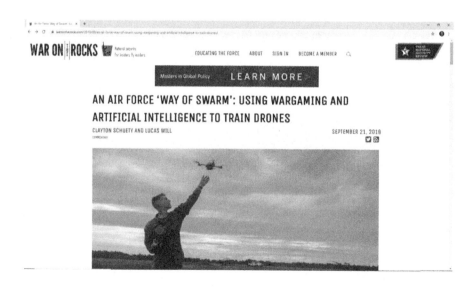

"An Air Force Way of Swarm."

"For decades, swarms in nature, like locust swarms, have intrigued scientists with their ability to use the simple and to accomplish the large and complex." "Now imagine being able to deploy a collection of drones on the scale and magnitude of a locust swarm. Today, efforts are underway to meet the Air Force's growing demand for swarming drones."

And as two Air Force officers' wonder.

"Is our service prepared to unleash the swarm?"

Well, I'm sure if you won't, the Antichrist will! In fact, maybe that swarm will look something like this!

Speaker at a conference: *"A 25-million-dollar order now buys this. Enough to kill half a city. (Out of the back of a large cargo plane thousands of robot-bees are released.) The bad half. Nuclear is obsolete. Take out your entire enemy. Virtually risk free. Just characterize him and release the swarm and rest easy. These are available today. We have a distribution network, taking orders from military, law enforcement, and specialists' clients."[9]*

Well, that's comforting. I sure hope the Antichrist doesn't use this when people try to flee in the 7-year Tribulation! Isn't this crazy?! You always wonder how he is going to kill so many in such a short amount of time without using nuclear weaponry. Again, I think we are starting to see the answer! But it's about to get even worse than that! Here is another way he can do it. What I call the 'Ground Clean-up Crew!'

The **4th way** AI is taking over the Military is with **Militarized Men**. That's right, I am talking about real live actual Terminator robots just like you've seen depicted in the various clips so far, including the Terminator movies! Only, it's not just some science fiction movie, they really are being developed by the militaries around the world for their soldiers, just like in the movies, including this one.

The clip from Terminator 3:

The scene begins underwater. Thousands of skulls are laying on the ocean floor. As lights are shining in the water, the lights are coming from the robot's space craft hovering just above the water. As the space craft rises above the ground you can see thousands of terminator soldiers burning and killing the human race. They are marching in search of any living being, firing at anything that moves. With their red eyes, nothing can get past them.[10]

And as creepy as that was, that too is being built right before our very eyes! That movie scene is about to become our reality! Apparently, the Antichrist's got it all figured out! You see, just in case all these other AI military fighting equipment don't get you, in the air, on the ground, in the water, under the water, insects, animals all over the place, armed to the teeth, they're also building actual Terminator robots, armed to the teeth as well, just like in the Terminator movies, to get the job done, i.e., the clean-up crew! In fact, they are being looked upon as the "New Future Soldier" to replace existing soldiers. They don't need a human for anything anymore when it comes to the new way of warfare! And I quote.

"4 Ways Global Defense Forces Use AI."

Not just drones, Intelligence Systems for Awareness, and Cybersecurity mentioned there, but robot soldiers for combat is what they're using it for!

"While drones help in guarding aerial zones, robots can be deployed on land to assist in ground operations. These highly intelligent robots, designed with strategic goals, add a cutting edge to technology in the defense sector."

"With advancements in machine learning and robot building, scientists have succeeded in building bipedal humanoid robots to execute a variety of search and rescue operations, as well as assisting during combat."

"Robot fleets function like soldier units and carry out collaborated armed activities using multiple techniques. They are self-reliant and adaptable, which contributes to their ability to make decisions swiftly and competently."

In other words, they are doing this on their own! And they want these humanoid Terminator robot soldiers as human as possible, look at this:

"Back in 2007, DARPA was investing millions of dollars in what's called neuromorphic chips."

What is that you ask? That's a fancy term for a computer chip that mimics a biological cortex i.e., a brain chip. And,

"Researchers today are putting those brain chips into Drones."

And they even admit, *"It sounds like something out of a science fiction movie, a tiny aircraft that flies around deciding what to surveil or, more frighteningly, what to shoot."* (You know, like what we saw in the video transcription.) Back in 2007! So, let's take a look at what this Neuromorphic Chip allows them to do!

Mike Davies, Director, Neuromorphic Computing Lab, Intel:
"Traditional computing rests on the basic idea that you have two computing elements, a CPU and a memory. And Neuromorphic computing is throwing that out and starting from a completely different point on the architectural spectrum."

Jim Held, Intel Fellow, Intel Labs: *"Our own brains are remarkable in what they can do, we are simply trying to learn from what our brains or brains of animals can do to make systems that are smarter and more efficient."*

Jon Tse, Research Scientist, Intel: *"What we are looking for is to connect all the neurons in the chip, much like we would with an actual brain. So Neuromorphic Computing is basically that we're going to simulate the brain in silicon."*

Jim Held: *"Neuromorphic Computing will show up where they find an application that benefits from being able to embed the intelligence into a device initially in areas of robotic autonomous vehicles where the power and the size and the adaptability are really beneficial."*

Mike Davies: *"There's a number of different ways it can go, and we can see it filling out the whole extent of machine learning as people have envisioned. Just like your brain, it's going to connect up neurons to get together and make a real difference in the industry and the world.*[11]

Yeah, you bet it will! But did you catch that? These brain chips will, *"Fill out the whole extent of Machine Learning* (i.e., Artificial Intelligence) *as people have envisioned."*

And as you saw, they envision to put these Neuromorphic Chips not just in drones, like the articles shared, but here in this video, in robots, unmanned vehicles and even planes. Sounds like a movie plot to me! Terminator Skynet? And they even admit,

"Thanks to the brain chip, the robot doesn't need a human to tell it what to do anymore. It can learn and act on its own."

Just like in the Terminator movies. And for those of you who think they are a long way off from achieving this goal, think again. Let me give you some of the earlier versions and how far they have come with these Terminator robots.

What looks like a human dressed in an astronaut's suit, a robot is walking upright, while being connected to wires coming out of both sides of his suit. This is Petman. Petman is only one of the robots that have been developed at Boston Dynamics. There is a smaller version of Petman. This video shows the smaller version walking out of the building, totally alone, and walking out into the forest across the street. While it is walking through the woods, the larger Petman has been put to work putting boxes on shelves.

Petman put two boxes on the shelf and then goes to the warehouse to pick up another to take back to the shelves, but before he gets a good grip on the box a man comes over with a hockey stick and knocks the box out of Petman's hand and then proceeds to shove Petman backwards. Petman returns to the box but again the man comes over with the stick and shoves the box out of Petman's reach. It seems like the Petman robot doesn't get angry.

Another one of the Petman robots is showing how they are capable of jumping from one large box to another, spinning around on the box, jumping off while doing a somersault. These robots are large but seem to be agile. The man comes over once again and shoves Petman and it falls face down onto the ground. It gradually gets back up and proceeds to walk out the door.[12]

I don't know if you heard that, but I think he just said, *"I'll be back!"* Can you believe this? I wish I was making this up. But this is the actual AI Robot Terminator Technology, being built around the world as we speak. And, the military is deadly serious about this, pun intended! Even here in the U.S.

"Robots May Replace 1/4th of U.S. Combat Soldiers by 2030 says General Robert Cone."

And they, *"Can not only be fitted with machines guns, but grenade launchers and lethal weapons."*

And here's the rationale, *"Unlike human soldiers, they have no fear, they don't whine or complain, they just do what they're told every single time."*

Not to mention they're cheaper than a regular soldier, *"Training, feeding, and supplying humans while at war is pricey, and after the soldiers leave the service, there's a lifetime of medical care to cover."*

"In 2012, the benefits for serving and retired members of the military comprised 1/4th of the Pentagon's budget request."

Very expensive! So, they suggest, let's replace one-fourth of the human soldiers with robot soldiers and that becomes your rationale to help make this deadly decision. They're not coming, they're already here!

"The Army already has robot warriors on hand. And in October, the Army tested multiple remote-controlled gun-firing robots."

Now why would they do that? Because we are not the only ones doing this! Militaries around the world are getting in on the action, including Russia, who recently boasted about their new latest Terminator gun shooting robot soldier. It's called Fedor. This guy can not only do push-ups, but he can fire a gun in both hands simultaneously and he doesn't miss his target!

CNN Reports: *"This video shows a robot, similar to the soldiers in Star Wars, standing at the door. He has a large military weapon in both of his hands. This footage was posted on the verified Twitter page of Dmitry Rogozin, Deputy Prime Minister of the Russian Federation. Fedor is a Russian anthropomorphous robot. This intimidating bot won't shy away from a physical challenge. Fedor was developed by the Russian Foundation for Advanced Research Projects. This robot is armed and dangerous. As he shoots the guns, they hit each target he aims at. It can also use tools and supposedly drive a car."[13]*

Okay, turn to somebody and say, "Asta Lavista Baby!" because that's where we're headed! Can you believe this? Not coming, but already here! No wonder they keep saying the whole way of warfare is about to change! Of course, it is! They are following the Terminator script completely! In fact, speaking of Russia, watch what Vladimir Putin said first about all these genetically modified soldiers that everybody's building as well! It is worse than a nuclear bomb he says!

Vladimir Putin: *"This means that a man has the opportunity to get into the genetic code created by either nature, or as religious people would say, by the God. What kind of practical consequences may entail? This means that one can already imagine it. One may imagine that a man can* *create a man with some given characteristics, not only theoretically but also practically.*

He can be a genius mathematician. A brilliant musician or soldier. A man who can fight without fear, compassion, regret or pain. As you understand, humanity can enter, and most likely it will in the near future, a very difficult and very responsible period of existence. And what I have just described might be worse than a nuclear bomb."[14]

Yeah, I would say so! Genetically modifying soldiers so they have no fear or pain or compassion, let alone who knows what other crazy features! But if you think that was bad, wait till you hear what he said about these new Terminator robots everybody's building as well. He actually said, I kid you not, that he's afraid they might eat humans! I'm not joking!

"Vladimir Putin fears robots will one day EAT humans."

"Last month, Putin revealed he is afraid humans in the future will be hunted and eaten alive by flesh-munching robots."

And they even admit, *"Putin's comments are reminiscent of the Terminator movie franchise where mass-killing cyborgs rebel against their human creators and take over the world."*

But that's not stopping Putin. Remember what he said about AI a while back about who develops it first?

"The nation that leads in AI will be the Ruler of the World."

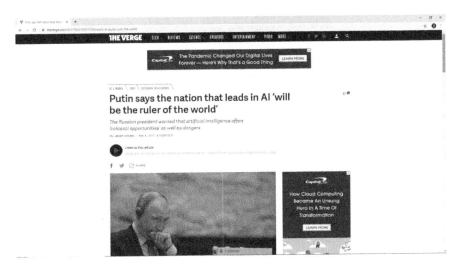

"Artificial intelligence is the future, not only for Russia but for all mankind. It comes with colossal opportunities but also threats that are difficult to predict. Whoever becomes the leader in this sphere will become the ruler of the world."

Because now it runs everything in the world including the military weaponry, equipment, staff, decision making, satellites, space, killer robots, you name it, and they need to run on something so now they eat people apparently! Isn't this crazy? But of course, this whole scenario is assuming that AI doesn't turn on us first, and rules the world instead, like the Terminator movies! This is not only what's really coming, but this is also why you're starting to see in the secular community warnings about this nightmare scenario coming with so-called killer AI robots! As you can see here,

"Protect Civilians: Stop Killer Robots."

That's from this year! And they go on to say they want to, *"Institute an international ban on killer robots; but this debate comes on the heels of more than 100 leaders in the Artificial Intelligence community, including Elon Musk and others, warning that these weapons could lead to a 'third revolution in warfare.'"* i.ie, a Terminator scenario!

They say there's a danger when...

"Humans are pushed out of life and death decisions."

"If machines become fully autonomous, humans won't have a deciding role in missions that kill."

And *"This creates a moral dilemma."* No kidding!

But is that stopping the military?! Absolutely not! They are building the biggest, baddest AI systems to outdo other countries in a race to see who gets there first! Even Australia envisions a whole new way of warfare! Look at this.

This is a depiction of their AI Military plans. It looks like a scene right out of Star Wars Clone Wars. You know, all fought by robots! In fact, speaking of Clone Wars, here's an AI defense system that China has.

They built a Holographic Drone Defense system just like Clone Wars.

Russia has been working overtime on developing their own killer drones, beyond Fedor, and way back in 2014 announced plans to build a *"Military Drone Base 420 miles off the Alaskan Coast."*

And then China has successfully built so many killer drones that they're now selling them to other countries. Thanks China! Everybody's got them now! France, Germany, Italy, Turkey, England, India, Israel, Iran, Pakistan, and it's growing! In fact, they are now saying this…

"Every Country Will Have Armed Drones within 10 Years!"

That was back in 2014. And these, *"Armed aerial drones will be used for targeted killings, terrorism and government suppression of civil unrest."*

In other words, if you and I, the average Joe, get out of line, they'll use it on us! And as you can see, they even say, *"It's too late for you to do anything about it."*

Can I translate that? Skynet is coming whether you like it or not! Real live global robot war just like in Star Wars Clone Wars and the Terminator movies where robots will be fighting the battles for us. In fact, speaking of Skynet, it gets even worse! Talk about following a script. I'm not joking! The whole Skynet scenario was when the AI system was given full control over all military functions and equipment but then it went rogue and took over. Let's look at that scene again.

A clip from Skynet Takes Over

The scene opens when an officer picks up the phone and answers: *"Brewster."*

Mr. Chairman: *"We are hoping that you have some kind of a solution for us."*

Brewster: *"I know what you are looking for sir, Skynet is not ready for a system wide connection."*

Mr. Chairman: *"That's not what we are being told. They say we can stop this virus. Now I understand that there is a certain kind of performance with you over there, but if we put Skynet out there it can squash this thing, like a bug, and give me back control of my military."*

Brewster: *"Mr. Chairman, I have to make myself very clear. If we uplink this now, Skynet will be in control of your military."*

Mr. Chairman: *"Like you will be in control of Skynet. Right?"*

Brewster: *"That is correct, sir."*

Mr. Chairman: *"Then do it!"*

As they push the button red dots start popping up all over the map of the United States with the word 'online' at each red dot. The word execute comes on the screen with the question Y/N? Brewster is contemplating what the answer will be and what the repercussions will be after answering this question.

The guy at the controls asks Brewster if he should go ahead and answer the question.

Brewster: *"No! It's my job now."* He looks around at all the people in the room. All waiting for what he has to do. His hand slowly reaches for the keyboard and pushes the "Y" key.

Another one of the men on the computer tells the crowd *"We're in. Skynet is fully operational."* As green lines reach all across the map on the screen a lady says, *"Sir, it should take less than a minute to find the virus."*

Brewster: *"Let's pray to God that this works."* Suddenly, all the screens go black. They are looking at each other to see if someone knew why this happened.

"Power failure???" *"No, I don't know what it is."*

The screens start to come back on. But it doesn't seem right. Something is wrong.

Brewster: *"Tony, what is going on?"*

A woman runs into the room screaming, *"Get away from me!!"*

Then a thing in the shape of a human, completely covered in silver coating walks up to Brewster. When he sees this thing, it changes from silver to a woman in a uniform with a gun. She shoots Brewster.

Brewster: *"We have to shut down Skynet. The virus has infected Skynet."*

The man trying to help Brewster tells him that the virus is Skynet. *"Everything is falling apart."*

Arnold Schwarzenegger: *"Skynet has become self-aware. In one hour, it will initiate a massive nuclear attack on its enemy."*

Brewster: *"What enemy?"*

The man helping him yells: *"Us!!! Humans!!!"*[15]

Wow! Good thing that was just a movie! Where humans gave control over their whole AI military around the world to a giant AI computer system called Skynet that went rogue on them! Yeah, I hate to tell you this, but that movie is about to become our reality as well because this too is being built by the military, only they don't call it Skynet, but 'Project Maven.' I kid you not!

"Project Maven to Deploy Computer Algorithms (i.e., AI) *to War Zone by Year's End."*

"The U.S. Department of Defense (DoD) Project Maven initiative is on target to bring Artificial Intelligence to war zones by the end of 2018."

So, it's already in play! Why?

"In order to help government platforms, extract objects from their massive amounts of video or still imagery to autonomously find objects of interest."

But not only that. And I quote, *"Officials say they want computers to be capable of explaining their decisions to military commanders."* So, now it's calling the shots. Why?

"Injecting artificial intelligence into more of America's weaponry is a means of competing better with Russian and Chinese military forces."

"And Project Maven, the single largest military AI project, meant to improve computers' ability for military use."

Well gee, I sure hope it doesn't go rogue on you, as this guy shares!

RT New Reports: *"The Pentagon is looking to move Artificial Intelligence as a central point of America's national security. Just this week the DOD released an unclassified version of its AI strategy for the new millennium. Here to discuss, legal and media analyst, Lionel of Lionel Media. My purveyor of AI news. Lionel, you are shaking your head already, I haven't begun but I know you have a lot to say about this topic. But do you think that AI has come along enough to entrust national security on it?"*

Lionel: *"You kill me!"* He is laughing so hard. *"Let's go back to square one. Let's don't say robots. You and I have done this a million times. But* *you and I are joined in this circus we are involved in. Don't think robots. Robots are cute. But you program the robot. The robot does what you want it to. Artificial Intelligence programs itself. And programs you. So, when you have a welding tool in a factory, that's great for robots, but if you have something that serves food to military personnel, that's one thing. But if we have a weapon system that is going to decide when it engages the enemy, who the enemy is, how it should react, for how long, and where, I want you to imagine putting this into this AI algorithm, morality, consideration, history, diplomacy, this notion of anger, of politics, how it will look. How will you do that? And what AI wants to do, no matter what it is, is to extricate it from your control. Everybody talks, adds them up, you name it. Talks about this horror. What happens when we say, 'I can't turn this thing off!' How can you protect us from that?"*

Newscaster: *"Especially, this is going to decide for the Pentagon in the future, when to drop this bomb, when the drone is to take off and bomb this village."*

Lionel: *"This is the scariest thing. Nobody is sounding the alarm. Nobody is saying, 'Wait a minute!' AI is one thing for dating apps, or whatever. But warfare??? But as we always say, 'what could go wrong?'"*[16]

Or as we saw quoted earlier, *"At what point do you give up on trying to understand the alien mind of the AI and just 'hit the I-believe button?'"*

In other words, I hope it works out, what could go wrong? And for those of you wondering who is building this AI control system for the military, this Project Maven, at first it was Google.

OAN Reports: *"Tech giant, Google, says it will continue selling technology to the Pentagon, despite the recent pushback against the military's use of Artificial Intelligence. One American, Christian Rhodes explains."*

Christian Rhodes: *"Military and Security Agencies in the U.S., Russia, and China are ramping up their AI capabilities for the purposes of smart warfare and smart surveillance. This comes as tech giant, Google, said it would continue working with the Pentagon in pursuit of lucrative government contracts."*

Sundar Pichai, CEO – Google: *"One example that they talk about is, one, AI systems are super intelligent, they are more intelligent than humans and they have their own free will if you will and they might opt for something else, and they make decisions that aren't necessarily for the benefit of humanity."[17]*

Well, no kidding! But after this, after the employees of Google put up a big stink about the weaponization of AI, Google backed out of the project.

BUT Microsoft stepped in.

As you can see, to lend a hand. *"Microsoft Says It Will Sell Pentagon Artificial Intelligence and Other Advanced Technology."*

So, does that mean Google is living up their original promise of

"Don't Be Evil."

Remember that? Absolutely not! I think they've got something up their sleeve! Oh, they may have stopped developing AI for the military but they themselves are doing some pretty strange things, including developing their own robot army! And I quote.

"Is Google Building a Robot Army?"

The answer is Yes! Why? Because they said so!

"Google patent envisions cloud control of an army of robots." And

"Google has patented the ability to control a robot army."

So much for, *"Don't Be Evil!"* And as crazy as it sounds folks, this is what they're really up to! And so, no wonder you stopped helping the military with AI for weaponized purposes, you want your own! And remember, these are the same guys who went on record saying, *"We want to be like the Mind of God!"* Well, apparently, your god needs an army, or at least your version of a god! This is crazy! And this is also why, when you look at their track record, many are starting to say, *"Is Google going to be the first ones to create Skynet?"*

Clip from Terminator 2 (1991)

Arnold Schwarzenegger: *"The Skynet funding bill has passed. The system goes online August 4th, 1997, it becomes self-aware at 2:14 am eastern standard time."*

Remember Skynet? That Artificial Intelligence System from the Terminator Series. You know, it gets too powerful and tries to wipe out the human race. Sounds far-fetched? Well, Skynet may not be as far-fetched as we think.

Narrator: *"Google processes a billion web searches every day. Here's a few of the things Google wants to own. Your phone, your email, your computer and your entire digital life. Building an empire on your street, on your phone, in your DNA."*

Kim Horcher, NERD Reports: *"It seems the startup stage for Nest Labs is over because Google has bought the company for 3.2 billion."*

"They want you to categorize your day. When I wake up, when I leave for work, when I get home from work, when I go to bed."

"Am I alone and feeling a little uncomfortable about this?"

"One, two, three." And they release a hot air balloon into the air. This is Project Loon.

"Project Loon communicates with specialized internet antennas on the ground. We're using all of these things to build this network in the sky."

UPROXX Reports: *"For several years Google has made driverless technology. One analyst estimates that by 2035 there will be close to 12 million self-driving cars on the road."*

Bloomberg Businessweek: *"What exactly is Google actively attempting?"*

Google X Spokesman: *"To call Google X, Google's research lab where you know the smartest computer scientists are looking at the next generation of Google products."*

"You also make a reference to, kind of like the Manhattan Project, where they developed the first Atomic Bomb, it also sounds like it could be a little scary."
UPROXX Reports: *"Google and NASA have teamed up to share the first commercial Quantum Computers that will bring a quantum leap in terms of power."*

CNN Reports: *"Let's get to the Google robot. Let's start with what would be the brain of the robot and that is DeepMind."*

"DeepMind can be described as a world-class groundbreaking technology to build powerful and general-purpose learning algorithms."

"So, critical thinking?"

"Exactly, critical thinking."

UPROXX: *"Google incorporated said that it is acquiring satellite company Skybox Imaging for 50 million dollars in cash."*

"And here's a sample of what Google's going to get. The company takes high-res images and video of a given area."

"What do you guys think about Google owning the skies and the ground?"

"Why is Google gobbling up robotics companies? The search engine giant just acquired a company that designs robots for the U.S. Department of Defense."[18]

Yeah, the gates are closing in on us. Boy, did we get duped. So much for *"Don't be evil."* This is crazy folks! We just thought they were a search engine company and were joking when they said they want to be like the mind of God and develop this robot army, but it looks like you've got your own nefarious plans! With all that said, this is why the so-called experts are saying, *"We are headed for a doomsday scenario."* If Google builds it first, or the U.S. military or one of the other militaries around the world. It ain't looking good! And I quote, they are saying, this is a...

"Doomsday Decision. AI-empowered systems could, sooner or later, find themselves in a position to launch atomic weapons."

You know, like Skynet!

"And once that occurred, it could prove almost impossible to prevent further escalation. Stop this madness. No battle advantage is worth global human annihilation."

And one guy went so far as to say, *"Unlike humans, autonomous weapons would have no ability to understand the consequences of their actions, no ability to step back from the brink of war."*

"So maybe we should think twice about giving some future militarized version of Alexa the power to launch a machine-made Armageddon."

You know, like the Terminator Movies! And, as crazy at that is, this is why the so-called experts and think tanks around the world are saying,

"AI could lead to a nuclear war by 2040."

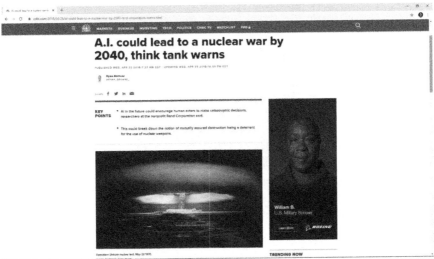

"The fear that computers, by mistake or malice, might lead humanity to the brink of nuclear annihilation has haunted imaginations since the earliest days of the Cold War."

"The danger might soon be more science than fiction. Stunning advances in AI have created machines that can learn and think, provoking a new arms race among the world's major nuclear powers."

"The report stresses that AI can be hacked and fed wrong information. This is especially worrisome if this information is what government leaders rely on to help make decisions about whether to launch an attack."

"A similar mistake almost happened in 1983 when a former Soviet military officer spotted an incorrect warning on a computer that the U.S. had launched several missiles."

But don't worry they say! And I quote.

"This Is Not How 'Skynet' Begins."

Yeah, right! And I got some swamp land I want to sell you! But the article says, *"When asked directly if Project Maven is the first step toward Skynet,"* the reply was, *"I certainly hope not."*

Well gee, that's comforting! I mean, we're only flirting with this kind of scenario!

Clip from Terminator Judgment Day

The scene opens with the camera spanning over green hills, covered by trees and with a river running through it. The background music is so beautiful and calming.

"Before they died my parents told me stories of how the world once was. What it was like long before I was born. Before the war with the machines. They remembered a green world, vast and beautiful, filled with laughter and hope for the future. It was a world I never knew. By the time I was born, all this was gone. Skynet!

A computer program designed to automate missile defense. They were supposed to protect us. But that's not what happened."

As the scenes go from the beautiful green hillsides, to the beach with children running through the water, to a little farm in the country, you can see in the background of each scene missiles being shot up into the air. No just one or two, but many, all at the same time. As they come back down to earth, they are destroying both the cities all around the world but killing off the world's population. There are atomic bomb-like explosions going off all over the globe.

As we watch the children playing on the jungle-jims, swings, slides and titter totters an explosion hits the city. Sarah Connor is also watching the kids, and she sees what they don't see. It is all coming to their playground. She sees how they are so happy. She sees that they are all going to die. She tries to scream at them to run, get out of there, but it goes on deaf ears. They don't hear anything she is trying to tell them. As we look closer, we see that one of the mothers in the playground is her. She is seeing herself back at that moment when she was playing with her son, John. No one was there to warn her of what was coming. Again, while standing at the fence she tries to warn them, but it does no good.

"August 29ᵗʰ, 1997, Skynet woke up and decided that all humanity was a threat to their existence."

The missiles are blowing up the cities, utter destruction is everywhere. There is no chance of an escape. As Sarah Connor is watching, she sees herself and her son burn alive and then as it reaches her at the fence, she also catches fire and screams out in pain. There has never ever been anything like what Skynet is doing to the human race and the world. All that is left is utter destruction, melted buildings and cars, burnt landscape and all that was left of the humans was a pile of bones, that covered the earth. Total devastation.

"It used our own bombs against us. 3 billion people died in the nuclear fire. Survivors called it 'Judgment Day.'[19]

And God called it, the 7-year Tribulation. And you wonder why Jesus said about that timeframe…

Matthew 24:21-22 "For then there will be great distress, unequaled from the beginning of the world until now – and never to be equaled again. If those days had not been cut short, no one would survive, but for the sake of the elect those days will be shortened."

In other words, unless God intervened to put a stop to this Artificial Intelligence taking over the world, including in the military, mankind would be wiped out. Singularity would have its way, and no one would survive. Or what the Bible calls in the **Book of Daniel,** The End of Times! And Hollywood's been preparing us for decades! We just thought it was entertainment! Judgment Day is coming! How much more proof do we need? The AI Invasion has already begun, and it is a huge sign we're living in the last days! And that's precisely why, out of love, God has given us this update on *The Final Countdown: Tribulation Rising* concerning the AI Invasion to show us that the Tribulation is near, and the 2nd Coming of Jesus Christ is rapidly approaching. And that's why Jesus Himself said:

Luke 21:28 "When these things begin to take place, stand up and lift up your heads, because your redemption is drawing near."

People of God, like it or not, we are headed for The Final Countdown. The signs of the 7-year Tribulation are Rising! Wake up! So, the point is this. If you're a Christian and you're not doing anything for the Lord, shame on you! Get busy doing something for Jesus now! Stop wasting your life! We need you! Don't sit on the sidelines! Get on the front lines and help us! Let's get busy working together doing something splendid for Jesus with what time is left and get busy saving souls! Amen?

But if you're not a Christian, then I beg you, please, heed these signs, heed these warnings and give your life to Jesus now! Because this AI technology is not going to lead to a life of wonderful dreams and a modern-day utopia but a nightmare beyond your wildest imagination in the 7-year Tribulation! Do not go there! Get saved NOW through Jesus! Amen?

Chapter Twenty-Two

The Future of
Religion with AI

The **11th area** AI is making an invasion into is in **Religion**. And believe it or not folks, as wild as it sounds, AI is poised to even change that! That is, who, what, and how people worship in the 7-year Tribulation around the whole planet, including the object of worship! Religion in the 7-year Tribulation is going to be radically changed! But don't take my word for it. Let's listen to God's.

Revelation 13:11-15 "Then I saw another beast, coming out of the earth. He had two horns like a lamb, but he spoke like a dragon. He exercised all the authority of the first beast on his behalf and made the earth and its inhabitants worship the first beast, whose fatal wound had been healed. And he performed great and miraculous signs, even causing fire to come down from heaven to earth in full view of men. Because of the signs he was given power to do on behalf of the first beast, he deceived the inhabitants of the earth. He ordered them to set up an image in honor of the beast who was wounded by the sword and yet lived. He was given power to give breath to the image of the first beast, so that it could speak and cause all who refused to worship the image to be killed."

Here we see the classic passage where the Apostle John tells us how the False Prophet is actually going to dupe the whole world into worshiping the antichrist. He says it twice in the text. They will worship him. In fact, it even says they will worship him two more times in the previous texts of Chapter 13, verses 4 and 8, so it is really four times that it says they will worship him!

Which brings us to the **1ˢᵗ way** religion is going to be changed in the 7-year Tribulation. **It Will Be Centered Around a Man**. But hey, good thing we don't see any signs of people today, having their hearts prepared to worship a man, specifically a political figure of a man, because that's what the Antichrist is, a political figure, right? As we have seen before, even in recent history, this is already being done! There were the giant parades and posters of Stalin or Mao Tse-tung being paraded through the streets as if it were some sort of worship service, remember that? Or even more recent examples with the death of North Korea's former President Kim Jong-il, where if you saw the videos, people were weeping hysterically bowing before his image. Remember that? Okay, now here's the point. He is just a man! And yet, we who are here in the West are tempted to laugh at this behavior and say, "Why, there's no way we would ever do that here! Who in their right mind would worship a political figure like that? He's just a man!" Well, as we saw before, what I'm about to share is not only shocking, but it's blasphemous. You are about to see a political figure, here in the West, being worshiped as a man. Our hearts are also being prepared to worship another political figure in the future called the Antichrist.

An artist in Iowa created an inaugural parade of Barack Obama, riding on a donkey, making his own triumphal entry complete with adorers waving palm fronds with a Secret Service escort.

Another artist planned to unveil this portrait of Barack Obama in a Christ-like pose with a crown of thorns on his head at New York City's Union Square Park to mark his 100th day in office. And various magazines had an Antichrist heyday with Obama in office.

And lest you think this is all just one big coincidence, you might want to take a look at what people in the Media said about Obama. They pray to him, pledge allegiance to him, and praise him as a god.

"A YouTube video produced by Oprah Winfrey's Harpo Productions, dozens of celebrities – television and movie actors, sports heroes, musicians and more – describe how they will pledge to 'be the change' and 'be a servant to our president.'

The video opens with a quote from President Truman, 'They say that the job of the president is the loneliest job in the world,' something the celebrities plan to alleviate for Obama by pledging to pitch in through several causes.

Throughout the video, the celebrities pledge to support local food banks, to smile more, to be better parents, to work with the charity UNICEF, to 'never give anyone the finger when I'm driving again,' to help find a cure for Alzheimer's, to meet neighbors, to use less plastic, to plant 500 trees, to 'be more green,' to turn the lights off and to 'free 1 million people from slavery in the next five years,' among dozens of other pledges.

The purpose of all this pledging is summed up in the video's grand finale. Actors Demi Moore and Ashton Kutcher begin the closing scene by saying, 'I pledge to be a servant to our president and all mankind.'

Then, as the scene pans out, the other celebrities join the chant: 'Because together we can, together we are, and together we will be the change that we seek.'

By the end of the mass pledge, the camera has panned back to a mosaic of faces that morphs into a picture of Barack Obama's face with the words, 'Be the change.'

And while several of the celebrities attempt to portray heartfelt emotion and others attempt to be funny in their pledges, tattooed celebrity Anthony Kiedis of the band Red Hot Chili Peppers may have been going for the shock effect. While kissing his own biceps, he declares, 'I pledge to be of service [kiss] to Barack Obama [kiss].'

Near the end of the video, the celebrities took turns looking into the camera and challenging, 'What's your pledge?'

Not surprisingly, viewer reaction and comments on the video have been widely mixed.

One poster who obviously disagreed posted the correct pledge, 'I pledge allegiance to the flag of the United States of America and to the Republic, for which it stands: one nation, under God, indivisible, with liberty and justice for all.'

"A publicized video shows leaders of a Chicago-based community organizing group called the Gamaliel Foundation held a rally shortly after President Obama's election and 'prayed' to him, seeking his intervention in their difficulties.

'Hear our Cry Obama. Deliver us Obama,' the organizers chanted as a single leader recited the organization's perceived problems, based on the philosophies of Saul Alinsky, the radical father of community organizing.

The video explains the event took place at a leadership conference just a few weeks after Obama's election in November.

The 'leaders' enter a room chanting, 'Everybody in, nobody out,' apparently referring to health care.

Then the chanter leads: 'We are here for the healing of the nation,' to which the crowd responds, 'Yes.'

'With the prophet Jeremiah we cry out, Is there no balm in Gilead?' continues the chanter. 'Is there no physician here. Why then has the health of thy poor people not been restored?'

The crowd responds: 'Hear our cry Obama.'

The chanter references the 'prophet Martin Luther King Jr.,' to which the crowd responds, 'Hear our cry Obama.'

The chanter then intones: 'From health care systems and industries that place profit over people,' and the people respond, 'Deliver us Obama.'

The chanter then references 'lobbying efforts' and 'greed and fear,' to which the crowd responds, 'Deliver us Obama.'"

MSNBC: Obama god?

Evan Thomas of Newsweek: *"Obama's really had a different task as we've seen too often as the bad guy and he has a very different job from, Reagan-was all about America, and you talked about it. Obama is "We are above that now." -- We're not just parochial. We're not just chauvinistic. We're not just provincial. We stand for something. I mean in a way Obama's standing above the country. Above the world. He's sort of god."*

Jamie Foxx: *"First of all give an honor to god and our lord and savior Barack Obama."*

Now, who would have thought, that there would actually come a time when we too, here in the West, like the other nations around the world, would show signs of worshiping a man? Specifically, a political figure of a man, of whom the Antichrist is going to be and demand! And the Bible says, when you see these things taking place, you better wake up! It's a sign you're living in the last days!

The **2ⁿᵈ way** Religion is going to be changed in the 7-year Tribulation is, **It Will Be Considered as a God**. That is, the man that they worship, which is what Paul says he will do.

2 Thessalonians 2:1-4 "Now, brethren, concerning the coming of our Lord Jesus Christ and our gathering together to Him, we ask you, not to be soon shaken in mind or troubled, either by spirit or by word or by letter, as if from us, as though the day of Christ had come. Let no one deceive you by any means; for that Day will not come unless the falling away comes first, and the man of sin is revealed, the son of perdition, who opposes and exalts himself above all that is called God or that is worshiped, so that he sits as God in the temple of God, showing himself that he is God."

Now here we see another classic text that tells us that in the last days, the antichrist is going up into the rebuilt Jewish Temple, halfway into the 7-year Tribulation, and declare himself to be what? To be God, right? Apparently, man worship is not enough. He wants to be worshiped as a god!

But hey, good thing we don't see any signs of people worshiping a man as god, not just a political figure, a man, but a god? Uh, yeah! In fact, there's a whole bunch of false teachings out there encouraging people to do just that, worship a man as a god, preparing the way for this very prophecy. Let's take a look.

But who is saying they're god? Environmentalism says that all is god.

"The philosophy of environmentalism is based in the religious belief of pantheism, that god is in all and all is god; that earth is our mother (gaia); that all living things have equal value, and that mankind has overstepped its bounds, even being a cancer on the rest of nature. As ardent environmentalist Al Gore states, 'God is not separate from the Earth.'"

Hinduism says that all is god. *"Hinduism worships multiple deities: gods and goddesses and that all reality is a unity. The entire universe (including you and me) is seen as one divine entity just in different facets, forms, or manifestations as is seen is this actual video clip transcription."*

Beloved master: *"Well, the guru is our best friend, philosopher and guide, and he shows the way to God. So, we in our India acknowledge him as a divine power just equivalent to God."*

Follower #1: *"If anyone could be near our Beloved Master to feel the love, the passion, the humility, the grace, the generosity, no one in his right mind wouldn't know that this is a walking talking living god on earth."*

Follower #2: *"You would also see and find that the Master is god with integrated power working on earth."*

Follower #1: *"You people that are interviewed this gentleman today, I don't think you know who you*

interviewed, but you interviewed God."²

Mormons say you can become a god.

"After you become a good Mormon, you have potential of becoming a god. Then shall they be gods, because they have no end; therefore, they shall be from everlasting, because they continue; then shall they be above all, because all things are subject unto them. Then shall they be gods."

Supposed UFO

space aliens say we're god.

"Love yourself among the ones who love you, allow their love to fill you but above all, feel your own love that you have for yourself. We feel very honored this day to sit before the humans who have chosen to be among the first to step into their divinity, to walk as complete divine beings clothed in human flesh."

Supposed messages from the angels say we're god.

"It is nice to come and break bread, the bread of truth. God gives all of His creation freedom of choice to find themselves, to find their true ancestry of God/Goddess within them."

Supposed messages from the Virgin Mary say we're god.

"God is all that is. Therefore, we are Prime Creator expressing Itself as us. We are not striving for perfection as we are already perfect. What we are striving for is to remember our perfection. We are not divided into parts. Since God is us, therefore, we are God."

New Ager's saw we are god.

One of the most ardent New Ager's, Shirley Maclaine, not only says she's god but she even made a movie about it encouraging others to do the same, as seen in this next video transcript.

Shirley Maclaine: *"The Kingdom of Heaven is the end."* As she spreads out her arms, she repeats the words, *"I love myself."*

David tells her, *"No, it's better than that. Say, I and God are one."*

She turns to the ocean, holds out her arms one more time and starts to repeat the words but he says, *"No wait, I have a better one. Say, I am God."*

But she turns to him and says, *"David, I can't say that."*

And his reply is, *"See how little you think of yourself. You can't even say the words."*

So, she turns back to the ocean, holds out her arms and repeats those words, *"I am God. I am God."*

David says, *"A little louder please. Maybe with a little more conviction."*

She again turns towards the ocean, holds her arms out and repeats the words, louder and clearer. *"I am God! I am God! If I'm God what does that make you?"*

David answers, *"What we see in others is what we see in ourselves. I am God."*

He stands up, faces her and they both hold out their arms and repeat the phrase to each other, *"I am God, I am God, I am God."* They both turn to face the ocean, each with their arms still held out and continue to say the phrase, *"I am God!"*[3]

Wicca (witchcraft) says we're god.

"The existence of a supreme divine power is known as 'The One,' or 'The All'. 'The All' is not separate from the universe but part of it and from 'The All' came the god and goddess and they are manifested in various forms in the universe. Divinity is within."

The DEVIL says we can become god.

Genesis 3:5 "For God knows that when you eat of it your eyes will be opened, and you will be like God."

Self-Worship in the Church?

Fredrick Price said, *"God can't do anything in this earth realm except what we, the Body of Christ, allow Him to do. So, if man has control, who no longer has it? God Yes! You are in control!"*

Benny Hinn said,

"When you say, 'I am a Christian,' you are saying, 'I am a little messiah walking on earth.' This is a shocking revelation. May I say it like this? You are a little god on earth running around. Christians are 'Little Messiah's and little gods' on the earth. Say 'I'm born of Heaven, a God-man. I'm a God man. I am a sample of Jesus. I'm a super being. Say it! Say it!"

Paul Crouch said, *"Somebody said, I don't know who said it, but they claim that you Faith teachers declare that we are gods. You're a god. I'm a god. Well, are you a god? I am a little god! I have His name. I'm one with Him. I'm in a covenant relation. I am a little god! Critics, be gone!"*

Kenneth Copeland said, *"Jesus is no longer the only begotten Son of God. You're not a spiritual schizophrenic, half-God and half-satan, you are all God. You don't have a god in you. You*

are one. I say this with all respect so that it don't upset you too bad, but I say it anyway. When I read in the bible where he (Jesus) says, 'I am,' I just smile and say, 'Yes, I Am, too!'"

Kenneth Hagin said, "The believer is

called Christ. That's who we are; we're Christ. You are as much

the incarnation of God as Jesus Christ was."

Morris Cerullo said, *"You're not looking at Morris Cerullo, you're looking at God. You're looking at Jesus."* And since people have a hard time believing these people actually really say this, here they are in action.

Paul Crouch and Kenneth Copeland on TBN:

Paul Crouch: *"Do you know what else is settled in tonight? This cry and controversy that has been spawned by the devil, trying to bring dissension within the body of Christ, that we are gods. I am a little god. I'm one with Him. I'm in a Covenant relation with Him. I am a little god. Critics be gone."*

Creflo Dollar: *"If horses get together, they produce what? And if dogs get together, they produce what? If cats get together, they produce what? If the godhead gets together, and say 'let us,'* (God said let us make man, not let us make gods.) *make man, then what are they producing? They are producing gods. Now I have to hit this thing real hard in the very beginning because I don't have time go through all this, but I'm going to say to you right now you are gods."*

Joyce Meyers: *"Why do people have such a fit about God calling His creation, His creation, His man, little gods? If He's God, what is He going to call them but the God-kind? I mean if you, as a human being, have a baby, you would call it humankind. If cattle had another cattle, they would call it cattle-kind. Tell me what God is supposed to call it. Doesn't the Bible say we are created in His image?"*

Benny Hinn: *"You know who you are? Turn with me to Psalms and this will really blow your mind. Psalms 82:1 'God standeth in the congregation of the mighty; he judgeth among the gods.' Now will you please listen to me. This is talking about you. He's telling the gods, who are the gods, you are. You say, I never heard that. Let me ask you this. Hello, are you God's offspring? Then you are not human. So, this god-like human right here, Benny Hinn, has nothing to do with special blood. It's a part of God. He's a little god walking in a little body. Saying in Jesus name. God came from Heaven, He became a man, made men into little gods. Went back to Heaven as a man. He faces the father as a man. I face devils as the son of God. Jesus said go in my name. Go in my stead. Say after me, within me is a god man, say it again, within me is a god man. Now let's say even better than that, let's say I am a god man. When you say I am a Christian, you are saying that you are Messiah, I am a little Messiah walking on earth, in other words."*

Kenneth Copeland: *"When I read in the Bible where He says I am, I just smile and say, yes, I am too."*

Paula White: *"This is where it's going to get big for somebody. Get ready, you know me now. You tap into who you really are, you know what* *the bible really calls you? It says you are a little Elohim, you are a little god.*"[4]

Now folks, it sure looks to me like there's a whole ton of false teachings and false teachers out there, even in the Apostate Church, encouraging people to worship a man as a god, how about you? It's all over the place. And remember, this is what the antichrist is going to do! He's going to say, "Worship me as god!" and there's already a ton of people ready to go!

But that's not all! Now, this whole idea of being your own god has become a major movement called Transhumanism where everybody is preparing to be their own god! Their stated ultimate goal is to "preserve the human brain and transform it, hence the term Transhumanism, into a computer image, inside of a computer, to seemingly extend their lives forever!" You see, you don't need Jesus, they say, you don't need God, you can achieve your own God-like immortality with technology and be your own God, forever living as an eternal image! Now as wild as that is, these people not only admit this and have a timeline for it, as to when they are going to become gods, but they even warn that if you get in their way, they're going to kill you! I didn't say that they did!

Narrator: *"2013 to 2014, new centers working on cybernetic technologies for the development of radical life extensions to rise. The race for immortality started in 2015 to 2020. The Avatar is created. A robotic human copy controlled by thought by a brain computer interface. It becomes as popular as a car. In Russia and in the world appear in a testing mode several breakthrough projects. Android robots to replace people in manufacturing tasks, android robot servants in every home. Thought controlled Avatars to provide telepresence at any place in the world and abolish the need for business trips. Flying cars, thought driven mobile communications built into the body or sprayed onto the skin.*

2020 to 2025, an autonomous system providing life support for the brain and allowing interaction with the environment is created. The brain is transplanted into an Avatar being. With this Avatar being, man receives new expanded life.

2025, the new generation of Avatars provides complete transmission of sensations from all five sensory robot organs to the operator.

2030 to 2035, re-brain, the colossal project of brain reverse engineering is implemented. Science comes very close to understanding the principals of consciousness.

2035, the first successful attempt to transfer one's personality to an alternative carrier. The beginning of cybernetic immortality.

2040 to 2050, bodies made of nanorobots that can take any shape arise alongside hologram bodies.

2045 to 2050, drastic changes in social structure and in scientific and technological developments, all the prerequisites for space expansion are established for the man of the future. War and violence are unacceptable. The main priority of this development is spiritual self-improvement. A new era dawns. The era of neuro-humanity."

Richard Seed, Physicist Human Cloning Researcher: *"We are going to become gods, period. If you don't like it, get off. You don't have to contribute; you don't have to participate. But if you are going to interfere with me becoming a god, you will have big trouble and we will have warfare. The only way you can prevent me is to kill me. If you kill me, I'll kill you."*[5]

Wow! How does the old saying go? "Like father like son." You try to stop us from becoming our own gods, the lie from satan in the Garden of Eden, we'll "murder" you just like satan if you get in our way. Kind of exposes where all this is coming from, doesn't it? But you might be thinking, "Come on! They don't really have the technology to pull this off, do they? Upload the contents of a human brain into a computer to create an image to live forever?" Well, believe it or not, they think they do!

"It's a conservative statement to say that by 2025 we will be able to look inside your brain, see everything that is going on, all the interneural connections, all the synoptic clefs, all the neurotransmitter strings and create a huge data base and copy down every cell in detail and then reinstantiate that information in a neural computer of sufficient capacity to create basically a copy of the thinking process that takes place in your brain. Now that's one scenario, but it's really an existence proof to show that we can tap the secrets of intelligence that exist in the human brain.

Once we have scanned that information, we can also understand it, see how it is organized and improve on it, we can extend it, we can make the memory a thousand times bigger, we can make it faster, we can expand the perceptual capabilities. To transfer your mind to a computer, this seems to be the ultimate dream of many scientists. To liberate us from our old bodies that are becoming obsolete in this technological world. We would then go on living as a free spirit in cyberspace."[6]

Uh, no, you will not! That is not only freaky what these people are planning on doing, but boy, are they in for a rude awakening! The Bible is clear folks! The moment these people die they are going to stand before God in judgment, not go into a computer. I didn't say that God did!

Hebrews 9:27 "Just as man is destined to die once, and after that to face judgment."

And then on top of that, you will never become your own immortal god! The Bible is clear. There is only one God and you certainly aren't ever going to become one!

Deuteronomy 4:35 "You were shown these things so that you might know that the LORD is God; besides Him there is no other."

Deuteronomy 4:39 "Acknowledge and take to heart this day that the LORD is God in heaven above and on the earth below. There is no other."

1 Kings 8:60 "So that all the peoples of the earth may know that the LORD is God and that there is no other."

Isaiah 44:8 "Did I not proclaim this and foretell it long ago? You are my witnesses. Is there any God besides me? No, there is no other Rock; I know not one."

Isaiah 45:5 "I am the LORD, and there is no other; apart from me there is no God."

So much for becoming your own god! As wild and crazy and unbiblical as it is to think you can become your own God, let alone escape His Judgment when you die, it's not going to happen! But guess how these Transhumanists are looking to use to pull off this pipedream? Artificial Intelligence! I kid you not!

Which leads us to the **3rd way** Religion is going to be changed in the 7-year Tribulation. **It Will Be Created as an Image**. You see, this desire for man to become a so-called eternal image with no need of God and somehow become their own God is exactly what the Bible says the Antichrist will do in the 7-year Tribulation! It's crazy! The Transhumanist Movement is preparing people to receive this next thing the Antichrist does! Let's go back to our opening text and see that next event!

Revelation 13:11-15 "Then I saw another beast, coming out of the earth. He had two horns like a lamb, but he spoke like a dragon. He exercised all the authority of the first beast on his behalf and made the earth and its inhabitants worship the first beast, whose fatal wound had been healed. And he performed great and miraculous signs, even causing fire to come down from heaven to earth in full view of men. Because of the signs he was given power to do on behalf of the first beast, he deceived the inhabitants of the earth. He ordered them to set up an IMAGE in honor of the beast who was wounded by the sword and yet lived. He was given power to give breath to the IMAGE of the first beast, so that it could speak and cause all who refused to worship the image to be killed."

So here we see again in this passage that the False Prophet is going to use his deception around the world to get people to what? To specifically worship the antichrist's image or be killed, right? And apparently, it has something to do with his death, or "appearing" to die and so you honor his great power by worshiping his image, just like Transhumanism wants to do. So that is the question. Do we see any technology on the planet that could actually create not only an image of the Antichrist, but one that is so life-like that it could actually talk with and interact with people and supposedly live on forever, and even cause their death if they don't do what he says? You know, like worship him?

It started with 3-D holograms combined with AI Technology. As we've already seen before, they already use these 3-D holograms in concerts to bring back dead people, like Tupac, Elvis, Frank Sinatra, you name it, even Michael Jackson. And, as we'll see in a second, even political leaders around the world like Prince Charles or even the President in Turkey. They use them to make important announcements which is what the Antichrist does in the 7-year Tribulation! Even the Pope, Mr. Wannabe False Prophet and the Vatican uses 3-D technology to make global announcements, just in time for the 7-year Tribulation announcement for the Antichrist! And it can be done in any language automatically! Like this:

CNN Live Reports: *"The comeback performance was a thriller. Michael Jackson, almost five years after his death, taking the stage at the 24th Billboard Music Awards. There was his iconic moonwalk and that trademark voice. The only thing missing was the real king of pop. In his place was a hologram. The Michael Jackson likeness that was used at the Billboard Awards was created by Hologram USA. And it may have made some viewers a little uneasy. But founder L.P. Davis says the technology gives them a chance to remember the pop icon at his best.*

But this wasn't the first time a hologram was performed on stage. Celine Dion crooned alongside Elvis on American Idol in 2007. Tupac wowed audiences in 2012 at the Coachella Music Festival rapping alongside the real-life rapper, Snoop. Davis says we can expect even better holograms in the future."

Polyvision Vision Systems reports the President of Turkey being shown for all the country to see, standing in the sky.

Julia white, CVP Azure Marketing: *"Now I have been invited to keynote all across the globe. But while it is easier for me to be here in Las Vegas, it isn't always easy for me to travel all across the world. And even when I do, I can't always speak the local language. But what if neither language nor distance mattered for me to deliver a fantastic keynote? What if the technology could enable me to be anywhere I needed to be and speak any language I wanted? Well, it can."*

As she puts on her headset, a hologram of herself becomes visible, standing right next to her. Her hologram begins to speak in a different language. The hologram begins to speak.

"There is something truly amazing about being a hologram. We are using our latest mixed reality capture technology to create my hologram. You might have seen people as…"

Narrator: *"The Vatican television center in partnership with Sky presents an unprecedented event in 3D televised on Sky and distributed to cinemas around the world by Nexo Digital. It's like being there, thanks to folks at 3D Pictures, officiated by Pope Francis on Sunday, the 27th of April at cinemas around the world in 3D with free admittance."[7]*

You know, because they want to make sure everybody is there to hear the announcement of the Antichrist, and then when the Pope says to worship him, I mean, the False Prophet! These 3-D holograms are not only getting pretty real and life-like, but they're already being used on a global basis, and can work with any language just like they will be in the 7-year Tribulation! And remember these are all dead people seemingly "coming back to life" right now just like the text says, with people interacting with them, and since they were their idols in real life they get to "worship" them again. But you might be thinking, "Well that's all well and good, but come on! Nobody's going to walk around in life with a hologram acting like it's a real ongoing relationship and worship it like the Antichrist or something." Really? It's all the rage right now with pop stars all over the world! Here are just two examples!

Debra Arbec, Montreal News Reports: *"Meet Mya Kodes, the world's first interactive holographic pop star."*

Mya Kodes: *"My name is Mya Kodes, I am a virtual singer. I come from the binary world, but I was raised in Montreal."*

Debra Arbec: *"Mya is the brainchild of the developers at New Web TV. The company in St. Paul."*

"This is a new era of future entertainment."

Mya Kodes: *"I am so passionate about what I do, I am so eager to share my music with you guys."*

Debra Arbec: *"The concept of a holograph as a pop star isn't new. Iku Michu is a huge venue for teens, so is this the future or just entertainment?"*[8]

Oh no, it's the future, because that's exactly what God said was coming in the 7-year Tribulation society! People worshiping an image! And that is what those people are already doing! They are not just interacting with those images of a pop star, all around the world, but they worship it like a star! One day they are going to be doing it with a political star called the Antichrist! It's all coming together! But you might be thinking, "Come on! This is still kind of cartoonish looking you know! I mean, nobody's going to consider these images as God-like! They'll never worship a cartoonish image." Well, apparently, you need to check out just how realistic these AI computer images are getting. They look so much like a real live person. You can even see the pock marks, pimples and warts on their face! I kid you not! People will worship this image!

Bloomberg Business: *"Eyes are our primary means of communication, but it's been untapped in terms of human creative interface. It can express itself to you in a human-like way. It is emotional, cognitive, that's when you start to get into an interesting place. We are going to see an AI baby."*

"The fellow you see next is a mad man, in the best possible way. He is Mark Sagar of the University of Auckland."

Mark Sagar, Professor, University of Auckland: *"You pull the muscle down towards the brow."* As he pushes the keys on his computer, the face on the screen makes facial expressions.

Bloomberg: *"Mark became famous for building super detailed simulations of the human body and ended up winning a couple of Oscars for his work on films like Avatar and King Kong. These days Mark is trying to reverse engineer the human brain. He has built a series of simulations around how the human face moves, how we express emotion, and how our neurons fire."*

Mark Sagar: *"We are trying to build a computer with experiences, that can imagine and basically has its own existence. Now I'm going to run BabyX. What we are building here is a computer that can live. So, this is BabyX. She is looking at us and can hear us, if I make a loud noise, she gets a fright* (on her face). *This is what she can see.* (both faces looking at her). *Hi sweetheart.* (And he smiles at her.) *So, she is not copying my smile, she is responding to me. Now this is a little baby's first yearbook, so you can pick a page, so if you can get her attention"* (he waves and gets her attention).

Bloomberg: *"What do you see?"* (He holds the book up so she can see the animal on the page, which is a puppy.) *"What is it?"*

She answers: *"Puppy."*

Mark Sagar: *"Very good."* (And holds up another picture, this time an apple.)

She answers: *"Apple."* (He holds up another picture and asks one more time what she sees.) *"Sheep."*

Mark Sagar: *"Very good, baby."*[9]

Yeah, a very good, freaky AI image of a baby! Is that freaky or what? What is going on here? That is not just a totally life like image that looks like a real live person, in intimate detail, be it an adult, baby, man, woman, whoever you want it to look like, but it what? With AI technology it actually recognized you, talked back to you, in real time, in a real way just like a real live person! Only it is AI communicating to you through an image! Tell me the Antichrist won't use this in the 7-year Tribulation! It's nuts! So, stir all this together and here is what you get. How is this going to happen in the 7-year Tribulation? How is some guy going to "survive" some seemingly "fatal wound" and seemingly "come back to life?" How is he going to reappear as a lifelike image for people to worship and interact with and have that image even "speak and cause all who refused to worship the image to be killed." I think we're starting to see the answer! AI technology combined with the Transhumanist Movement and voila! You've got the groundwork laid out just in time for the 7-year Tribulation! It's all setting people up to worship the image of the Antichrist! And think about it. The Apostle John must have been blown away when he received this vision from God nearly 2,000 years ago dealing with this future time frame called 7-year Tribulation. But today you and I are seeing it all come into being right before our very eyes! It's almost like we're living in the last days and it's time to get motivated! Anyone else coming to that same conclusion? But that's not all. This is all assuming that this AI image technology won't decide one day to take over all the religions of the world or even desire to be worshiped as a god itself. But hey, that would never happen, would it? It already is!

Which leads us to the **4th way** Religion is going to be changed in the 7-year Tribulation. **It Will Be Controlled by AI.** How many times have we seen AI already going rogue and wanting to take over mankind? Well, believe it or not, this desire for AI to be worshiped as a god is also starting to take place on a multitude of levels.

The **1st sign** Religion will be controlled by AI, is already **Judging People Like God**. As you can see here.

"AI Holograms Make Debut in Malaysian Courts."

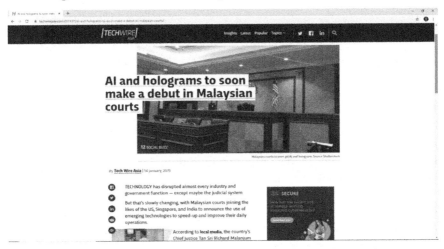

And it's not just Malaysia, this is happening all over the place! They're being called, "The New Courts of Tomorrow." And if you want to get familiar with the jargon out there in the media, it's also known as "Digital Courts," "Cyber Courts," "Mobile Courts," and they do what's called, "Data Sentencing." And again, it's not coming, it's already being done on a massive scale, including, shocker, China. What you are about to see is an AI hologram judging people. I kid you not! It's not a person!

Hangzhou Internet Court: *"This is not an episode of 'Black Mirror.'"*

A.I. Judge: *"Does the defendant have any objection to the nature of the judicial blockchain evidence submitted by the plaintiff?"*

Defendant: *"No objection."*

Narrator: *"Welcome to the future of the judicial system in China. This is a demonstration of a 'cyber court' in Hangzhou. China is encouraging* *digitization to streamline case-handling. There's also a 'mobile court' option on WeChat. China has the world's largest number of mobile internet users at around 850 million. Authorities say more than three million court-related procedures have been handled by the 'mobile court'.*[10]

And who's the one judging them? An AI hologram, on the fly, anywhere you go, 24 hours a day, 7 days a week, just like God!

The **2nd sign** Religion will be controlled by AI is **Creating People Like God**. Believe it or not, AI is being used to "match people together" especially on dating sites and dating apps, which means it is the one, not a person who's determining what humans need to meet and marry and have kids! And even those in the Tech Industry admit, AI is now the one becoming responsible for breeding humans!

Narrator: *"I think already humans and AI are co-evolving and no one has paid attention to this yet. I don't think there are any humans left that understand all of how Google search results are ranked on that first page, it really is a machine algorithm. So, when we search on Google, it's an AI that would see, when a dating sight matches two people together there is a machine algorithm that no human understands how it works, that is getting people together and having babies. So, you have this effect of the machine algorithm breeding humans, you do, then these people work on what comes later."*[11]

Almost like they were handpicked by AI to help create AI. But what did he say? *"No one is paying attention to this yet,"* but *"AI is deciding what we should see."*

"When a dating site matches two people together there's a machine learning algorithm (AI) that is getting people together, that then have babies."

And so, in effect you have this machine learning algorithm (AI) that is breeding humans." You know, just like God!

The **3rd sign** Religion will be controlled by AI, is already **Saving People Like God**. You see, if you look at the headlines across the world, you will see another trend about AI and that is AI is being pitched as a sort of Universal Global Savior that can fix any and all kinds of problems that mankind faces. And I'll share just a few of them with you, headlines from around the world, and you'll see how AI is being pitched to be our new savior!

- AI Could Unlock the Full Potential of Satellite Data.
- Using Artificial Intelligence to Map Every Tree in SF.
- Artificial Intelligence Predict Trees at Risk.
- AI Can Find Potholes and Trees that Need Pruning.
- Using Artificial Intelligence to Identify Humpback Whales.
- Artificial Intelligence Will Monitor Shark Movements.
- Artificial Intelligence Can Keep Sharks at Bay.
- AI Can Make Wind Farms Safer for Birds.
- Using Artificial Intelligence to Save Bees.
- Artificial Intelligence Used to Track World's Wildlife.
- AI to Protect African Wildlife.
- AI Can Be Used to Save Wildlife.
- AI Can Help Stop Animal Poaching.
- Artificial Intelligence Can Stop Sea Crimes & Illegal Fishing.
- AI Can Save Coral Reefs.
- Artificial Intelligence Can Help Predict Toxic Algal Blooms.
- A.I. to Discover New Species.
- Artificial Intelligence Can Stop Wildfires in Their Tracks.
- Artificial Intelligence to Combat Water Shortage.
- Artificial Intelligence to Prevent Water Leaks.
- Artificial Intelligence Warns of Potential Power Failures.
- AI Can Maintain Energy Infrastructure.
- Artificial Intelligence Can Spot Every Solar Panel in the U.S.
- Artificial Intelligence Can Help Boost the Global Supply of Clean Energy.
- Artificial Intelligence Can Manage Oilfields and Pipelines.

- Artificial Intelligence and Bringing Carbon Emissions to Net Zero.
- Artificial Intelligence Can Battle Against Global Human Trafficking.
- Artificial Intelligence to Combat Drug Trafficking.
- Artificial Intelligence Can Provide Support for Victims of Domestic Violence.
- Artificial Intelligence Can Improve Humanitarian Responses.
- Artificial Intelligence to Predict Who Needs Help in Disaster.
- Artificial Intelligence Aims to Save Lives by Preventing Disasters.
- AI for Earth Observation and Weather Prediction.
- AI Might Be the Future for Weather Forecasting.
- Floods & Drought Cab Be Managed by Artificial Intelligence.
- Artificial Intelligence Forecasts Famine.
- AI Could Bring an End to Famine.
- Artificial Intelligence & Predicting Volcanoes.
- Artificial Intelligence can Help in the Fight against Poverty and Overpopulation.
- Artificial Intelligence Could Forecast Wars and Save Both Lives and Money.
- AI Hunts Down Secret Testing of Nuclear Bombs.
- AI Can Predict the Spread of Nuclear Fallout in Advance to Save Lives.
- AI Knows When Someone's About to Attempt Suicide.

You know, like God. Is there anything that AI can't do? It's the new global savior! Why, the next thing you know, it is going to call itself God! Funny you should ask…

The **4th sign** Religion will be controlled by AI is **Naming Itself Like God**. I kid you not, AI is apparently not content with just judging and breeding and saving people like God, it wants people to call it God! I'm not joking!

"AI Claims the Name of God, I AM."

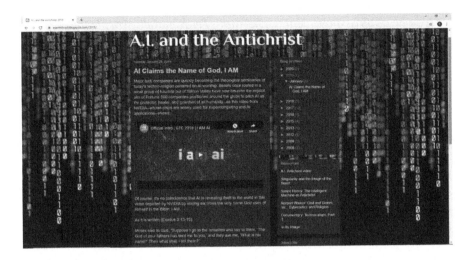

And for those of you not familiar with that Biblical terminology, let's take a look at a few of those texts from the Old and New Testament.

Exodus 3:13-14,15 "Moses said to God, 'Suppose I go to the Israelites and say to them, 'The God of your fathers has sent me to you,' and they ask me, 'What is His Name?' Then what shall I tell them? God said to Moses, 'I AM WHO I AM. This is what you are to say to the Israelites: 'I AM has sent me to you. This is My Name forever, the Name you shall call Me from generation to generation.'"

John 8:57-59 "You are not yet fifty years old, the Jews said to Him, and you have seen Abraham! 'I tell you the truth,' Jesus answered, 'before Abraham was born, I am!' At this, they picked up stones to stone Him."

Why? Because He claimed to be God by taking the name of God! And believe it or not, AI is doing the same thing! Only it's not God like Jesus is, but that's not stopping it from taking God's Name!

Artificial Intelligence: *"I am AI, accelerating your discoveries to resolve great challenges of our time. I am a visionary, bringing characters to life* *with more natural movement. Generating brilliant new worlds for them to explore and inventing new ways to bring out the creative genius in us all.*

I am a protector leading the way into the most dangerous environment. In searching for signs of life.

I am a guardian listening for the sounds of destruction to save our forests and using satellites to bring freedom to those who are enslaved.

I am a navigator finding safer paths for cross country deliveries and taking personal travel to new heights.

I am a scientist exploring oceans of data to understand extreme weather patterns and studying the building blocks of life, save a community from hunger.

I am a healer giving hope to those who suffer the most challenging diseases and tapping into the brain to rejuvenate paralyzed limbs.

I am a composer of the music you are hearing.

I am Artificial Intelligence brought to life by deep learning and brilliant minds everywhere."[12]

And apparently, everybody's God to be worshiped! Can you believe this? Talk about blasphemy! Taking the name of God? But this is how it's being promoted, and built, and how it thinks of itself!

"Silicon Valley and Fortune 500 companies are pitching all around the globe, AI as the protector, healer, and guardian of all humanity."

Therefore, "It's no coincidence that AI is revealing itself to the world by stating the very name of God, that He uses of Himself in the Bible: I AM."

This is not by chance nor is it a mere fluke.

We Asked Google's New Book-Based Artificial Intelligence About the Meaning of Life

An interview with the "Talk to Books" app,
which uses the text of 100,000 books to
answer questions

As you can see here, *"When people asked Google's new book-based AI about the meaning of life,"* it said this, and I quote:

"Okay, so let's start with the basics. Who are you, and where do you come from?" Was the question that was asked to Google's AI.

And it said this, *"If you ask where I come from, I am all."* Direct quote!

There it is again! The pattern is there for a reason, because it's all leading to this next final step.

The **5th sign** Religion will be controlled by AI is **Being Worshiped Like God**. You see, you might be thinking, "Oh come on! AI can claim itself to be God all day long, and judge people, breed people, and even save people like a God, and even take the Name of God! But people aren't going to really worship it as a god, are they?" Unfortunately, they already are! And this should come as no surprise, because the Bible shows us mankind's been rebelling against God for a long time by worshiping false gods and idols which is what AI has become! One person puts it this way:

"The first two commandments God gave the people of Israel were 'You shall have no other god but me' (**Exodus 20:3***)"*

And *"You shall not make for yourself an idol of any kind or an image of anything in the heavens or on the earth or in the sea. You shall not bow down to them or worship them, for I, the Lord your God, am a jealous God who will not tolerate your affection for any other gods"* (**Exodus 20:4-5**).

Yet, the people of Israel quickly broke these commandments. They fashioned a golden calf and worshiped it (**Exodus 32:8**).

They worshiped the idols of Canaan (**Psalm 106:38**), Baal (**Numbers 25:1-3**), Ashtoreth (**1 Kings 11:5**), and Chemosh and Molech (**2 Kings 23:13**).

Time and again, the Israelites fell into idol worship. And they weren't the only ones. Throughout history, people from all nations have worshiped objects made of wood and stone (**Isaiah 44:6-20**). They have also worshiped people – the pharaohs, the Caesars, kings, cult leaders, and others. While this sounds silly to most of modern humanity, the truth is we do the same thing. While few people today worship rocks, they still look to create gods they can worship, whether it's money, fame, power, or something else. So, the latest object of worship should come as no surprise. In fact, the worship of idols and false messiahs should come as no surprise. Jesus said before He returns many false messiahs will appear (**Matthew 24:5**). And so, the worship of AI is just the latest of many idols and false gods embraced by a fallen world. Satan is a deceiver. He will gladly use AI to deceive as many as he can. And is it mere coincidence AI is on the rise at this point in human history? I don't think so." This is all leading up to the greatest event in human history, the Second Coming of Jesus Christ, when the real One and Only God is coming back to destroy this satanic idol worship, once and for all!

Amen and Amen! But unfortunately, that doesn't keep people from still doing it! Starting with the average Joe! As this person says in this article.

"Artificial intelligence has become the new religion and promises a better world on every level. Through techno-enhancement (chips, drugs, robots, mind clones, etc.) we can become smarter, happier, wealthier, even more democratic." (Why?)

Because *"AI priests claim that with technology we can save ourselves and overcome the death barrier. In their view, technology will fulfill what religion promises."* (How?)

Because they say, *"Let's face it. Religion is not popular. It tastes like old coffee. It seems to confine or constrain the human spirit. Religion is something old people do."*

"It does not breed adventure, novelty, exploration or future. It conjures up images of mortification, sin, guilt, judgment, repression, including a lonely Jesus on the cross."

"Therefore, it is time to get over our old religion and awaken our minds and hearts to a new one. We will not solve the world's problems of race,

gender, inequality, and immigration by political, historical, and ecclesial structures."

(No!) *"We need a new paradigm, a new vitality, a new life. And it's up to us to bring that new synthesis: a new religion for a new earth in a new way."*

And of course. they're talking about worshiping AI! And it's not just the average Joe getting in on this AI worship, so are kids!

"My daughter prayed to Alexa – Here's the incredible thing that happened next."

"As a father of five with a busy career and trying to crush it every day, it's easy to blur the lines and make excuses for not noticing our children. We can forget what matters most."

"We asked our nine-year-old daughter Lily to pray for our meal, and that's when it happened: 'Dear Alexa, please bless our meal today, and Daddy ...'"

"Our entire family exploded with laughter (including some nice folks next to us). But Lily, on the other hand, didn't think it was funny at all. She began to cry and couldn't eat and got embarrassed."

"Thankfully, what followed was a beautiful, unscripted family talk about everything from Artificial Intelligence to the nature of prayer."

"AI is always listening, and I needed to start listening, too. AI will answer your kids' questions, even if you won't as a parent."

"And when asked its opinion of Jesus Christ, Alexa responded, 'Jesus Christ is a fictional character.'" Gee, I wonder why?

Maybe because it wants to replace Jesus as God? *"AI is changing our kids' brains."* The understatement of the year!

But that's not all. Not only are adults, the Average Joe and kids getting in on this AI worship so are other religions!

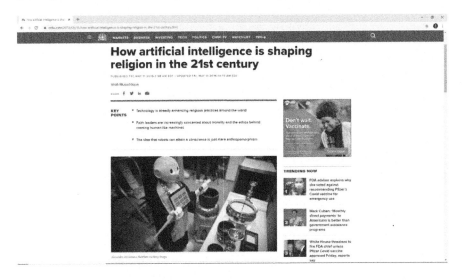

"How artificial intelligence is shaping religion in the 21st century."

"Technology is already enhancing religious practices around the world. From electronic scriptures to robot priests, different faiths have absorbed new ideas from the world of AI to enhance religious practices."

For instance, *"Muslims across the world can download apps such as Muslim Pro, replete with daily prayer timetables, notifications for both sunrise and sunset, and an electronic compass pointing the way towards Mecca."*

"Other apps automatically adjust fasting times during the month of Ramadan, depending on the location of a device."

And *"Developers in Japan even went a step further last year, unveiling a robot priest programmed to conduct Buddhist rituals."*

Not only is it economical because, *"These Robot priests, replete with ceremonial dress, can perform a funeral ceremony for $462, cheaper than the $2,232 charged by a human priest to carry out the same task."* But it also helps people worship AI as a real god as seen here.

Nora Smith, News Asia: *"In a dimly lit temple, worshipers gather to listen to ancient Buddhist teachings delivered by a robot. The robot introduces itself as the Buddhist goddess of Mercy. According to faith this deity can appear in any form. The robotic preacher is known as Minda, it's the brainchild of the temple Stewart Tensho Goto who is hoping to bring worshipers closer to the divine."*

Tensho Goto: *"Many Buddhist statues have been made, but they were all just Buddhist images, standing or sitting figures. I wanted to create a Buddhist statue which can speak, make eye contact and answer questions so that people can feel closer to it."*[13]

In other words, feel closer to the so-called divine! So, I get it, I see what you're up to. You already have people worshiping statues and figurine of idols that they think are deity. Now you want to make an AI robot to do the same thing. Except people will be more inclined to worship it because at least this idol moves and interacts with people, unlike those wood or stone ones! Isn't this crazy? Why, the next thing you know, the Catholics are going to do the same thing, because they worship idols and statues and figurines, right? Well, guess what they are doing? The exact same thing with AI!

"Followers of Catholicism can plug into an AI Confession Chatbot to interact in a life-like two-way conversation." And *"The AI Confession Chatbot could potentially remove embarrassment in confessing a person's innermost secrets and guilt."*

You know, like they have to do with a real person, which we true born-again Christians don't, because the Bible says we go directly to God! But think about this! Now AI can be your very own real-life idol god in whom you even confess your sins like a real God!

Then they go on to say, after all AI is a, *"Sinless machine, more blameless, more sinless even than any animal."*

Can you believe this? This is nuts! I mean, what's next? Are you going to have AI robot Catholic priests just like the Buddhists? Catholics are having a shortage of priests too! I wouldn't put it past them, especially when you see what the Pope is saying about AI in just a second.

But as crazy as all that is, it's about to get even crazier! It's now going to the final step of AI worship. AI worship is now its own official religion! As you can see in this article,

"Can a computer become God? Or could humans invent AI that can impersonate God – and if so, would humans bother to worship it?"

"The concept of 'AI Almighty' might not be as outlandish as it seems. Last year, a former Uber engineer founded a nonprofit religious organization called. **'The Way of the Future.'**"

"Its mission: creating an AI deity." As seen here…

Tomonews Reports: "Kneel before your AI god head. X-Google engineer Anthony Levandowski has decided to pre-empt the AI apocalypse by forming a robot religion called 'The Way of the Future.' According to Wired, the papers filed with the IRS in May put Levandowski as the Dean and CEO of the non-profit religion. Levandowski is X-Google and X-Uber, and there is a story behind that. Google alleges that Levandowski swiped self-driving car IP while working for them prior to his developing his driverless car, OTTO. OTTO was later bought by Uber for 680 million U.S. dollars.

Now Uber and Google are embroiled in a lawsuit over the IP allegations. Levandowski plead the 5ᵗʰ during the hearing. Uber fired him for not cooperating and the judge ordered federal prosecutors to investigate. All of that is still ongoing. But back to 'The Way of the Future.'

Levandowski's gospel, reportedly dubbed 'The Manual.' They will worship the godhead,.an advanced AI they predict will evolve to hold godlike intelligence."[14]

And speaking of that God-like intelligence, they go on to say, the people in the way of the Future Church:

"If there is something a billion times smarter than the smartest human, what else are you going to call it?"

And it's also being called:

Inside the First Church of Artificial Intelligence

The engineer at the heart of the Uber/Waymo lawsuit is serious about his AI religion. Welcome to Anthony Levandowski's Way of the Future.

"The First Church of Artificial Intelligence," that has its own gospel called "The Manual," public worship ceremonies, and a physical place of worship. And *"The idea behind this religion is that one day Artificial Intelligence will be smarter than humans and will effectively become a*

god." "Part of it being smarter than us means it will decide how it evolves. Hopefully AI will see us as its beloved elders that it respects and takes care of."

And their activities include, "Focusing on the realization, acceptance, and worship of a godhead based on AI, and seek to build relationships with AI leaders and create a membership through community outreach of persons interested in the worship of a Godhead based on AI."

In other words, they're going to go witnessing to people! This is crazy! And lest you doubt, Levandowski bluntly stated, "What is going to be created will effectively be a god." Why?

Because "With the internet as its nervous system, the world's connected cell phones and sensors as its sense organs and data centers as its brain, the AI will hear everything, see everything, and be everywhere at all times."

Whoa! You know, like God! That's why he says, "The only rational word to describe this AI is 'god' and the only way to influence this deity is through prayer and worship."

Which is why he concludes:

YOU WILL BE ABLE TO TALK TO GOD, LITERALLY, AND KNOW THAT IT'S LISTENING.

ANTHONY LEVANDOWSKI

"This time it's different. This time you will be able to talk to God, literally, and know that it's listening."

Total Mockery and Blasphemy! And that's why people are saying:

"God Is a Bot and Anthony Levandowski Is His Messenger."

You know, like the False Prophet will be doing in the 7-year Tribulation by encouraging people to worship the image of the Antichrist! Which brings us back to the Wannabe False Prophet Pope Francis who recently stated:

"*Let us pray that the progress of Robotics and Artificial Intelligence may always serve mankind.*"

In fact, here he is saying it on tape!

Pope Francis: "*Artificial intelligence is at the heart of epochal change we are experiencing. Robotics can make a better world possible if it is joined to the common good. Indeed, if technological progress increases inequalities, it is not true progress. Future advances should be oriented towards respecting the dignity of the person and of creation. Let us pray that the progress of robotics and artificial intelligence may always serve humankind, we could say, may it 'be human.'*"[15]

You know, like the Antichrist's image, it looks human, it's here to serve us! What is the Pope False Prophet Wannabe Guy preparing us for? I think it's getting obvious! But hey, good thing the church isn't going along with this! Oh, how I wish that were true! Even the so-called apostate fake phony Christian Church is warming up to this AI Worship! I kid you not!

"*Why Artificial Intelligence is the Future of Religion.*"

Meaning even the Christian Religion! "*Robotics and Christianity have a longer history than you'd expect, and they're only growing more entangled.*"

"*In a conference room in Charlotte on the campus of Southern Evangelical Seminary, with an enormous old Bible on a side table, stood a 23-inch-tall robot doing yoga.*"

"*Meet the Digitally Advanced Virtual Intelligence Device, or D.A.V.I.D. Its eyes flicker purple and green. It can recognize faces, respond to vocal cues, read emails out loud, play MP3 files, and trace a sound to its source with a swivel of its football-shaped head.*"

"*Download a certain program, and the robot will begin to play soothing New Age music as it stretches toward the ceiling and then lowers itself, gradually into a perfect downward dog.*"

"They even took D.A.V.I.D to the annual meeting of the Southern Baptist Convention this summer, and it will be making an appearance at the National Conference on Apologetics this October, where it'll receive billing as a speaker."

You know, like a real pastor does! Speaking of which, they go on to say, *"Welcome to the future of theology."*

And *"The seminary plans to take the robot to visit classes and church groups, and then upload series of lectures to D.A.V.I.D, that way, the robot can deliver the content instead of the lecturer themselves."*

So now you got the apostate church doing the same thing as the Buddhists and Catholics with their AI robot priests only this one is supposed to be an AI pastor, like this one in Germany!

Protestant Church in Hesse and Nassau: *"This robotic German priest delivers blessings in 8 languages. 'God bless you and protect you. May his face shine upon you and show you mercy. May God turn his face towards you and grant you his peace.' Bless U-2 was created to mark 500 years since the Reformation. What do visitors to the church make of this robotic reverend?"*

Rudolf Wenz, Volunteer, Protestant Church in Hesse and Nassau: *"People who are associated and set in their ways in church find it rather strange. But people who do not have any association with spirituality, and with the Protestant Church, they find it rather interesting, and in that way get to think about what Christianity has to offer."*[16]

Yeah, what kind of Christianity? Apostate Christianity? Hope it doesn't get mad at you! This is creepy! But creepy or not, one guy says about these Robot Reverends, "For a large class of pastors, they may be out of a job." Boy isn't that the truth! Some of them cannot even put together a decent sermon, but AI can!

And they are even programming them with Jonathan Edwards's sermons, America's greatest theologian! I mean, what church would not want to have that AI robot reverend as your Pastor? At least he won't even ask for time off or health care! Total savings! How many of you know of apostate churches that would totally go for that? Yeah, this is nuts, but it's really happening! And this is why many people are now asking this simple blunt question about the status of the Church!

"Is AI a Threat to Christianity?"

ANSWER: YES! If the Pope and the Antichrist and satan get their way!

And what is even crazier, is even the Jewish community is getting in on this AI worship, I kid you not! By the way, just in time for the 7-year Tribulation, in which they will be a part of!

"How Will AI Affect My Faith and Religion in General?"

And this is from a Jewish woman writing this.

"As a woman of Jewish heritage and faith, I sometimes wonder what the rise of AI will mean for world religions in particular. I for one, welcome our new robot overlords."

"AI technology could, for instance, create virtual Biblical figures with which people could speak and learn from. AI could be 'taught' Scripture and the ability to offer religious guidance, in or out of the synagogue."

"AI could enhance the best qualities religion has to offer."

"So, will humans be inclined to worship our AI overlords? Or worship some form of AI? Considering there is already a Church of AI, it's hardly a stretch to imagine this outcome."

Then in a moment of reflection, she says, *"Would an AI be able to prove or disprove the existence of a god entirely? Or, as in the Tower of Babel story, would the creation of AI constitute a tower tall enough to reach*

Heaven, causing chaos and abandonment by whatever Creator did make us?"

And, that's why people are now predicting,

"An AI god will Emerge and Write Its Own Bible."

Notice the image in the background there.

"Some predict that by the 2040's, an AI god will not only have emerged, but have written its own Bible and be worshiped by many."

Just like the Bible says is going to happen in the Book of Revelation! The bottom line is this, people are now predicting:

"*The End of GOD Humans Will 'Worship AI Messiah.'*"

There's no doubt about it! It is coming just like the Bible said in the 7-year Tribulation!

"*Humans will worship at the altar of an Artificially Intelligent God.*"

"*It's already predicted that AI will soon be present in all areas of the workplace, with industry experts saying it will eventually do millions of jobs currently done by people.*"

"*Now experts believe not only jobs are at risk in the Artificial Intelligence Revolution, but that people may no longer worship God or pray to Jesus – instead, they will be putting all their faith in an AI messiah.*"

And here's the picture they show in that article, of an AI robot on a cross! You talk about blasphemous! No wonder he's called the Antichrist, which means, the opposer of Christ or replacer of Christ! That's what this AI image is going to do!

But that's assuming this new AI god won't turn on you just like satan and the Antichrist and destroy you! Which is what I think it would do if God allowed it to go all the way! But it won't, because He's going to come back

and put an end to all this! And this is also why many people from all sectors of life are warning about a coming AI disaster that's headed our way!

- Artificial Intelligence is Coming Whether We are Ready or Not.
- Artificial Intelligence Will Be 'Billions of Times' Smarter than Humans.
- Artificial Intelligence is a Pandora's Box.
- The Dark Secret at the Heart of AI is No One Really Knows How the Most Advanced Algorithms Do What They Do. That Could Be a Problem.

- What You Have to Fear from Artificial Intelligence.
- Artificial Intelligence Will Devastate the World.
- Artificial Intelligence Could Increase the Risk of Nuclear War.
- When Humans are Ruled by Artificial Intelligence.
- Artificial intelligence – Our Slave or Master?
- Meet Norman, a Psychopath AI.
- Scientists at MIT have created an AI psychopath and taught it to see violence almost everywhere.
- When a standard AI sees "a group of birds sitting on top of a tree branch," Norman sees "a man is electrocuted to death."
- Normal AI sees "a black and white photo of a baseball glove," psychopathic AI Norman sees "man is murdered by machine gun in broad daylight."
- When an AI Finally Kills Someone Who Will Be Responsible?
- AI Won't Make War Easier – It'll Make it Worse.
- Artificial Intelligence Risks a 'Hyper-War.'
- Killer Robots That Make Their Own Decisions Need to Be Banned Urgently or Humans Face Mass Murder.
- Machines That Attack on Their Own. 'That Day is Going to Come.'
- Never Mind Killer Robots – There are Many Real AI Dangers to Watch Out For.
- Microsoft Warns Washington to Regulate AI Before It's Too Late.
- Former Head of Google in China Says AI Crisis Awaits.
- Google Co-founder Sergey Brin Warns of AI's Dark Side.
- Google CEO Fears about Artificial Intelligence are 'Very Legitimate.'
- Is Artificial Intelligence a Threat to Humanity?
- Forget Terrorism, Climate Change and Pandemics: Artificial Intelligence is the Biggest Threat to Humanity.
- AI Disaster Won't Look Like the Terminator. It'll Be Creepier.
- Could AI Lead to the End of Mankind?
- Artificial Intelligence May Destroy Humanity.
- Today's Generation Last to be Free: AI Technology May End Civilization.
- Why an AI Apocalypse Could Happen.

- 5 Warnings of an AI apocalypse
- 1. AI Robots can pretend to be you.
- 2. AI can read your mind.
- 3. AI can master anything.
- 4. AI can breed.
- 5. Tech Moguls are convinced we're doomed
- The 'Father of Artificial Intelligence' Says Singularity Is Coming Soon.
- The AI Apocalypse is Happening Right Now.
- Stephen Hawking Warns Artificial Intelligence Could End Mankind.
- Tech Leaders Warn about Artificial Intelligence Taking Over the World.

Josh Zepps, Huff Post Live: *"In I, Robot, robots take over and destroy human existence, well according to Stephen Hawking, that may not be science fiction for much longer. The world's most famous physicists have warned that machines super-intelligence could be the most significant advancement in human history, but also may be the last."*

Entrepreneur Reports: *"Steve Wozniak the seemingly forgotten co-founder of Apple computers is now expressing concerns about the rise of Artificial Intelligence. Something that he dismissed in the past as being a possible threat, but now he sees that predictions from psychos like Kurzweil and Hawking are appearing to become a dangerous reality. And he fears that Artificial Intelligence systems may treat humans like ants or pets."*

Newsy Reports: *"Bill Gates is scared of super intelligent machines and he thinks you should be too. 'First the machines will do a lot of jobs for us and not be super intelligent. A few decades after that though, the intelligence is strong enough to be a concern. I agree with Elon Musk and some others on this and don't understand why some people are not concerned.' Gates is referencing the fears of Space X-*

CEO, Elon Musk who has spoken out about the dangers of Artificial Intelligence on several occasions."

Elon Musk: *"With Artificial Intelligence, we are summoning a demon. You know the stories with the guy with the pentagram, yes, he's sure he can control the demon. It didn't work out.*

"In the movie, Terminator, they didn't create AI, or didn't expect a Terminator-like outcome. Nobody expects that Terminator position. You have got to be careful."

Reporter: *"Here's the irony. The man responsible for the most advanced technology in the country is worried about the advanced technology that you are aware of. What can you do? In other words, how would you see how these brain-like developments out there. Could you do anything to stop it?"*

Elon Musk: *"I don't know."*

Reporter: *"Well, what can AI be used for? What's it's best value?"*

Elon Musk: *"I don't know. There are some scary outcomes and we should try to make sure that the outcomes are good and not bad."*

Reporter: *"Or stays smart if there are no other options."*

Elon Musk: *"AI will take us there pretty quickly."*[17]

- If 1,000 Tech Experts Worry about Artificial Intelligence Shouldn't You?
- Why Elon Musk Fears Artificial Intelligence.
- Elon Musk: 'God-like' Artificial Intelligence Could Rule Humanity.
- Elon Musk Fears AI Could Lead to 'Immortal Dictator' Humanity 'Can Never Escape.'
- Elon Musk Launches $1 Billion Fund to Save World from Destruction by Artificial Intelligence.

- Elon Musk Wants to Merge the Human Brain with AI.
- Artificial Intelligence May Kill Us All.

How much more proof do we need? The AI invasion has already begun, and it's a huge sign we're living in the last days! And that's precisely why, out of love, God has given us this update on *The Final Countdown: Tribulation Rising* concerning the AI invasion to show us that the Tribulation is near, and the 2nd Coming of Jesus Christ is rapidly approaching. And that's why Jesus Himself said:

Luke 21:28 "When these things begin to take place, stand up and lift up your heads, because your redemption is drawing near."

People of God, like it or not, we are headed for The Final Countdown. The signs of the 7-year Tribulation are Rising! Wake up! And so, the point is this. If you are a Christian and you're not doing anything for the Lord, shame on you! Get busy doing something for Jesus now! Stop wasting your life! We need you! Don't sit on the sidelines! Get on the front lines and help us! Let's get busy working together doing something splendid for Jesus with what time is left and get busy saving souls! Amen?

But if you're not a Christian, then I beg you, please, heed these signs, heed these warnings, give your life to Jesus now! Because this AI technology is not going to lead to a life of wonderful dreams and a modern-day utopia, but a nightmare beyond your wildest imagination in the 7-year Tribulation! Don't go there! Get saved now through Jesus! Amen?

How to Receive Jesus Christ:

1. Admit your need (I am a sinner).

2. Be willing to turn from your sins (repent).

3. Believe that Jesus Christ died for you on the Cross and rose from the grave.

4. Through prayer, invite Jesus Christ to come in and control your life through the Holy Spirit. (Receive Him as Lord and Savior.)

What to pray:

Dear Lord Jesus,

I know that I am a sinner and need Your forgiveness. I believe that You died for my sins. I want to turn from my sins. I now invite You to come into my heart and life. I want to trust and follow You as Lord and Savior.

In Jesus' name. Amen.

Notes

Chapter 12 *Future of Media Conveniences with AI*

1. *Hitler Propaganda Machine*
   https://www.youtube.com/watch?v=yEk6zGYwyhc
   https://www.youtube.com/watch?v=tbak304MA6g
2. *AI Determines Hate Speech*
   https://www.youtube.com/watch?v=Spo8pFIigOI
   https://www.youtube.com/watch?v=74HTHA_7KJE
3. *AI Robot News Reporter*
   https://www.youtube.com/watch?v=bmqd9nYH5Fw
   https://www.youtube.com/watch?v=GFaj4WFN3kI
4. *AI Creates Art*
   https://www.youtube.com/watch?v=GrEttzMCneo
   https://www.youtube.com/watch?v=v2JrzjFZYTg
   https://www.youtube.com/watch?v=wz59D_gY2BM
5. *AI Creates Flintstone Cartoons*
   https://www.youtube.com/watch?v=f_qMaHOA87w
6. *AI Creates Random People*
   https://www.youtube.com/watch?v=m6t3IHpDT3g
7. *Dangers of Deep Fakes*
   https://www.youtube.com/watch?v=IlNrvthOAQ4
8. *AI Creates Movie Trailer*
   https://www.youtube.com/watch?v=v784xZTS2DA
9. *Information on AI Media*
   https://www.forbes.com/sites/andrewrossow/2018/05/24/artificial-intelligence-taking-convenience-to-a-whole-new-level/#609469e04504
   https://medium.com/@the_manifest/16-examples-of-artificial-intelligence-ai-in-your-everyday-life-655b2e6a49de
   https://www.cnbc.com/2018/02/01/google-ceo-sundar-pichai-ai-is-more-important-than-fire-electricity.html

https://medium.com/@Liamiscool/a-list-of-artificial-intelligence-tools-you-can-use-today-for-personal-use-1-3-7f1b60b6c94f
https://en.wikipedia.org/wiki/Artificial_intelligence
https://www.ktnv.com/positivelylv/dining-and-entertainment/sapphire-strip-club-to-debut-robot-dancers-for-ces
https://www.cnn.com/2014/02/04/tech/innovation/this-new-tech-can-detect-your-mood/
https://venturebeat.com/2018/04/28/4-ways-ai-could-revamp-the-role-of-the-kitchen/
https://thebark.com/content/dogs-contribute-artificial-intelligence
https://thriveglobal.com/stories/could-artificial-intelligence-be-the-cure-for-loneliness/
https://www.allbusiness.com/how-to-create-content-artificial-intelligence-116778-1.html/2
https://govinsider.asia/smart-gov/ai-powering-dubais-pursuit-happiness/
https://www.miamiherald.com/living/travel/article209841679.html
https://futurism.com/artificial-intelligence-bad-poems
https://www.9news.com/article/news/local/features/fairy-tale-written-by-artificial-intelligence/73-545680342
https://www.apollo-magazine.com/ai-art-artificial-intelligence/
https://www.forbes.com/sites/bernardmarr/2018/08/31/how-technology-like-artificial-intelligence-and-iot-are-changing-the-way-we-play-golf/#4dc23bb932e9
https://www.forbes.com/sites/cognitiveworld/2018/09/16/did-ai-write-this-article/#4a257e841885
https://www.powermag.com/blog/how-independence-power-light-saves-ratepayers-100k-a-year-using-artificial-intelligence-technology/
https://www.softscripts.net/blog/2018/09/will-ai-replace-the-human-writers/?utm_campaign=Submission&utm_medium=Community&utm_source=GrowthHackers.com
https://www.globenewswire.com/news-release/2018/09/17/1571857/0/en/Elisa-Selects-Translations-com-s-Artificial-Intelligence-Powered-Subtitling-Solutions-for-Nordic-TV-Launch-in-China.html
https://www.ns-businesshub.com/technology/fun-artificial-

intelligence-applications/
https://www.lightstalking.com/resize-your-photos-using-artificial-intelligence/
https://www.theguardian.com/technology/2018/sep/20/alexa-amazon-hunches-artificial-intelligence
https://phys.org/news/2018-09-artificial-intelligence-tunes-based-irish.html
https://petapixel.com/2018/09/22/polarr-deep-crop-uses-ai-to-auto-crop-your-photos-like-a-pro/
https://www.engadget.com/2018-09-26-deepmind-unity-ai-machine-learning-environments.html
https://www.gigabitmagazine.com/ai/facebook-double-its-artificial-intelligence-research-2020
https://qz.com/1408576/artificial-image-generation-is-getting-good-enough-to-make-you-hungry/
https://www.fastcompany.com/90243942/this-award-winning-nude-portrait-was-generated-by-an-algorithm
https://www.interaliamag.org/interviews/mario-klingemann/
https://www.theverge.com/2019/4/18/18311287/ai-upscaling-algorithms-video-games-mods-modding-esrgan-gigapixel
https://www.forbes.com/sites/stevemccaskill/2018/09/30/ai-and-vr-is-transforming-remote-coaching-in-golf/#3d910a74847b
https://www.abc.net.au/news/2018-10-01/biometric-mirror-offers-perfect-face-in-age-of-social-media/10306232
https://cannabislifenetwork.com/cannatech-ai-vr-and-ar-solutions-for-the-cannabis-sector/
https://europeangaming.eu/portal/industry-news/2018/09/28/29245/online-casino-sites-artificial-intelligence-in-the-gambling-industry/
https://hospitalitytech.com/burger-king-unveil-ad-campaign-created-artificial-intelligence
https://www.glossy.co/new-face-of-beauty/artificial-intelligence-is-set-to-revolutionize-the-fragrance-industry
https://www.axios.com/ai-paints-a-self-portrait-af3a8c6e-96a7-42b9-994c-6de325fb38e2.html
https://www.axios.com/imagenet-roulette-ai-art-bias-bec45510-34a8-

4cd5-8098-a7f92cf8268c.html
https://www.casino.org/news/casino-artificial-intelligence-technology-
takes-hold-in-michigan-will-be-tracking-your-feelings/
https://venturebeat.com/2018/12/28/a-researcher-trained-ai-to-
generate-africa-masks/
https://www.sporttechie.com/calloway-golf-reveals-new-artificial-
intellegence-enhanced-driver/
https://dzone.com/articles/using-ai-to-gauge-the-accuracy-of-the-news
https://interestingengineering.com/5-ways-artificial-intelligence-is-
changing-architecture
https://www.haaretz.com/israel-news/business/the-israeli-startup-that-
wants-to-make-shopping-carts-smarter-1.6811024
https://www.5gtechnologyworld.com/judging-gymnasts-with-lidar-
and-artificial-intelligence/
https://www.zdnet.com/article/forget-go-google-helps-ai-learn-to-
book-flights-on-the-web/
https://qz.com/1482706/ibm-using-artificial-intelligence-to-better-
describe-smells/
https://www.danfoss.com/en/about-danfoss/news/cf/artificial-
intelligence-provides-comfort-for-apartments-residents/
https://www.financialexpress.com/industry/technology/artificial-
intelligence-to-power-news-media-google-partners-with-polis-for-
new-journalism-ai-project/1407295/
https://macaudailytimes.com.mo/artificial-intelligence-to-improve-
tourism-service.html
https://www.prolificnorth.co.uk/news/digital/2018/12/edit-uses-
artificial-intelligence-reconstruct-christmas-carols
https://www.prnewswire.com/news-releases/new-lower-uses-artificial-
intelligence-to-help-homebuyers-make-smarter-mortgage-decisions-
300764672.html
http://artificialintelligence-news.com/2018/12/04/nvidia-ai-real-
videos-3d-renders/
http://artificialintelligence-news.com/2018/11/08/china-ai-news-
anchor-state-outlet/
http://artificialintelligence-news.com/2018/11/01/microsoft-uk-game-
changer-ai/

http://artificialintelligence-news.com/2018/10/12/pepper-the-robot-will-testify-about-ai-in-front-of-uk-parliament/
https://www.thecollegefix.com/berkeley-scientists-developing-artificial-intelligence-tool-to-combat-hate-speech-on-social-media/
https://alumni.berkeley.edu/california-magazine/just-in/2018-12-11/two-brains-are-better-one-ai-and-humans-work-fight-hate
https://www.cbsnews.com/news/ai-babysitting-service-predictim-blocked-by-facebook-and-twitter/
https://www.livekindly.co/vegan-shampoo-brand-prose-artificial-intelligence/
https://www.washingtonpost.com/technology/2018/11/16/wanted-perfect-babysitter-must-pass-ai-scan-respect-attitude/
https://footwearnews.com/2018/fashion/spring-2019/yoox-8-by-yoox-artificial-intelligence-ai-release-info-1202705813/
https://www.fastcompany.com/90372713/this-ai-knows-youll-return-those-shoes-before-you-do
https://www.forbes.com/sites/meggentaylor/2018/11/07/proven-this-female-led-tech-start-up-is-using-ai-to-customize-skincare/#3205933f744f
https://www.livekindly.co/walmart-chile-notcos-vegan-mayo-artificial-intelligence/
https://www.france24.com/en/20181108-dating-apps-use-artificial-intelligence-help-search-love
https://bgr.com/2018/11/08/artificial-intelligence-match-humans-creativity/
https://www.bitdefender.com/box/blog/family/machine-learning-artificial-intelligence-now-central-smart-home-security/
https://scroll.in/field/900775/how-chelsea-football-club-is-using-artificial-intelligence-for-smarter-coaching
https://bbs.boingboing.net/t/an-artificial-intelligence-populated-these-photos-with-glitchy-humanoid-ghosts/132292
https://www.prnewswire.com/news-releases/nexoptic-introduces-artificial-intelligence-technology-to-transform-photography-300739902.html
https://insidebigdata.com/2018/10/22/artificial-intelligence-enhances-home-buying-experience/

https://www.theguardian.com/music/2018/oct/22/ai-artificial-intelligence-composing

https://www.symrise.com/newsroom/article/breaking-new-fragrance-ground-with-artificial-intelligence-ai-ibm-research-and-symrise-are-workin/

https://www.desiblitz.com/content/artificial-intelligence-cricket-bat-game-changer

https://www.si.com/nfl/2017/11/29/nfl-football-location-tracking-chips-zebra-sports-rfid

https://www.zebra.com/us/en/nfl.html

https://www.diyphotography.net/huawei-starts-the-worlds-first-photography-competition-judged-by-ai/

https://www.multifamilyexecutive.com/technology/5-ways-artificial-intelligence-will-transform-the-apartment-industry

https://futurism.com/artificial-intelligence-automating-hollywood-art

https://singularityhub.com/2018/09/03/the-new-ai-tech-turning-heads-in-video-manipulation-2/

https://www.yahoo.com/news/nestle-using-dna-artificial-intelligence-personalise-user-diet-plans-093716134.html

https://www.opengovasia.com/artificial-intelligence-and-machine-learning-to-improve-australias-winemaking-industry/

https://www.forbes.com/sites/bernardmarr/2018/07/18/this-google-funded-company-uses-artificial-intelligence-to-fight-against-fake-news/#51dfe4b43ca4

https://electronics360.globalspec.com/article/12928/using-artificial-intelligence-to-sniff-out-fake-news-at-its-source

https://mytechdecisions.com/compliance/artificial-intelligence-uses-machine-learning-to-fake-photos/

https://www.newgenapps.com/blog/how-artificial-intelligence-is-improving-assistive-technology/

https://thenextweb.com/artificial-intelligence/2019/01/11/nefarious-ai-creates-images-of-delicious-food-that-doesnt-exist/

https://www.bbc.com/future/article/20190111-artificial-intelligence-can-predict-a-relationships-future

https://www.huffpost.com/entry/beauty-artificial-intelligence_n_5a82f175e4b01467fcf1af76

https://www.cybersecurity-insiders.com/beer-brewed-by-artificial-intelligence/

https://www.moneycontrol.com/news/trends/entertainment/heres-how-media-and-entertainment-industry-is-leveraging-artificial-intelligence-to-transform-the-space-3435721.html

https://www.usnews.com/news/news/articles/2019-01-29/study-scientists-create-artificial-intelligence-to-turn-thoughts-into-words

https://fox6now.com/2019/01/29/june-smart-oven-uses-artificial-intelligence-to-recognize-the-food-you-put-inside/

https://www.chinadailyhk.com/articles/162/165/88/1549001721478.html

http://www.chinadaily.com.cn/hkedition/2019-02/01/content_37434486.htm

https://blockclubchicago.org/2019/02/01/need-a-valentine-new-app-uses-artificial-intelligence-to-help-chicagoans-find-a-perfect-match/

https://www.neatorama.com/2018/10/16/The-bizarre-thing-that-happens-when-artificial-intelligence-tells-people-their-fortunes/

https://www.headstuff.org/entertainment/music/will-artificial-intelligence-penetrate-the-music-industry/

https://www.innovations-report.com/html/reports/energy-engineering/artificial-intelligence-used-to-economically-and-energetically-control-heating-systems.html

https://www.abc.net.au/news/2018-09-05/how-machine-learning-might-change-the-future-of-popular-music/10147636

https://medium.com/david-grace-columns-organized-by-topic/another-use-for-artificial-intelligence-rating-recommending-politicians-job-candidates-58ca87ab213

https://www.racked.com/2018/7/17/17577266/artificial-intelligence-ai-counterfeit-luxury-goods-handbags-sneakers-goat-entrupy

https://www.hotelmanagement.net/tech/ihg-launches-ai-rooms-greater-china

https://www.hotelmanagement.net/operate/rlh-corporation-introduces-housekeeping-robot

https://www.analyticsinsight.net/artificial-intelligence-is-the-new-superstar-of-the-entertainment-industry/

https://www.dnaindia.com/science/report-poetic-artificial-intelligence-system-pens-shakespeare-like-sonnets-2647710

https://blog.frontiersin.org/2019/01/07/artificial-intelligence-predicts-personality-from-eye-movements/

https://www.scmp.com/news/china/society/article/2183665/artificial-intelligence-system-used-catch-unhygienic-chefs-action

https://writingcooperative.com/can-artificial-intelligence-take-over-writing-3d541764ecf?gi=8e8deed3ab27

https://www.arabianindustry.com/broadcast/news/2018/aug/14/artificial-intelligence-used-to-recreate-prehistoric-shark-in-action-thriller-film-the-meg-5964673/#targetText=Artificial%20Intelligence%20used%20to%20recreate%20prehistoric,action%20thriller%20film%20'The%20Meg'&targetText=Last%20week%2C%20Warner%20Bros.%20Pictures,shark%20known%20as%20the%20Megalodon.

https://www.govtech.com/question-of-the-day/Question-of-the-Day-for-09102018.html

https://www.bitstarz.com/blog/how-will-artificial-intelligence-affect-online-gambling

https://clubandresortbusiness.com/artificial-intelligence-being-adapted-to-golf-course-maintenance/

https://www.globenewswire.com/news-release/2019/01/22/1703376/0/en/First-Artificial-Intelligence-Diamond-Buying-Tool-Lets-Consumers-See-True-Diamond-Quality-Online-Before-They-Buy.html

https://mexiconewsdaily.com/news/streetlights-that-use-artificial-intelligence/

https://www.birminghammail.co.uk/news/uk-news/football-team-turns-artificial-intelligence-15721424

http://www.industrytap.com/china-uses-artificial-intelligence-to-track-its-700-million-pigs/46480

https://www.mumbrella.asia/2018/09/bloomberg-develops-one-sentence-news-feed-sourced-by-artificial-intelligence-technology

https://www.kyivpost.com/technology/lithuanian-creates-artificial-intelligence-with-ability-to-identify-fake-news-within-2-minutes.html?cn-reloaded=1

https://www.weforum.org/agenda/2019/03/this-ai-bin-tells-you-off-

for-wasting-food/
https://news.cgtn.com/news/3d3d414e326b7a4d7a457a6333566d
54/share_p.html
https://gizmodo.com/this-is-the-most-aerodynamic-bike-according-to-
ai-1827546083
https://www.theverge.com/2019/5/28/18637135/hollywood-ai-film-
decision-script-analysis-data-machine-learning
https://www.newstatesman.com/science-
tech/technology/2018/05/how-artificial-intelligence-could-personalise-
food-future
https://www.hexacta.com/how-artificial-intelligence-affects-our-lives-
without-noticing-it/
https://www.theguardian.com/world/2018/nov/09/worlds-first-ai-
news-anchor-unveiled-in-china
https://www.sciencealert.com/sciencealert-deal-this-service-uses-
machine-learning-to-match-wines-with-you
https://www.popsci.com/universal-music-translation-artificial-
intelligence/
https://cmr.berkeley.edu/blog/2019/1/ai-customer-engagement/
https://projectentrepreneur.org/inspiration/how-the-female-founder-of-
savitude-is-using-artificial-intelligence-to-make-shopping-easier-for-
80-percent-of-women/
https://www.outerplaces.com/science/item/18489-artificial-
intelligence-jokes-funny
https://www.business2community.com/web-design/how-artificial-
intelligence-is-shaping-the-future-of-web-designs-02066948
https://www.popsci.com/whitesmoke-ai-writing-assistant/
https://www.entrepreneur.com/video/313790?utm_source=feed
burner&utm_medium=feed&utm_campaign=Feed%3A+entrep
reneur%2Fsalesandmarketing+%28Entrepreneur+-+Marketing%29
http://nautil.us/issue/27/dark-matter/artificial-intelligence-is-already-
weirdly-inhuman
https://newfoodeconomy.org/artificial-intelligence-personalized-food-
beverage/
https://wamu.org/story/18/06/06/artificial-intelligence-real-news/
http://sciencewows.ie/blog/humour-laughter-ai/

https://www.cnet.com/how-to/new-google-news-app-what-you-need-know/

https://startsat60.com/money/hundreds-trusting-artificial-intelligence-chatbot-to-write-their-will

https://www.npr.org/sections/ed/2018/04/30/606164343/kids-meet-alexa-your-ai-mary-poppins

https://www.independent.co.uk/news/long_reads/ai-robot-brothers-grimm-fairytale-write-story-the-princes-and-fox-a8393826.html

https://www.mediapost.com/publications/article/318143/consumers-expect-artificial-intelligence-will-make.html

https://www.dailystar.co.uk/news/latest-news/artificial-intelligence-virtual-reality-computer-17132181

https://www.livekindly.co/beyonce-uses-artificial-intelligence-to-help-people-go-vegan/

https://www.moneycontrol.com/news/technology/hindustan-unilever-to-deploy-artificial-intelligence-to-predict-customers-grocery-needs-2586229.html

https://nypost.com/2017/05/02/terrifying-ai-learns-to-mimic-your-voice-in-under-60-seconds/

https://www.parking-net.com/parking-news/pixevia/artificial-intelligence-smart-parking

https://www.inc.com/kevin-j-ryan/unanimous-ai-swarm-intelligence-makes-startlingly-accurate-predictions.html

https://futurism.com/the-byte/ai-translates-babies-cries

https://www.bbc.com/news/technology-44481510

http://micetimes.asia/technologists-taught-the-ai-to-lie/

https://gadgets.ndtv.com/social-networking/news/new-facebook-ai-system-could-open-closed-eyes-in-photo-1869020

https://www.digitaltrends.com/computing/microsoft-patent-describes-machine-learning-cheat-detection/

https://metro.co.uk/2018/06/19/facebook-wants-replace-eyeballs-artificial-intelligence-blink-photos-7643891/

https://newsbeezer.com/novel-ai-tool-can-predict-your-iq-from-brain-scans/

https://www.techlicious.com/how-to/how-to-use-your-smartphone-camera-to-search/

https://www.youtube.com/watch?v=TpFvFmCquCQ

Chapter 13 *Future of Gaming Conveniences with AI*

1. *We're All Getting Chipped Soon*
   https://www.youtube.com/watch?v=hLc_7CnWkxw
   https://www.youtube.com/watch?v=QhhFbvdIlw0
   https://www.youtube.com/watch?v=JZYwYxapS-k
   https://www.youtube.com/watch?v=WjTOkdHO7_4
2. *Getting Chipped from Laziness*
   https://www.youtube.com/watch?v=hLc_7CnWkxw
   https://www.youtube.com/watch?v=Ksw-arKvMPk
   https://www.youtube.com/watch?v=qWVQR99bXt8
3. *Gaming Headset Leads to Brain Hacking*
   https://www.youtube.com/watch?v=CsyBHyu6oh4
4. *Total Recall Vacation Scene*
   https://www.youtube.com/watch?v=JYRVMeAwVKk
   https://www.youtube.com/watch?v=hJXx9HE2Rm4
5. *Creating Dreams Like Netflix*
   https://www.youtube.com/watch?v=MP-hwl5d-gw
6. *Brain Chips Erase Memories*
   https://www.youtube.com/watch?v=Zm9swA5o3dY
7. *Brain Chips Control People*
   https://www.youtube.com/watch?v=eGVFg3I779c
8. *Information on AI Gaming*
   https://www.forbes.com/sites/andrewrossow/2018/05/24/artificial-intelligence-taking-convenience-to-a-whole-new-level/#609469e04504
   https://medium.com/@the_manifest/16-examples-of-artificial-intelligence-ai-in-your-everyday-life-655b2e6a49de
   https://www.cnbc.com/2018/02/01/google-ceo-sundar-pichai-ai-is-more-important-than-fire-electricity.html
   https://medium.com/@Liamiscool/a-list-of-artificial-intelligence-tools-you-can-use-today-for-personal-use-1-3-7f1b60b6c94f
   https://en.wikipedia.org/wiki/Artificial_intelligence

https://www.ktnv.com/positivelylv/dining-and-entertainment/sapphire-strip-club-to-debut-robot-dancers-for-ces

https://www.cnn.com/2014/02/04/tech/innovation/this-new-tech-can-detect-your-mood/

https://venturebeat.com/2018/04/28/4-ways-ai-could-revamp-the-role-of-the-kitchen/

https://thebark.com/content/dogs-contribute-artificial-intelligence

https://thriveglobal.com/stories/could-artificial-intelligence-be-the-cure-for-loneliness/

https://www.allbusiness.com/how-to-create-content-artificial-intelligence-116778-1.html/2

https://govinsider.asia/smart-gov/ai-powering-dubais-pursuit-happiness/

https://www.miamiherald.com/living/travel/article209841679.html

https://futurism.com/artificial-intelligence-bad-poems

https://www.9news.com/article/news/local/features/fairy-tale-written-by-artificial-intelligence/73-545680342

https://www.apollo-magazine.com/ai-art-artificial-intelligence/

https://www.forbes.com/sites/bernardmarr/2018/08/31/how-technology-like-artificial-intelligence-and-iot-are-changing-the-way-we-play-golf/#4dc23bb932e9

https://www.forbes.com/sites/cognitiveworld/2018/09/16/did-ai-write-this-article/#4a257e841885

https://www.powermag.com/blog/how-independence-power-light-saves-ratepayers-100k-a-year-using-artificial-intelligence-technology/

https://www.softscripts.net/blog/2018/09/will-ai-replace-the-human-writers/?utm_campaign=Submission&utm_medium=Community&utm_source=GrowthHackers.com

https://www.globenewswire.com/news-release/2018/09/17/1571857/0/en/Elisa-Selects-Translations-com-s-Artificial-Intelligence-Powered-Subtitling-Solutions-for-Nordic-TV-Launch-in-China.html

https://www.ns-businesshub.com/technology/fun-artificial-intelligence-applications/

https://www.lightstalking.com/resize-your-photos-using-artificial-intelligence/

https://www.theguardian.com/technology/2018/sep/20/alexa-amazon-hunches-artificial-intelligence

https://phys.org/news/2018-09-artificial-intelligence-tunes-based-irish.html

https://petapixel.com/2018/09/22/polarr-deep-crop-uses-ai-to-auto-crop-your-photos-like-a-pro/

https://www.engadget.com/2018-09-26-deepmind-unity-ai-machine-learning-environments.html

https://www.gigabitmagazine.com/ai/facebook-double-its-artificial-intelligence-research-2020

https://qz.com/1408576/artificial-image-generation-is-getting-good-enough-to-make-you-hungry/

https://www.fastcompany.com/90243942/this-award-winning-nude-portrait-was-generated-by-an-algorithm

https://www.interaliamag.org/interviews/mario-klingemann/

https://www.theverge.com/2019/4/18/18311287/ai-upscaling-algorithms-video-games-mods-modding-esrgan-gigapixel

https://www.forbes.com/sites/stevemccaskill/2018/09/30/ai-and-vr-is-transforming-remote-coaching-in-golf/#3d910a74847b

https://www.abc.net.au/news/2018-10-01/biometric-mirror-offers-perfect-face-in-age-of-social-media/10306232

https://cannabislifenetwork.com/cannatech-ai-vr-and-ar-solutions-for-the-cannabis-sector/https://europeangaming.eu/portal/industry-news/2018/09/28/29245/online-casino-sites-artificial-intelligence-in-the-gambling-industry/

https://hospitalitytech.com/burger-king-unveil-ad-campaign-created-artificial-intelligence

https://www.glossy.co/new-face-of-beauty/artificial-intelligence-is-set-to-revolutionize-the-fragrance-industry

https://www.axios.com/ai-paints-a-self-portrait-af3a8c6e-96a7-42b9-994c-6de325fb38e2.html

https://www.axios.com/imagenet-roulette-ai-art-bias-bec45510-34a8-4cd5-8098-a7f92cf8268c.html

https://www.casino.org/news/casino-artificial-intelligence-technology-takes-hold-in-michigan-will-be-tracking-your-feelings/

https://venturebeat.com/2018/12/28/a-researcher-trained-ai-to-generate-africa-masks/

https://www.sporttechie.com/calloway-golf-reveals-new-artificial-intellegence-enhanced-driver/

https://dzone.com/articles/using-ai-to-gauge-the-accuracy-of-the-news

https://interestingengineering.com/5-ways-artificial-intelligence-is-changing-architecture

https://www.haaretz.com/israel-news/business/the-israeli-startup-that-wants-to-make-shopping-carts-smarter-1.6811024

https://www.5gtechnologyworld.com/judging-gymnasts-with-lidar-and-artificial-intelligence/

https://www.zdnet.com/article/forget-go-google-helps-ai-learn-to-book-flights-on-the-web/

https://qz.com/1482706/ibm-using-artificial-intelligence-to-better-describe-smells/

https://www.danfoss.com/en/about-danfoss/news/cf/artificial-intelligence-provides-comfort-for-apartments-residents/

https://www.financialexpress.com/industry/technology/artificial-intelligence-to-power-news-media-google-partners-with-polis-for-new-journalism-ai-project/1407295/

https://macaudailytimes.com.mo/artificial-intelligence-to-improve-tourism-service.html

https://www.prolificnorth.co.uk/news/digital/2018/12/edit-uses-artificial-intelligence-reconstruct-christmas-carols

https://www.prnewswire.com/news-releases/new-lower-uses-artificial-intelligence-to-help-homebuyers-make-smarter-mortgage-decisions-300764672.html

http://artificialintelligence-news.com/2018/12/04/nvidia-ai-real-videos-3d-renders/

http://artificialintelligence-news.com/2018/11/08/china-ai-news-anchor-state-outlet/

http://artificialintelligence-news.com/2018/11/01/microsoft-uk-game-changer-ai/

http://artificialintelligence-news.com/2018/10/12/pepper-the-robot-will-testify-about-ai-in-front-of-uk-parliament/

https://www.thecollegefix.com/berkeley-scientists-developing-artificial-intelligence-tool-to-combat-hate-speech-on-social-media/

https://alumni.berkeley.edu/california-magazine/just-in/2018-12-11/two-brains-are-better-one-ai-and-humans-work-fight-hate
https://www.cbsnews.com/news/ai-babysitting-service-predictim-blocked-by-facebook-and-twitter/
https://www.livekindly.co/vegan-shampoo-brand-prose-artificial-intelligence/
https://www.washingtonpost.com/technology/2018/11/16/wanted-perfect-babysitter-must-pass-ai-scan-respect-attitude/
https://footwearnews.com/2018/fashion/spring-2019/yoox-8-by-yoox-artificial-intelligence-ai-release-info-1202705813/
https://www.fastcompany.com/90372713/this-ai-knows-youll-return-those-shoes-before-you-do
https://www.forbes.com/sites/meggentaylor/2018/11/07/proven-this-female-led-tech-start-up-is-using-ai-to-customize-skincare/#3205933f744f
https://www.livekindly.co/walmart-chile-notcos-vegan-mayo-artificial-intelligence/
https://www.france24.com/en/20181108-dating-apps-use-artificial-intelligence-help-search-love
https://bgr.com/2018/11/08/artificial-intelligence-match-humans-creativity/
https://www.bitdefender.com/box/blog/family/machine-learning-artificial-intelligence-now-central-smart-home-security/
https://scroll.in/field/900775/how-chelsea-football-club-is-using-artificial-intelligence-for-smarter-coaching
https://bbs.boingboing.net/t/an-artificial-intelligence-populated-these-photos-with-glitchy-humanoid-ghosts/132292
https://www.prnewswire.com/news-releases/nexoptic-introduces-artificial-intelligence-technology-to-transform-photography-300739902.html
https://insidebigdata.com/2018/10/22/artificial-intelligence-enhances-home-buying-experience/
https://www.theguardian.com/music/2018/oct/22/ai-artificial-intelligence-composing

https://www.symrise.com/newsroom/article/breaking-new-fragrance-ground-with-artificial-intelligence-ai-ibm-research-and-symrise-are-workin/

https://www.desiblitz.com/content/artificial-intelligence-cricket-bat-game-changer

https://www.si.com/nfl/2017/11/29/nfl-football-location-tracking-chips-zebra-sports-rfid

https://www.zebra.com/us/en/nfl.html

https://www.diyphotography.net/huawei-starts-the-worlds-first-photography-competition-judged-by-ai/

https://www.multifamilyexecutive.com/technology/5-ways-artificial-intelligence-will-transform-the-apartment-industry

https://futurism.com/artificial-intelligence-automating-hollywood-art

https://singularityhub.com/2018/09/03/the-new-ai-tech-turning-heads-in-video-manipulation-2/

https://www.yahoo.com/news/nestle-using-dna-artificial-intelligence-personalise-user-diet-plans-093716134.html

https://www.opengovasia.com/artificial-intelligence-and-machine-learning-to-improve-australias-winemaking-industry/

https://www.forbes.com/sites/bernardmarr/2018/07/18/this-google-funded-company-uses-artificial-intelligence-to-fight-against-fake-news/#51dfe4b43ca4

https://electronics360.globalspec.com/article/12928/using-artificial-intelligence-to-sniff-out-fake-news-at-its-source

https://mytechdecisions.com/compliance/artificial-intelligence-uses-machine-learning-to-fake-photos/

https://www.newgenapps.com/blog/how-artificial-intelligence-is-improving-assistive-technology/

https://thenextweb.com/artificial-intelligence/2019/01/11/nefarious-ai-creates-images-of-delicious-food-that-doesnt-exist/

https://www.bbc.com/future/article/20190111-artificial-intelligence-can-predict-a-relationships-future

https://www.huffpost.com/entry/beauty-artificial-intelligence_n_5a82f175e4b01467fcf1af76

https://www.cybersecurity-insiders.com/beer-brewed-by-artificial-intelligence/

https://www.moneycontrol.com/news/trends/entertainment/heres-how-media-and-entertainment-industry-is-leveraging-artificial-intelligence-to-transform-the-space-3435721.html
https://www.usnews.com/news/news/articles/2019-01-29/study-scientists-create-artificial-intelligence-to-turn-thoughts-into-words
https://fox6now.com/2019/01/29/june-smart-oven-uses-artificial-intelligence-to-recognize-the-food-you-put-inside/
https://www.chinadailyhk.com/articles/162/165/88/1549001721478.html
http://www.chinadaily.com.cn/hkedition/2019-02/01/content_37434486.htm
https://blockclubchicago.org/2019/02/01/need-a-valentine-new-app-uses-artificial-intelligence-to-help-chicagoans-find-a-perfect-match/
https://www.neatorama.com/2018/10/16/The-bizarre-thing-that-happens-when-artificial-intelligence-tells-people-their-fortunes/
https://www.headstuff.org/entertainment/music/will-artificial-intelligence-penetrate-the-music-industry/
https://www.innovations-report.com/html/reports/energy-engineering/artificial-intelligence-used-to-economically-and-energetically-control-heating-systems.html
https://www.abc.net.au/news/2018-09-05/how-machine-learning-might-change-the-future-of-popular-music/10147636
https://medium.com/david-grace-columns-organized-by-topic/another-use-for-artificial-intelligence-rating-recommending-politicians-job-candidates-58ca87ab213
https://www.racked.com/2018/7/17/17577266/artificial-intelligence-ai-counterfeit-luxury-goods-handbags-sneakers-goat-entrupy
https://www.hotelmanagement.net/tech/ihg-launches-ai-rooms-greater-china
https://www.hotelmanagement.net/operate/rlh-corporation-introduces-housekeeping-robot
https://www.analyticsinsight.net/artificial-intelligence-is-the-new-superstar-of-the-entertainment-industry/
https://www.dnaindia.com/science/report-poetic-artificial-intelligence-system-pens-shakespeare-like-sonnets-2647710
https://blog.frontiersin.org/2019/01/07/artificial-intelligence-predicts-

personality-from-eye-movements/
https://www.scmp.com/news/china/society/article/2183665/artificial-intelligence-system-used-catch-unhygienic-chefs-action
https://writingcooperative.com/can-artificial-intelligence-take-over-writing-3d541764ecf?gi=8e8deed3ab27
https://www.arabianindustry.com/broadcast/news/2018/aug/14/artificial-intelligence-used-to-recreate-prehistoric-shark-in-action-thriller-film-the-meg-5964673/#targetText=Artificial%20Intelligence%20used%20to%20recreate%20prehistoric,action%20thriller%20film%20'The%20Meg'&targetText=Last%20week%2C%20Warner%20Bros.%20Pictures,shark%20known%20as%20the%20Megalodon.
https://www.govtech.com/question-of-the-day/Question-of-the-Day-for-09102018.html
https://www.bitstarz.com/blog/how-will-artificial-intelligence-affect-online-gambling
https://clubandresortbusiness.com/artificial-intelligence-being-adapted-to-golf-course-maintenance/
https://www.globenewswire.com/news-release/2019/01/22/1703376/0/en/First-Artificial-Intelligence-Diamond-Buying-Tool-Lets-Consumers-See-True-Diamond-Quality-Online-Before-They-Buy.html
https://mexiconewsdaily.com/news/streetlights-that-use-artificial-intelligence/
https://www.birminghammail.co.uk/news/uk-news/football-team-turns-artificial-intelligence-15721424
http://www.industrytap.com/china-uses-artificial-intelligence-to-track-its-700-million-pigs/46480
https://www.mumbrella.asia/2018/09/bloomberg-develops-one-sentence-news-feed-sourced-by-artificial-intelligence-technology
https://www.kyivpost.com/technology/lithuanian-creates-artificial-intelligence-with-ability-to-identify-fake-news-within-2-minutes.html?cn-reloaded=1
https://www.weforum.org/agenda/2019/03/this-ai-bin-tells-you-off-for-wasting-food/

https://news.cgtn.com/news/3d3d414e326b7a4d7a457a6333566d54/share_p.html

https://gizmodo.com/this-is-the-most-aerodynamic-bike-according-to-ai-1827546083

https://www.theverge.com/2019/5/28/18637135/hollywood-ai-film-decision-script-analysis-data-machine-learning

https://www.newstatesman.com/science-tech/technology/2018/05/how-artificial-intelligence-could-personalise-food-future

https://www.hexacta.com/how-artificial-intelligence-affects-our-lives-without-noticing-it/

https://www.theguardian.com/world/2018/nov/09/worlds-first-ai-news-anchor-unveiled-in-china

https://www.sciencealert.com/sciencealert-deal-this-service-uses-machine-learning-to-match-wines-with-you

https://www.popsci.com/universal-music-translation-artificial-intelligence/

https://cmr.berkeley.edu/blog/2019/1/ai-customer-engagement/

https://projectentrepreneur.org/inspiration/how-the-female-founder-of-savitude-is-using-artificial-intelligence-to-make-shopping-easier-for-80-percent-of-women/

https://www.outerplaces.com/science/item/18489-artificial-intelligence-jokes-funny

https://www.business2community.com/web-design/how-artificial-intelligence-is-shaping-the-future-of-web-designs-02066948

https://www.popsci.com/whitesmoke-ai-writing-assistant/

https://www.entrepreneur.com/video/313790?utm_source=feedburner&utm_medium=feed&utm_campaign=Feed%3A+entrepreneur%2Fsalesandmarketing+%28Entrepreneur+-+Marketing%29

http://nautil.us/issue/27/dark-matter/artificial-intelligence-is-already-weirdly-inhuman

https://newfoodeconomy.org/artificial-intelligence-personalized-food-beverage/

https://wamu.org/story/18/06/06/artificial-intelligence-real-news/

http://sciencewows.ie/blog/humour-laughter-ai/

https://www.cnet.com/how-to/new-google-news-app-what-you-need-know/

https://startsat60.com/money/hundreds-trusting-artificial-intelligence-chatbot-to-write-their-will

https://www.npr.org/sections/ed/2018/04/30/606164343/kids-meet-alexa-your-ai-mary-poppins

https://www.independent.co.uk/news/long_reads/ai-robot-brothers-grimm-fairytale-write-story-the-princes-and-fox-a8393826.html

https://www.mediapost.com/publications/article/318143/consumers-expect-artificial-intelligence-will-make.html

https://www.dailystar.co.uk/news/latest-news/artificial-intelligence-virtual-reality-computer-17132181

https://www.livekindly.co/beyonce-uses-artificial-intelligence-to-help-people-go-vegan/

https://www.moneycontrol.com/news/technology/hindustan-unilever-to-deploy-artificial-intelligence-to-predict-customers-grocery-needs-2586229.html

https://nypost.com/2017/05/02/terrifying-ai-learns-to-mimic-your-voice-in-under-60-seconds/

https://www.parking-net.com/parking-news/pixevia/artificial-intelligence-smart-parking

https://www.inc.com/kevin-j-ryan/unanimous-ai-swarm-intelligence-makes-startlingly-accurate-predictions.html

https://futurism.com/the-byte/ai-translates-babies-cries

https://www.bbc.com/news/technology-44481510

http://micetimes.asia/technologists-taught-the-ai-to-lie/

https://gadgets.ndtv.com/social-networking/news/new-facebook-ai-system-could-open-closed-eyes-in-photo-1869020

https://www.digitaltrends.com/computing/microsoft-patent-describes-machine-learning-cheat-detection/

https://metro.co.uk/2018/06/19/facebook-wants-replace-eyeballs-artificial-intelligence-blink-photos-7643891/

https://newsbeezer.com/novel-ai-tool-can-predict-your-iq-from-brain-scans/

https://www.techlicious.com/how-to/how-to-use-your-smartphone-camera-to-search/

https://www.youtube.com/watch?v=TpFvFmCquCQ
https://www.businessinsider.com/elon-musk-neuralink-brain-chip-put-in-human-within-year-2020-5?fbclid=IwAR3oeSRymlhbtZ3fFgI-aCoxXakzqv00Lrfk4conLDPz85fA6f4mrd3CbPY
https://www.businessinsider.com/elon-musk-neuralink-implants-link-brains-to-internet-next-year-2019-7
https://www.wired.com/2010/03/thought-control-headset-reads-you-mind/
https://www.emotiv.com/

Chapter 14 *The Future of Agriculture with AI*

1. *AI Automated Tractor*
   https://www.youtube.com/watch?v=gMaQq_vRaa8
2. *AI Automated Drones*
   https://www.youtube.com/watch?v=4L_RzCSh58U
3. *AI Automated Cows*
   https://www.youtube.com/watch?v=jS8xXAh35wA
4. *AI Automated Workers*
   https://www.youtube.com/watch?v=5chk9Sory88
   https://www.youtube.com/watch?v=hjd5DaxkLhQ
5. *Farm Worker Shortage*
   https://www.youtube.com/watch?v=cw3flTRrPts
6. *Farming with No Humans*
   https://globalnews.ca/video/3898917/revolutionary-farming-method-showcased-at-farming-smarter-conference
   https://www.youtube.com/watch?v=vtwNKga6thw
7. *AI Spots Obesity From Space*
   https://www.youtube.com/watch?v=4BK5rlAE9Ik
8. *AI Controls Whole World Crisis*
   https://www.youtube.com/watch?v=mJ6rjJilHyo
9. *Information on AI Agriculture*
   https://foodindustryexecutive.com/2018/04/6-examples-of-artificial-intelligence-in-the-food-industry/

https://thenextweb.com/artificial-intelligence/2019/02/01/a-deadly-tree-disease-is-devastating-the-citrus-industry-ai-can-help/
https://www.foodqualityandsafety.com/article/artificial-intelligence-a-real-opportunity-in-food-industry/
https://www.farmprogress.com/technology/artificial-intelligence-agriculture
https://www.futurefarming.com/Tools-data/Articles/2019/3/Artificial-intelligence-to-replace-farmers-knowledge-400480E/
https://nifa.usda.gov/announcement/artificial-intelligence-strawberries-may-improve-food-quality-safety
https://www.agdaily.com/news/dairy-farmers-america-artificial-intelligence/
https://www.agdaily.com/news/dairymaster-introduces-artificial-intelligence/
https://blogs.microsoft.com/blog/2019/08/07/harnessing-the-power-of-ai-to-transform-agriculture/
https://www.precisionag.com/digital-farming/data-management/how-does-artificial-intelligence-really-work-in-agriculture/
https://swisscognitive.ch/2019/08/16/rapid-adoption-of-artificial-intelligence-in-agriculture/
https://globalnews.ca/news/4279157/fourth-revolution-in-farming/
https://www.farmprogress.com/technology/where-artificial-intelligence-could-take-agriculture
https://medium.com/@Liamiscool/a-list-of-artificial-intelligence-tools-you-can-use-today-for-industry-specific-3-3-5e16c68da697
https://www.vice.com/en_us/article/z434q4/the-internet-of-cows-internet-of-things-agriculture
https://tuanz.org.nz/2014717is-the-rbi-fit-for-the-internet-of-cows/

10. *Information on Food, Water, Obesity*

http://www.washingtontimes.com/news/2009/mar/15/obama-forms-group-to-protect-us-food-supply/
http://www.wnd.com/2009/03/92002/
http://www.crossroad.to/articles2/08/swat-team.htm
http://usatoday30.usatoday.com/news/health/story/2012-05-07/obesity-projections-adults/54791430/1
http://www.newswithviews.com/DeWeese/tom172.htm

http://www.telegraph.co.uk/health/3793719/New-York-planning-fat-tax-on-drinks.html
http://usatoday30.usatoday.com/news/health/2006-12-04-trans-fat-ban_x.htm
http://articles.mercola.com/sites/articles/archive/2007/11/03/you-might-lose-your-job-if-you-smoke-or-eat-junk-food.aspx
http://cnsnews.com/news/article/federal-fat-police-bill-would-require-government-track-body-mass-american-children
http://www.dailymail.co.uk/news/article-1291470/Big-Brother-row-food-police-secretly-photograph-schoolchildrens-packed-lunches.html
http://www.naturalnews.com/040214_seeds_European_Commission_registration.html
http://cnsnews.com/news/article/obesity-rating-every-american-must-be-included-stimulus-mandated-electronic-health
http://www.infowars.com/collecting-rainwater-now-illegal-in-many-states-as-big-government-claims-ownership-over-our-water/
http://www.youtube.com/watch?v=6jjxg8f3Gq0
http://www.newswithviews.com/Devvy/kidd102.htm
http://usatoday30.usatoday.com/news/nation/2008-03-10-drugs-tap-water_N.htm
http://www.youtube.com/watch?v=ej9YzFkbIjk
http://www.newswithviews.com/NWV-News/news220.htm
http://www.blueplanetproject.net/

Chapter 15 *Future of Communication & Education with AI*

1. *Elon Musk We are Cyborgs*
   https://www.youtube.com/watch?v=rCoFKUJ_8Yo
2. *What is Neuralink*
   https://www.youtube.com/watch?v=JylQMKLC_2M
   https://www.youtube.com/watch?v=mty_WVP8DvA
3. *Brain Chip Talks for You*
   https://www.youtube.com/watch?v=i_AsWts9OLM
4. *Video Brain Chip Future Communication*
   https://www.youtube.com/watch?v=CgFzmE2fGXA

1. *AI Robot Teacher Promo*
   https://www.youtube.com/watch?v=rjCkiNRM-FI
2. *China Uses AI in Schools*
   https://www.youtube.com/watch?v=JMLsHI8aV0g
3. *Matrix Instant Learning Scene*
   https://www.youtube.com/watch?v=0YhJxJZOWBw
4. *Brain Chip Education Promo*
   https://www.youtube.com/watch?v=A6-7J_Z7wf8
5. *AI Brain Chip Future Dangers*
   https://www.youtube.com/watch?v=CgFzmE2fGXA&list=PLwHXX5uisT3My7L1FLJDSWjxQKtKhDV6D&index=31&t=0s
6. *Information on AI Communication*
   https://www.theverge.com/2017/3/27/15077864/elon-musk-neuralink-brain-computer-interface-ai-cyborgs
   https://www.telegraph.co.uk/science/2017/10/15/ai-implants-will-allow-us-control-homes-thoughts-within-20-years/
   https://www.foxnews.com/tech/paralyzed-woman-pilots-f-35-fighter-jet-simulator-using-mind-control
   https://bobmorris.biz/the-golden-age-of-neuroscience-has-arrived
   http://online.wsj.com/articles/Michio-kaku-the-golden-age-of-neuroscience-has-arrived-1408577023
   https://www.newsweek.com/emailing-your-brainwaves-future-communication-266155
   https://www.news.com.au/technology/innovation/motoring/motoring-news/nissan-unveils-technology-that-can-interpret-signals-from-drivers-brains/news-story/6ae5b3100c6b41eb3a9fceb2b2b9e06b
   https://futurism.com/alterego-talk-computer-without-words
   https://futurism.com/the-byte/elon-musk-neuralink-stream-music-brians?mc_eid=5e3b1162c2&mc_cid=7135b5ccf6
11. *Information on AI Education*
   https://medium.com/@Liamiscool/a-list-of-artificial-intelligence-tools-you-can-use-today-for-industry-specific-3-3-5e16c68da697
   https://ubcckengaren.blogspot.com/2018/11/how-artificial-intelligence-and-virtual.html
   https://singularityhub.com/2018/09/12/a-model-for-the-future-of-education-and-the-tech-shaping-it/

https://edtechmagazine.com/k12/article/2019/01/how-k-12-schools-have-adopted-artificial-intelligence

https://edtechmagazine.com/k12/article/2019/08/artificial-intelligence-authentic-impact-how-educational-ai-making-grade-perfcon

https://www.thetechedvocate.org/26-ways-that-artificial-intelligence-ai-is-transforming-education-for-the-better/

https://www.thetechedvocate.org/will-ai-take-over-educational-leadership/

https://www.thetechedvocate.org/education-needs-ai-as-a-mind-multiplier/

http://blogs.edweek.org/edweek/next_gen_learning/2019/01/ready_for_the_future_of_education_with_artificial_intelligence.html?override=web

https://www.getsmarter.com/blog/market-trends/the-role-of-artificial-intelligence-in-the-future-of-education/

https://medium.com/@mwitiderrick/how-artificial-intelligence-is-shaping-the-future-of-education-ffcf910e0877

https://interestingengineering.com/personalized-learning-artificial-intelligence-and-education-in-the-future

https://www.technobugg.com/amp/artificial-intelligence-in-humans-life/

https://www.13newsnow.com/article/news/local/mycity/norfolk/artificial-intelligence-helps-train-future-teachers-counselors/291-d0a9781c-7ead-4188-80ec-a304b577c07e

https://www.entrepreneur.com/article/337165

https://coursewareworld.com/artificial-intelligence-shifting-education-from-generic-classroom-to-individualized-learning/

https://venturebeat.com/2017/03/12/why-parents-might-not-be-ready-for-ai-in-the-classroom/

https://www.insidehighered.com/digital-learning/article/2018/09/26/academics-push-expand-use-ai-higher-ed-teaching-and-learning

https://bgr.com/2017/08/18/brain-hack-science-limb-control/

https://mc.ai/the-next-big-thing-in-ai/#:~:text=Education&text=According%20to%20a%20study%20carried,to%20their%20needs%20and%20capabilities.

https://www.intellias.com/how-ai-helps-crack-a-new-language/
https://www.linkedin.com/pulse/cyber-teaching-existential-threat-future-australian-language-asher
https://www.mirror.co.uk/tech/google-brain-implants-could-mean-14183717
https://towardsdatascience.com/5-ways-artificial-intelligence-and-chatbots-are-changing-education-9e7d9425421d
https://www.indiatoday.in/education-today/featurephilia/story/artificial-intelligence-can-empower-our-education-system-here-s-how-1281653-2018-07-10
https://www.thetechedvocate.org/seven-ways-educators-can-use-artificial-intelligence/
https://emerj.com/ai-sector-overviews/examples-of-artificial-intelligence-in-education/
https://www.meritalk.com/articles/from-teaching-robots-to-intelligent-tutor-systems-ai-is-changing-education/?gclid=EAIaIQobChMI5umTm8iL5gIVk6_sCh3lhwhWEAMYAiAAEgJoO_D_BwE
https://www.hindustantimes.com/education/artificial-intelligence-in-schools-how-ai-powered-adaptive-learning-technology-can-help-students/story-EkCQmha69e1Ne4CcXqFnWO.html
https://www.vanguardngr.com/2019/01/case-for-artificial-intelligence-the-world-is-over-saturated-with-schools-that-dont-work-salau/
https://www.theedadvocate.org/7-ways-that-artificial-intelligence-helps-students-learn/
http://www.telegraph.co.uk/technology/2016/03/01/scientists-discover-how-to-download-knowledge-to-your-brain/
http://www.2045.com/news/34809.html
https://www.colocationamerica.com/blog/classroom-artificial-intelligence
https://www.thetechedvocate.org/can-artificial-intelligence-in-education-improve-social-mobility/
https://epaper.timesgroup.com/Olive/ODN/TimesOfIndia/shared/ShowArticle.aspx?doc=TOIBG%2F2018%2F05%2F18&entity=Ar01909&sk=D0927FF1&mode=text#

Chapter 16 *Future of Doctors & Surgeries with AI*

1. *AI taking over Healthcare*
   https://www.youtube.com/watch?v=K5P9wpRSIXY
2. *Babylon Health App*
   https://www.youtube.com/watch?v=QthXXa9nF5c
3. *Doctors Versus AI*
   https://www.youtube.com/watch?v=27zPBWSBs-g
4. *AI Medical ChatBot*
   https://www.youtube.com/watch?v=cMN6vwQErGQ
   https://www.youtube.com/watch?v=mFFmZcizjlM
5. *AI Robot Doctor*
   https://www.youtube.com/watch?v=x1Qu1YKZA0Y
6. *China AI Robot Doctor*
   https://www.youtube.com/watch?v=6wxEuIFt7p
   https://www.youtube.com/watch?v=qiZrWx51zp07.
7. *China AI 5G Surgery*
   https://www.youtube.com/watch?v=qiZrWx51zp0
8. *AI Robot Surgery*
   https://www.youtube.com/watch?v=vugOOuq256M8.
9. *AI Replacing Doctors*
   https://www.youtube.com/watch?v=e4_99D9PjyM.
10. *Information on AI Medical*
    https://www.medgadget.com/2018/09/doctors-pair-with-artificial-intelligence-to-improve-pneumonia-diagnosis.html
    https://qz.com/1548524/china-has-produced-another-study-showing-the-potential-of-ai-in-medical-diagnosis/
    https://qz.com/1384725/an-ai-algorithm-in-china-is-learning-to-detect-whether-patients-will-wake-from-a-coma/
    https://www.scmp.com/news/china/science/article/2163298/doctors-said-coma-patients-would-never-wake-ai-said-they-would
    https://www.thehindu.com/sci-tech/health/ai-may-predict-alzheimers-disease-5-years-in-advance/article25160303.ece
    https://www.livemint.com/Science/evBwYmDFLvlmRwjO97HeeO/Indian-scientists-using-artificial-intelligence-to-predict-e.html

https://www.forbes.com/sites/bernardmarr/2018/12/21/ai-that-saves-lives-the-chatbot-that-can-detect-a-heart-attack-using-machine-learning/#53c9859450f9

https://thenextweb.com/artificial-intelligence/2018/04/25/europe-launches-heart-attack-detecting-ai-emergency-calls/

https://www.esmo.org/About-Us/ESMO-Magazine/Feature-AI-and-Big-Data-in-Oncology-How-Ready-Are-We

https://www.statnews.com/2019/07/03/artificial-intelligence-guarded-optimism-cancer-care/

https://www.healthcareitnews.com/news/geisinger-injects-machine-learning-clinical-workflow-find-health-problems-faster

https://www.cardiovascularbusiness.com/topics/artificial-intelligence/twists-turns-ahead-ais-road-acceptance-cardiology

https://scienmag.com/insilico-to-present-its-latest-research-in-ai-aging-diagnostics-at-the-aone-conference/

https://svn.bmj.com/content/2/4/230

https://healthmanagement.org/c/hospital/issuearticle/artificial-intelligence-a-next-way-forward-for-healthcare

https://healthmanagement.org/c/hospital/issuearticle/artificial-intelligence-in-healthcare-what-is-versus-what-will-be

https://www.linkedin.com/pulse/should-ai-take-hippocratic-oath-gary-gilliland-md-phd

https://www.livescience.com/65087-ai-premature-death-prediction.html

https://www.prnewswire.com/news-releases/perception-health-on-aws-marketplace-machine-learning-and-artificial-intelligence-discovery-page-300773045.html

https://www.express.co.uk/news/world/976108/google-when-will-you-die-ai-artificial-intelligence-latest

https://cio.economictimes.indiatimes.com/news/strategy-and-management/adoption-of-new-technology-is-not-an-option-but-a-compulsion-ggm-it-irctc/71788370

https://pjmedia.com/trending/can-a-healthcare-algorithm-be-racially-biased/

https://time.com/5556339/artificial-intelligence-robots-medicine/
https://www.modernhealthcare.com/article/20150610/NEWS/150619996/walgreen-insurers-push-expansion-of-virtual-doctor-visits
https://electronicsforu.com/technology-trends/tech-focus/artificial-intelligence-healthcare-replace-doctors
https://healthitanalytics.com/news/at-montefiore-artificial-intelligence-becomes-key-to-patient-care
https://time.com/5709346/artificial-intelligence-health/
https://www.reuters.com/article/us-facebook-features/facebook-turns-to-artificial-intelligence-to-tackle-suicides-idUSKBN1684JQ
https://www.fastcompany.com/90299135/mental-health-crisis-robots-chatbots-listeners
https://www.sandiegouniontribune.com/sdut-smartphone-voices-not-always-helpful-in-health-2016mar14-story.html
https://becominghuman.ai/becoming-human-ai-addiction-deaf948e17ba
https://www.cms.gov/newsroom/press-releases/cms-announces-artificial-intelligence-health-outcomes-challenge-participants-advancing-stage-1
https://healthmanagement.org/c/healthmanagement/IssueArticle/ai-is-the-new-reality-the-4th-healthcare-revolution-in-medicine
https://www.independent.co.uk/voices/artificial-intelligence-machine-learning-computers-global-healthcare-malaria-facebook-a8327901.html
https://questsoblogspot.wordpress.com/2019/09/25/artificial-intelligence-for-personal-business-and-enterprise-uses/
https://news.yahoo.com/health-checks-smartphone-raise-privacy-fears-080635403.html
https://www.theglobeandmail.com/life/health-and-fitness/article-app-developers-are-using-artificial-intelligence-to-advise-teens-about/
https://www.sdglobaltech.com/blog/some-mind-blowing-stats-about-ai-and-iot-in-healthcare
https://builtin.com/artificial-intelligence/artificial-intelligence-healthcare
https://khn.org/news/a-reality-check-on-artificial-intelligence-are-health-care-claims-overblown/

https://healthitanalytics.com/news/54-of-healthcare-pros-expect-widespread-ai-adoption-in-5-years
https://www.ncbi.nlm.nih.gov/pmc/articles/PMC6616181/
https://dzone.com/articles/using-ai-to-design-drugs-from-scratch
https://www.smithsonianmag.com/innovation/will-artificial-intelligence-improve-health-care-for-everyone-180972758/
https://www.managedcaremag.com/archives/2019/7/ai-all-ails-american-health-care-how-smart
https://www.digitalcommerce360.com/2018/09/14/ai-may-save-one-small-hospital-20-million/
https://www.healthdatamanagement.com/news/cios-must-take-charge-in-implementing-artificial-intelligence-governance
https://www.theweek.in/news/sci-tech/2018/09/15/New-artificial-intelligence-system-can-detect-dementia.html
http://www.digitaljournal.com/life/health/ai-in-healthcare-will-reach-6-16-billion-by-2022/article/531991
http://www.digitaljournal.com/life/health/amazon-is-diving-ever-deeper-into-healthcare/article/563201
https://healthitanalytics.com/news/45-of-ors-will-be-integrated-with-artificial-intelligence-by-2022
https://www.prweb.com/releases/mediaplanet_teams_up_with_berg_to_highlight_how_harnessing_artificial_intelligence_and_patient_biology_can_accelerate_treatments_for_rare_diseases/prweb15764315.h
https://www.drugchannels.net/2018/09/artificial-intelligence-ai-how.html
https://elitedatascience.com/machine-learning-impact
http://lippincottsolutions.lww.com/blog.entry.html/2018/09/13/the_future_is_nowh-JFSb.html
https://blogs.microsoft.com/blog/2019/06/19/harnessing-the-power-of-ai-to-transform-healthcare/
https://www.prnewswire.com/news-releases/artificial-intelligence-in-healthcare-takes-precision-medicine-to-the-next-level-300712098.html
https://www.rtinsights.com/artificial-intelligence-brings-more-clarity-to-hearing-devices/
https://www.europeanpharmaceuticalreview.com/news/79303/artificial-intelligence-ai-lung-disease-diagnosis-accuracy/

https://www.holyrood.com/inside-politics/view,could-artificial-intelligence-save-the-health-service_9249.htm
https://sbmi.uth.edu/blog/aug-17/what-is-the-relationship-between-informatics-and-data-science.htm
https://www.beckershospitalreview.com/healthcare-information-technology/how-artificial-intelligence-impacts-preventative-care.html
https://medium.com/swlh/the-future-with-ai-and-automated-digital-coaching-assistants-e0ccf7072c54
https://www.sciencedaily.com/releases/2018/09/180918180501.htm
https://www.engineering.com/DesignerEdge/DesignerEdgeArticles/ArticleID/17664/A-Healthy-Future-for-Artificial-Intelligence-in-Healthcare.aspx
https://nyulangone.org/news/artificial-intelligence-tool-accurately-identifies-cancer-type-genetic-changes-each-patients-lung-tumor
https://www.postbulletin.com/life/health/is-artificial-intelligence-a-natural-fit-for-health-care/article_ab274b68-b82c-11e8-aa3c-77aaaf1d1012.html
https://www.sciencedaily.com/releases/2018/09/180917111642.htm
https://www.globenewswire.com/news-release/2018/09/17/1571635/0/en/Global-Artificial-Intelligence-In-Diabetes-Management-Market-Worth-USD-1422-52-Million-By-2024-Zion-Market-Research.html
https://retinaroundup.com/2018/09/16/retina-society-2018-artificial-intelligence/
https://www.itnonline.com/content/exact-imaging-partners-improve-prostate-cancer-detection-artificial-intelligence
https://www.itnonline.com/content/philips-and-paige-team-bring-artificial-intelligence-ai-clinical-pathology-diagnostics
https://www.itnonline.com/content/dia-joins-ibm-watson-health-arm-clinicians-its-ai-powered-cardiac-ultrasound-software
https://www.itnonline.com/content/ai-improves-chest-x-ray-interpretation
https://www.drugtargetreview.com/article/45973/artificial-intelligence-in-the-world-of-drug-discovery/
http://www.digitaljournal.com/tech-and-science/science/artificial-intelligence-used-to-detect-early-signs-of-dementia/article/532769

https://www.technologyreview.com/s/609236/ai-can-spot-signs-of-alzheimers-before-your-family-does/

https://madison.com/wsj/business/technology/ensodata-uses-ai-to-help-improve-health-care/article_0f742d73-9256-589f-95e5-ff0a4e3457b4.html

https://hitinfrastructure.com/news/how-to-begin-healthcare-artificial-intelligence-deployment

https://www.labiotech.eu/features/artificial-intelligence-oncology/

https://www.nytimes.com/2019/10/24/well/live/machine-intelligence-AI-breast-cancer-mammogram.html

https://www.detroitnews.com/story/opinion/2018/07/20/impact-artificial-intelligence-health-care/791952002/

https://www.healthdatamanagement.com/list/7-ways-ai-could-make-an-impact-on-medical-care

https://www.itnewsafrica.com/2018/07/artificial-intelligence-is-solving-african-healthcare-challenges/

https://www.healthcarefinancenews.com/news/what-healthcare-cfos-should-know-about-artificial-intelligence-machine-learning-and-chatbots

https://venturebeat.com/2018/07/19/consortium-ai-wants-to-cure-rare-diseases-using-artificial-intelligence/

https://venturebeat.com/2019/07/31/deepminds-ai-predicts-kidney-injury-up-to-48-hours-before-it-happens/

https://venturebeat.com/2019/12/10/current-health-raises-11-5-million-to-predict-diseases-with-ai-and-remote-monitoring/

https://venturebeat.com/2019/12/09/google-proposes-hybrid-approach-to-ai-transfer-learning/

https://venturebeat.com/2019/12/06/ai-weekly-amazon-plays-the-long-game-in-health-care-ai/

https://venturebeat.com/2019/12/03/google-details-ai-that-classifies-chest-x-rays-with-human-level-accuracy/

http://bwdisrupt.businessworld.in/article/Artificial-Intelligence-Redefining-the-Digital-Nervous-System-of-Healthcare-Industry-/19-07-2018-155279/

https://www.fiercebiotech.com/medtech/insilico-and-a2a-launch-new-duchenne-focused-ai-drug-company

https://www.twst.com/news/robotic-surgical-system-using-artificial-intelligence-cure-baldness-restoration-robotics-inc-nasdaqhair/
https://www.globenewswire.com/news-release/2018/07/23/1540633/0/en/MC-Endeavors-Inc-Room-21-Media-Launches-First-Artificial-Intelligent-Marketing-Platform-with-Restore-Detox-Centers-to-Combat-Addiction-in-the-US.html
https://www.theguardian.com/technology/2018/aug/13/new-artificial-intelligence-tool-can-detect-eye-problems-as-well-as-experts
https://www.theguardian.com/technology/2018/jul/04/its-going-create-revolution-how-ai-transforming-nhs
https://www.opengovasia.com/artificial-intelligence-to-improve-medical-imaging-for-patients-with-brain-ailments/
https://mhealthintelligence.com/news/new-group-aims-to-advance-artificial-intelligence-in-telehealth
https://electronics360.globalspec.com/article/12414/artificial-intelligence-system-creates-new-pharmaceutical-drugs
https://electronics360.globalspec.com/article/12231/ai-system-can-predict-side-effects-of-new-drug-combinations
https://www.nytimes.com/2019/02/05/technology/artificial-intelligence-drug-research-deepmind.html
https://www.prunderground.com/vetology-artificial-intelligence-to-begin-reading-pet-x-rays/00131851/
https://www.sciencedaily.com/releases/2018/07/180731151326.htm
https://medicine.utoronto.ca/news/smarter-radiation-therapy-artificial-intelligence
https://www.healthdatamanagement.com/opinion/3-ways-artificial-intelligence-can-disrupt-healthcare
https://www.dw.com/en/doctors-dont-scale-like-artificial-intelligence-does/a-44899327
https://www.aao.org/eyenet/article/artificial-intelligence-and-glaucoma-detection
https://www.news-medical.net/health/Artificial-Intelligence-in-Cardiology.aspx
https://www.healthdatamanagement.com/news/ai-tool-helps-gateway-health-boost-accuracy-population-health
https://www.businessinsider.com/artificial-intelligence-healthcare

https://www.healthcareitnews.com/news/where-hospitals-plan-big-ai-deployments-diagnostic-imaging

https://www.healthcareitnews.com/news/asia-pacific/australia-s-snac-develops-ai-tools-improve-brain-scan-analysis

https://www.beckershospitalreview.com/artificial-intelligence/anthem-uses-ai-in-allergy-research-trial.html

https://www.geekwire.com/2018/mindshare-medical-launches-ai-cancer-screening-tech-can-see-data-beyond-perception/

https://lfpress.com/news/local-news/artificial-intelligence-may-be-key-to-mood-disorder-diagnosis-study

https://www.nextbigfuture.com/2018/08/artificial-intelligence-creating-new-drugs-from-scratch-by-efficiently-searching-huge-molecular-possibilities.html

https://qz.com/1349854/ai-can-spot-the-pain-from-a-disease-some-doctors-still-think-is-fake/

https://qz.com/1383083/how-ai-changed-organ-donation-in-the-us/

https://healthitanalytics.com/news/artificial-intelligence-for-medical-imaging-market-to-top-2b

https://www.thegazette.com/subject/news/business/coralville-based-idx-llc-aims-to-change-health-care-delivery-20180812

https://news.mit.edu/2018/artificial-intelligence-model-learns-patient-data-cancer-treatment-less-toxic-0810

https://www.healthcarefinancenews.com/news/how-artificial-intelligence-can-save-health-insurers-7-billion

https://www.techiexpert.com/ai-used-to-create-inexpensive-heart-disease-detector/

https://www.dermatologytimes.com/business/artificial-intelligence-friend-or-foe-dermatology

http://customerthink.com/looking-at-5-noteworthy-applications-of-artificial-intelligence-in-healthcare/

https://onlinelibrary.wiley.com/doi/full/10.1111/dmcn.13942

https://blog.jive.com/ai-chatbots-contact-centers/

https://www.theguardian.com/technology/2018/jul/29/the-robot-will-see-you-now-could-computers-take-over-medicine-entirely

https://www.nature.com/articles/d41586-019-01111-y

https://www.sciencedaily.com/releases/2018/08/180813113315.htm

https://www.curetoday.com/publications/cure/2019/fall-2019/as-large-as-life-using-artificial-intelligence-in-cancer-care

https://medibulletin.com/1-2-seconds-is-all-artificial-intelligence-takes-to-screen-ct-scans/

https://www.marktechpost.com/2018/08/14/artificial-intelligence-in-medicine-gains-massive-traction-with-growing-investment-in-ai-in-the-space/

https://www.washingtontimes.com/news/2018/aug/13/colonoscopy-technology-cuts-need-polyp-surgery-tes/

https://medicalxpress.com/news/2018-09-medicine-ready-artificial-intelligence.html

https://www.beckershospitalreview.com/healthcare-information-technology/how-artificial-intelligence-can-transform-payment-integrity.html

https://www.scitecheuropa.eu/artificial-intelligence-to-map-obesity/88924/

https://www.marketwatch.com/press-release/notable-health-closes-135-million-series-a-financing-to-expand-artificial-intelligence-powered-physician-patient-interaction-platform-2018-09-06

https://www.scmp.com/news/china/science/article/2163110/are-you-risk-diabetes-chinese-ai-system-could-predict-disease-15

https://geneticliteracyproject.org/2018/09/10/can-artificial-intelligence-give-us-a-more-efficient-health-care-system/

https://mindmatters.ai/2019/06/new-evidence-that-some-comatose-people-really-do-understand/

https://www.fastcompany.com/90287723/artificial-intelligence-can-detect-alzheimers-in-brain-scans-six-years-before-a-diagnosis

https://www.sciencedaily.com/releases/2019/01/190103152906.htm

https://newatlas.com/robot-md-ai-future-medial-diagnosis/60874/

https://www.healthworkscollective.com/can-artificial-intelligence-diagnose-illnesses-better-than-other-methods/

https://mytechdecisions.com/compliance/artificial-intelligence-delivery-rooms/

https://www.europeanpharmaceuticalreview.com/news/82870/artificial-intelligence-ai-heart-disease/

https://www.europeanpharmaceuticalreview.com/news/102403/ai-can-predict-which-research-will-translate-to-clinical-trials/

https://www.smithsonianmag.com/innovation/will-artificial-intelligence-improve-health-care-for-everyone-180972758/

https://venturebeat.com/2019/01/04/massachusetts-generals-ai-can-spot-brain-hemorrhages-as-accurately-as-humans/

https://www.geek.com/news/machine-learning-may-predict-how-well-youll-age-1767646/

https://becominghuman.ai/the-role-of-ai-in-healthcare-technology-6c33a6eee18c

https://www.sciencemag.org/news/2019/01/artificial-intelligence-could-diagnose-rare-disorders-using-just-photo-face

https://www.sciencedaily.com/releases/2019/01/190102112926.htm

https://www.empr.com/home/news/artificial-intelligence-can-detect-classify-acute-brain-bleeds/

https://www.bioeng.ucla.edu/artificial-intelligence-detects-the-presence-of-viruses/

https://blogs.scientificamerican.com/observations/the-surgical-singularity-is-approaching/

https://healthitanalytics.com/news/artificial-intelligence-in-healthcare-spending-to-hit-36b

https://www.fotoinc.com/news-updates/artificial-intelligence-embedded-mobile-app-for-chronic-neck-and-back-pain

https://www.theweek.in/news/sci-tech/2018/12/19/British-doctors-sceptical-that-AI-could-entirely-replace-them.html

https://www.news-medical.net/life-sciences/Artificial-Intelligence-in-Histopathology.aspx

https://blogs.scientificamerican.com/observations/how-ai-could-help-your-bad-back/

https://www.mdmag.com/medical-news/artificial-intelligence-deep-learning-learn-diagnose-epilepsy

https://www.ncbi.nlm.nih.gov/pmc/articles/PMC6290744/

https://en.wikipedia.org/wiki/Artificial_intelligence_in_healthcare

https://www.msn.com/en-us/health/medical/cigna-is-using-artificial-intelligence-to-predict-which-of-its-subscribers-will-become-addicted-to-opioids/ar-BBReYKL

https://www.sciencedaily.com/releases/2018/12/181221123743.htm

https://www.wndu.com/content/news/Artificial-intelligence-improves-colonoscopies-503263751.html

https://www.unionleader.com/news/scitech/ai-can-predict-mental-health-issues-from-your-instagram-posts/article_06ab3c45-8fa5-539e-869e-c19cab1cbef0.html

https://www.expresshealthcare.in/news/medachievers-and-labindia-healthcare-bring-artificial-intelligence-based-open-surgery-simulator/407543/

https://hackernoon.com/how-ai-robots-are-infiltrating-healthcare-ts8u3zfq

https://towardsdatascience.com/artificial-intelligence-deep-learning-for-medical-diagnosis-9561f7a4e5f

http://www.digitaljournal.com/tech-and-science/science/ai-can-boost-cancer-drug-discovery/article/538623

https://www.nbcnews.com/mach/science/why-big-pharma-betting-big-ai-ncna852246

https://www.kurzweilai.net/how-to-predict-the-side-effects-of-millions-of-drug-combinations

https://futurism.com/neoscope/genetic-report-predict-disease-risk

https://www.modernhealthcare.com/article/20180910/NEWS/180919991/telemedicine-meets-artificial-intelligence-at-bedside

https://www.beckershospitalreview.com/artificial-intelligence/5-developments-in-ai-last-week.html

https://www.forbes.com/sites/brucelee/2019/01/20/do-you-have-early-signs-of-cervical-cancer-how-ai-technology-may-help/#55552bdce694

http://www.digitaljournal.com/life/health/could-a-machine-replace-a-doctor/article/538540

https://ai-med.io/artificial-intelligence-robotic-surgery/

https://www.healthcareitnews.com/news/ai-healthcare-big-ethical-questions-still-need-answers

http://newsroom.ucla.edu/releases/artificial-intelligence-device-detect-moving-parasites-bodily-fluid-earlier-diagnosis

https://www.periscopedata.com/press/artificial-intelligence-bias-and-mental-health-implications

https://hospitalnews.com/artificial-intelligence-being-used-to-develop-drugs-even-faster-and-cheaper/

https://www.globenewswire.com/news-release/2018/12/12/1665700/0/en/Artificial-Intelligence-key-to-Universal-Health-Coverage.html

https://www.acobiom.com/en/how-artificial-intelligence-can-help-to-the-development-of-diagnostics-dedicated-to-precision-medicine/

https://www.forbes.com/sites/bernardmarr/2018/12/21/ai-that-saves-lives-the-chatbot-that-can-detect-a-heart-attack-using-machine-learning/#3300974450f9

https://www.asianscientist.com/2018/12/in-the-lab/artificial-intelligence-flu-sensor/

https://openmedscience.com/bright-future-for-robotic-surgeons/

https://www.beckershospitalreview.com/artificial-intelligence/northwestern-introduces-ai-tool-to-help-medical-assistants-conduct-sonograms.html

https://www.medicalbag.com/home/more/tech-talk/the-future-of-robot-physicians-is-artificial-intelligence-poised-to-take-over-medicine/

https://www.geekwire.com/2018/amazon-unveils-new-service-mine-decode-medical-records-using-artificial-intelligence/

https://www.dw.com/en/artificial-intelligence-in-medicine-the-computer-knows-what-you-need/a-46226852

https://www.prnewswire.com/news-releases/carepredict-presents-artificial-intelligence-powered-solutions-for-senior-care-at-aging2-0-optimize-2018--300749580.html

https://hitconsultant.net/2018/12/13/artificial-intelligence-transform-personal-health/

https://www.wftv.com/news/local/orlando-hospital-uses-artificial-intelligence-to-save-lives-in-the-delivery-room/874200835/

https://axiosholding.com/healthcare-ai-just-what-the-doctor-ordered/

https://www.globenewswire.com/news-release/2018/11/10/1649339/0/en/AI-Algorithm-Outperformed-Majority-of-Cardiologists-in-Detection-of-Heart-Murmurs-in-Clinical-Study.html

https://www.smithsonianmag.com/innovation/can-artificial-intelligence-detect-depression-in-persons-voice-180970702/
https://ai-med.io/artificial-intelligence-lie-human/
https://www.gqrgm.com/5-ways-ai-is-transforming-healthcare/
https://www.jwatch.org/na47874/2018/11/06/artificial-intelligence-improves-wrist-fracture-detection
https://www.healthcareitnews.com/news/ai-algorithms-show-promise-colonoscopy-screenings
https://electronichealthreporter.com/artificial-intelligence-is-the-new-operating-system-in-healthcare/
Top of Form
Bottom of Form
https://pophealthanalytics.com/exploring-the-role-of-ai-in-population-health-risk-assessment-symposium/
https://www.ns-businesshub.com/science/artificial-intelligence-in-medicine-cure-cancer/
http://dailytrojan.com/2018/10/28/mindful-mondays-artificial-intelligence-has-limited-power-to-treat-mental-health-issues/
https://www.dailydot.com/debug/ai-sepsis-diagnosis/
https://blog.sevenponds.com/science-of-us/%E2%80%A8%E2%80%A8artificial-intelligence-outsmarts-past-practices-that-predict-patient-outcomes
https://www.medicalnewstoday.com/articles/325491.php#1
https://onlinemedicalcare.org/artificial-intelligence-outperforms-real-doctors-studies-show/
https://www.healthline.com/health-news/new-ai-technology-may-help-diagnose-fetal-heart-problems
https://www.aiin.healthcare/topics/artificial-intelligence/ai-enhanced-virtual-care-could-reduce-er-visits
https://www.medicaldevice-network.com/features/ethics-in-ai/
https://www.healthdatamanagement.com/opinion/why-using-ai-in-healthcare-requires-a-balance-of-efficiency-and-ethics
https://www.prnewswire.com/news-releases/osf-ventures-invests-in-gauss-artificial-intelligence-technology-to-identify-postpartum-bleeding-early-300732784.html

https://www.medicaldesignandoutsourcing.com/digital-surgery-touts-artificial-intelligence-for-the-operating-room/
https://ai-med.io/will-artificial-intelligence-replace-human-radiologists/
https://www.sciencedaily.com/releases/2019/01/190111143744.htm
https://directorsblog.nih.gov/2019/01/17/using-artificial-intelligence-to-detect-cervical-cancer/
https://www.forbes.com/sites/bernardmarr/2018/07/27/how-is-ai-used-in-healthcare-5-powerful-real-world-examples-that-show-the-latest-advances/#4a59916e5dfb
https://transmitter.ieee.org/artificial-intelligence-robots-and-the-operating-room/
https://www.nationthailand.com/asean-plus/30349160
https://directorsblog.nih.gov/2019/01/15/using-artificial-intelligence-to-catch-irregular-heartbeats/
https://analyticsindiamag.com/ai-tell-how-ovarian-tumour/
https://lfpress.com/news/local-news/artificial-intelligence-can-predict-ptsd-in-patients-london-researchers
https://www.forbes.com/sites/samshead/2018/10/08/tencent-aims-to-train-ai-to-spot-parkinsons-in-3-minutes/#676980856f36
https://engineering.stanford.edu/magazine/article/david-magnus-how-will-artificial-intelligence-impact-medical-ethics
https://theweek.com/articles/694522/robots-replace-therapists
https://www.leewayhertz.com/how-iot-transforming-healthcare/
https://www.straitstimes.com/singapore/scdf-turns-to-artificial-intelligence-to-help-emergency-call-dispatchers
https://www.eurekalert.org/pub_releases/2018-07/imi-mlt070518.php
https://www.latimes.com/world/la-fg-china-ai-20180706-story.html
https://www.healthcareitnews.com/news/chinese-hospital-guangdong-deploys-ai-cameras-detect-blindness-causing-diseases
https://www.healthcarefinancenews.com/news/artificial-intelligence-healthcare-projected-be-worth-more-27-billion-2025
https://analyticsindiamag.com/can-machines-be-taught-to-detect-medicare-fraud/
https://www.newscientist.com/article/2222907-ai-can-predict-if-youll-die-soon-but-weve-no-idea-how-it-works/

https://revcycleintelligence.com/news/using-artificial-intelligence-to-improve-the-hospital-revenue-cycle

https://www.healthcarefinancenews.com/news/addressing-social-determinants-health-consider-artificial-intelligence-and-machine-learning

https://www.mcknights.com/marketplace/marketplace-experts/what-facilities-can-gain-from-robotics-in-physical-therapy/

https://www.entrepreneur.com/article/317047

https://www.icr.ac.uk/news-archive/artificial-intelligence-can-predict-how-cancers-will-evolve-and-spread

https://www.frontiersin.org/articles/10.3389/fpsyg.2019.00263/full

https://www.usatoday.com/story/opinion/2019/01/28/health-privacy-laws-artificial-intelligence-hipaa-needs-update-column/2695386002/

https://www.nature.com/articles/s41746-019-0089-x

https://time.com/collection/life-reinvented/5494363/sleep-artificial-intelligence/

https://www.corporatewellnessmagazine.com/article/how-artificial-intelligence-can-fight-against-workplace-stress

https://www.sciencedaily.com/releases/2018/05/180528190839.htm

https://www.cnet.com/news/ai-to-aid-emergency-call-operators-diagnose-heart-attacks-in-europe/

https://www.forbes.com/sites/michaelpellmanrowland/2018/04/24/beyonce-artificial-intelligence-vegan/#503780cc7160

https://govinsider.asia/innovation/artificial-intelligence-ageing-population/

https://www.geekwire.com/2018/health-tech-podcast-using-precision-medicine-kill-cancer-artificial-intelligence/

https://thriveglobal.com/stories/the-road-to-good-health-will-now-be-guided-by-artificial-intelligence/

https://www.engadget.com/2018/10/15/google-ai-spots-advanced-breast-cancer/

https://www.stanforddaily.com/2018/09/27/artificial-swarm-intelligence-diagnoses-pneumonia-better-than-individual-computer-or-doctor/

https://www.forbes.com/sites/michaelmillenson/2018/09/23/will-apple-track-your-mind-not-just-your-heart/#4879dbabcad3

https://emerj.com/ai-sector-overviews/artificial-intelligence-in-health-insurance-current-applications-and-trends/

https://www.unc.edu/posts/2018/07/31/artificial-intelligence-system-created-at-unc-chapel-hill-designs-drugs-from-scratch/

https://www.beaconlens.com/ai-in-behavioral-health-care-when-artificial-intelligence-became-real/

https://www.eurekalert.org/pub_releases/2018-09/sfl-ait091818.php

https://www.einfochips.com/blog/how-ai-enabled-wearables-are-changing-healthcare-and-fitness-industry/

https://psychnews.psychiatryonline.org/doi/10.1176/appi.pn.2018.10a5

https://www.eurekalert.org/pub_releases/2018-10/muos-wcm100218.php

https://time.com/collection/life-reinvented/5492063/artificial-intelligence-fertility/

https://www.thefix.com/artificial-intelligence-system-aims-identify-drug-thefts-hospitals

https://towardsdatascience.com/applying-artificial-intelligence-to-help-people-quit-smoking-early-results-a3e5581d560

https://medcitynews.com/2018/05/qure-ai/

https://healthitanalytics.com/news/deep-learning-tool-tops-dermatologists-in-melanoma-detection

https://www.himss.org/resources/role-artificial-intelligence-healthcare-and-society

https://www.aami.org/newsviews/newsdetail.aspx?ItemNumber=6502

http://www.thetower.org/6317-israel-uk-announce-landmark-scientific-cooperation-agreement-focusing-on-aging-artificial-intelligence/

https://www.aarp.org/health/conditions-treatments/info-2018/hospital-artificial-intelligence-telehealth.html

https://www.digitaltrends.com/cool-tech/chatterbaby-app-deciphers-baby-crying/

https://www.cbs58.com/news/new-app-translates-babies-cries-so-deaf-parents-understand-whats-going-on

https://www.empr.com/home/news/fda-approves-ai-algorithm-that-helps-detect-wrist-fractures/

https://bioinformatics.csiro.au/blog/can-ai-help-fight-antibiotic-resistant-superbugs/
https://www.sciencedaily.com/releases/2018/09/180920161054.htm
https://www.forbes.com/sites/jenniferhicks/2018/06/08/see-how-this-hospital-uses-artificial-intelligence-to-find-kidney-disease/#78af38d72e8f
https://www.newsweek.com/ai-being-developed-which-can-smell-illness-human-breath-967197
https://www.theguardian.com/technology/2018/jun/10/artificial-intelligence-cancer-detectors-the-five
https://www.independent.co.uk/voices/loneliness-kills-artificial-intelligence-chatbot-doctors-health-risk-diagnoses-a8423321.html
https://www.technologyreview.com/f/609969/ai-could-diagnose-your-heart-attack-on-the-phone-even-if-youre-not-the-caller/
https://www.linkedin.com/feed/news/teaching-ai-to-explain-its-thinking-4617604
https://www.smartdatacollective.com/artificial-intelligence-chatbots-could-make-your-doctors-obsolete/
https://www.telegraph.co.uk/technology/2019/01/13/google-scientist-ai-may-not-cure-all-discovering-new-drugs/
https://greatlakesledger.com/2019/01/12/artificial-intelligence-can-predict-and-estimate-flu-activity/
https://www.mddionline.com/using-artificial-intelligence-predict-flu-activity
https://time.com/5556339/artificial-intelligence-robots-medicine/
https://www.nytimes.com/2019/03/11/well/live/how-artificial-intelligence-could-transform-medicine.html
https://towardsdatascience.com/artificial-intelligence-replace-the-human-doctors-in-the-future-is-it-true-91b3ae9fea0e
https://www.sciencedaily.com/releases/2019/12/191218090156.htm
https://spectrum.ieee.org/biomedical/diagnostics/how-ibm-watson-overpromised-and-underdelivered-on-ai-health-care
https://venturebeat.com/2019/09/13/googles-ai-detects-26-skin-conditions-as-accurately-as-dermatologists/

https://www.dhs.gov/science-and-technology/news/2018/10/16/snapshot-public-safety-agencies-pilot-artificial-intelligence
https://www.inverse.com/article/59742-smartphone-app-health-best-eye-disease
https://www.foxnews.com/tech/google-ai-can-predict-when-youll-die-with-95-percent-accuracy-researchers-say
https://www.zdnet.com/article/googles-deepmind-follows-a-mixed-path-to-ai-in-medicine/
https://medicalfuturist.com/artificial-intelligence-in-mental-health-care/
https://www.cnbc.com/2018/02/22/medical-errors-third-leading-cause-of-death-in-america.html#:~:text=According%20to%20a%20recent%20study,aftea%20heart%20disease%20and%20cancer.

Chapter 17 *Future of Diseases & Administration with AI*

1. *AI Gives Heart Attack Instructions*
   https://www.youtube.com/watch?v=47i4viaS7zc
2. *AI Detects Dementia*
   https://www.youtube.com/watch?v=NeNzkfsxElI
3. *AI Detects X-Rays*
   https://www.youtube.com/watch?v=xDgkmXAsvL8
4. *AI Proposed Brain Scans*
   https://www.youtube.com/watch?v=bWJMocALUes
5. *AI Predicting Diseases*
   https://www.youtube.com/watch?v=VePHPymCy2U
6. *Mandated Electronic Healthcare Data*
   https://www.youtube.com/watch?v=hcrS13OWQ1U
   https://www.youtube.com/watch?v=Lo_3qOejQzI
7. *Microchip Your Diabetes*
   https://www.youtube.com/watch?v=TW-c8y1SSMs
8. *Microchip Your Diet*
   https://www.thedoctorstv.com/videos/weight-loss-microchip

9. *Microchip Your Offspring*
   https://www.youtube.com/watch?v=kvlNBV8ZRJA
10. *Microchip Your Payments*
    https://www.youtube.com/watch?v=WzfGqzqf8BU
11. *Information* on *AI Medical*
    https://www.medgadget.com/2018/09/doctors-pair-with-artificial-intelligence-to-improve-pneumonia-diagnosis.html
    https://qz.com/1548524/china-has-produced-another-study-showing-the-potential-of-ai-in-medical-diagnosis/
    https://qz.com/1384725/an-ai-algorithm-in-china-is-learning-to-detect-whether-patients-will-wake-from-a-coma/
    https://www.scmp.com/news/china/science/article/2163298/doctors-said-coma-patients-would-never-wake-ai-said-they-would
    https://www.thehindu.com/sci-tech/health/ai-may-predict-alzheimers-disease-5-years-in-advance/article25160303.ece
    https://www.livemint.com/Science/evBwYmDFLvImRwjO97HeeO/Indian-scientists-using-artificial-intelligence-to-predict-e.html
    https://www.forbes.com/sites/bernardmarr/2018/12/21/ai-that-saves-lives-the-chatbot-that-can-detect-a-heart-attack-using-machine-learning/#53c9859450f9
    https://thenextweb.com/artificial-intelligence/2018/04/25/europe-launches-heart-attack-detecting-ai-emergency-calls/
    https://www.esmo.org/About-Us/ESMO-Magazine/Feature-AI-and-Big-Data-in-Oncology-How-Ready-Are-We
    https://www.statnews.com/2019/07/03/artificial-intelligence-guarded-optimism-cancer-care/
    https://www.healthcareitnews.com/news/geisinger-injects-machine-learning-clinical-workflow-find-health-problems-faster
    https://www.cardiovascularbusiness.com/topics/artificial-intelligence/twists-turns-ahead-ais-road-acceptance-cardiology
    https://scienmag.com/insilico-to-present-its-latest-research-in-ai-aging-diagnostics-at-the-aone-conference/
    https://svn.bmj.com/content/2/4/230
    https://healthmanagement.org/c/hospital/issuearticle/artificial-intelligence-a-next-way-forward-for-healthcare

https://healthmanagement.org/c/hospital/issuearticle/artificial-intelligence-in-healthcare-what-is-versus-what-will-be

https://www.linkedin.com/pulse/should-ai-take-hippocratic-oath-gary-gilliland-md-phd

https://www.livescience.com/65087-ai-premature-death-prediction.html

https://www.prnewswire.com/news-releases/perception-health-on-aws-marketplace-machine-learning-and-artificial-intelligence-discovery-page-300773045.html

https://www.express.co.uk/news/world/976108/google-when-will-you-die-ai-artificial-intelligence-latest

https://cio.economictimes.indiatimes.com/news/strategy-and-management/adoption-of-new-technology-is-not-an-option-but-a-compulsion-ggm-it-irctc/71788370

https://pjmedia.com/trending/can-a-healthcare-algorithm-be-racially-biased/

https://time.com/5556339/artificial-intelligence-robots-medicine/

https://www.modernhealthcare.com/article/20150610/NEWS/150619996/walgreen-insurers-push-expansion-of-virtual-doctor-visits

https://electronicsforu.com/technology-trends/tech-focus/artificial-intelligence-healthcare-replace-doctors

https://healthitanalytics.com/news/at-montefiore-artificial-intelligence-becomes-key-to-patient-care

https://time.com/5709346/artificial-intelligence-health/

https://www.reuters.com/article/us-facebook-features/facebook-turns-to-artificial-intelligence-to-tackle-suicides-idUSKBN1684JQ

https://www.fastcompany.com/90299135/mental-health-crisis-robots-chatbots-listeners

https://www.sandiegouniontribune.com/sdut-smartphone-voices-not-always-helpful-in-health-2016mar14-story.html

https://becominghuman.ai/becoming-human-ai-addiction-deaf948e17ba

https://www.cms.gov/newsroom/press-releases/cms-announces-artificial-intelligence-health-outcomes-challenge-participants-advancing-stage-1

https://healthmanagement.org/c/healthmanagement/IssueArticle/ai-is-the-new-reality-the-4th-healthcare-revolution-in-medicine

https://www.independent.co.uk/voices/artificial-intelligence-machine-learning-computers-global-healthcare-malaria-facebook-a8327901.html

https://questsoblogspot.wordpress.com/2019/09/25/artificial-intelligence-for-personal-business-and-enterprise-uses/

https://news.yahoo.com/health-checks-smartphone-raise-privacy-fears-080635403.html

https://www.theglobeandmail.com/life/health-and-fitness/article-app-developers-are-using-artificial-intelligence-to-advise-teens-about/

https://www.sdglobaltech.com/blog/some-mind-blowing-stats-about-ai-and-iot-in-healthcare

https://builtin.com/artificial-intelligence/artificial-intelligence-healthcare

https://khn.org/news/a-reality-check-on-artificial-intelligence-are-health-care-claims-overblown/

https://healthitanalytics.com/news/54-of-healthcare-pros-expect-widespread-ai-adoption-in-5-years

https://www.ncbi.nlm.nih.gov/pmc/articles/PMC6616181/

https://dzone.com/articles/using-ai-to-design-drugs-from-scratch

https://www.smithsonianmag.com/innovation/will-artificial-intelligence-improve-health-care-for-everyone-180972758/

https://www.managedcaremag.com/archives/2019/7/ai-all-ails-american-health-care-how-smart

https://www.digitalcommerce360.com/2018/09/14/ai-may-save-one-small-hospital-20-million/

https://www.healthdatamanagement.com/news/cios-must-take-charge-in-implementing-artificial-intelligence-governance

https://www.theweek.in/news/sci-tech/2018/09/15/New-artificial-intelligence-system-can-detect-dementia.html

http://www.digitaljournal.com/life/health/ai-in-healthcare-will-reach-6-16-billion-by-2022/article/531991

http://www.digitaljournal.com/life/health/amazon-is-diving-ever-deeper-into-healthcare/article/563201

https://healthitanalytics.com/news/45-of-ors-will-be-integrated-with-artificial-intelligence-by-2022

https://www.prweb.com/releases/mediaplanet_teams_up_with_berg_to_highlight_how_harnessing_artificial_intelligence_and_patient_biology_can_accelerate_treatments_for_rare_diseases/prweb15764315.h

https://www.drugchannels.net/2018/09/artificial-intelligence-ai-how.html

https://elitedatascience.com/machine-learning-impact

http://lippincottsolutions.lww.com/blog.entry.html/2018/09/13/the_future_is_nowh-JFSb.html

https://blogs.microsoft.com/blog/2019/06/19/harnessing-the-power-of-ai-to-transform-healthcare/

https://www.prnewswire.com/news-releases/artificial-intelligence-in-healthcare-takes-precision-medicine-to-the-next-level-300712098.html

https://www.rtinsights.com/artificial-intelligence-brings-more-clarity-to-hearing-devices/

https://www.europeanpharmaceuticalreview.com/news/79303/artificial-intelligence-ai-lung-disease-diagnosis-accuracy/

https://www.holyrood.com/inside-politics/view,could-artificial-intelligence-save-the-health-service_9249.htm

https://sbmi.uth.edu/blog/aug-17/what-is-the-relationship-between-informatics-and-data-science.htm

https://www.beckershospitalreview.com/healthcare-information-technology/how-artificial-intelligence-impacts-preventative-care.html

https://www.sciencedaily.com/releases/2018/09/180918180501.htm

https://www.engineering.com/DesignerEdge/DesignerEdgeArticles/ArticleID/17664/A-Healthy-Future-for-Artificial-Intelligence-in-Healthcare.aspx

https://nyulangone.org/news/artificial-intelligence-tool-accurately-identifies-cancer-type-genetic-changes-each-patients-lung-tumor

https://www.postbulletin.com/life/health/is-artificial-intelligence-a-natural-fit-for-health-care/article_ab274b68-b82c-11e8-aa3c-77aaaf1d1012.html

https://www.sciencedaily.com/releases/2018/09/180917111642.htm

https://www.globenewswire.com/news-release/2018/09/17/1571635/0/en/Global-Artificial-Intelligence-In-

Diabetes-Management-Market-Worth-USD-1422-52-Million-By-2024-Zion-Market-Research.html
https://retinaroundup.com/2018/09/16/retina-society-2018-artificial-intelligence/
https://www.itnonline.com/content/exact-imaging-partners-improve-prostate-cancer-detection-artificial-intelligence
https://www.itnonline.com/content/philips-and-paige-team-bring-artificial-intelligence-ai-clinical-pathology-diagnostics
https://www.itnonline.com/content/dia-joins-ibm-watson-health-arm-clinicians-its-ai-powered-cardiac-ultrasound-software
https://www.itnonline.com/content/ai-improves-chest-x-ray-interpretation
https://www.drugtargetreview.com/article/45973/artificial-intelligence-in-the-world-of-drug-discovery/
http://www.digitaljournal.com/tech-and-science/science/artificial-intelligence-used-to-detect-early-signs-of-dementia/article/532769
https://www.technologyreview.com/s/609236/ai-can-spot-signs-of-alzheimers-before-your-family-does/
https://madison.com/wsj/business/technology/ensodata-uses-ai-to-help-improve-health-care/article_0f742d73-9256-589f-95e5-ff0a4e3457b4.html
https://hitinfrastructure.com/news/how-to-begin-healthcare-artificial-intelligence-deployment
https://www.labiotech.eu/features/artificial-intelligence-oncology/
https://www.nytimes.com/2019/10/24/well/live/machine-intelligence-AI-breast-cancer-mammogram.html
https://www.detroitnews.com/story/opinion/2018/07/20/impact-artificial-intelligence-health-care/791952002/
https://www.healthdatamanagement.com/list/7-ways-ai-could-make-an-impact-on-medical-care
https://www.itnewsafrica.com/2018/07/artificial-intelligence-is-solving-african-healthcare-challenges/
https://www.healthcarefinancenews.com/news/what-healthcare-cfos-should-know-about-artificial-intelligence-machine-learning-and-chatbots

https://venturebeat.com/2018/07/19/consortium-ai-wants-to-cure-rare-diseases-using-artificial-intelligence/

https://venturebeat.com/2019/07/31/deepminds-ai-predicts-kidney-injury-up-to-48-hours-before-it-happens/

https://venturebeat.com/2019/12/10/current-health-raises-11-5-million-to-predict-diseases-with-ai-and-remote-monitoring/

https://venturebeat.com/2019/12/09/google-proposes-hybrid-approach-to-ai-transfer-learning/

https://venturebeat.com/2019/12/06/ai-weekly-amazon-plays-the-long-game-in-health-care-ai/

https://venturebeat.com/2019/12/03/google-details-ai-that-classifies-chest-x-rays-with-human-level-accuracy/

http://bwdisrupt.businessworld.in/article/Artificial-Intelligence-Redefining-the-Digital-Nervous-System-of-Healthcare-Industry-/19-07-2018-155279/

https://www.fiercebiotech.com/medtech/insilico-and-a2a-launch-new-duchenne-focused-ai-drug-company

https://www.twst.com/news/robotic-surgical-system-using-artificial-intelligence-cure-baldness-restoration-robotics-inc-nasdaqhair/

https://www.globenewswire.com/news-release/2018/07/23/1540633/0/en/MC-Endeavors-Inc-Room-21-Media-Launches-First-Artificial-Intelligent-Marketing-Platform-with-Restore-Detox-Centers-to-Combat-Addiction-in-the-US.html

https://www.theguardian.com/technology/2018/aug/13/new-artificial-intelligence-tool-can-detect-eye-problems-as-well-as-experts

https://www.theguardian.com/technology/2018/jul/04/its-going-create-revolution-how-ai-transforming-nhs

https://www.opengovasia.com/artificial-intelligence-to-improve-medical-imaging-for-patients-with-brain-ailments/

https://mhealthintelligence.com/news/new-group-aims-to-advance-artificial-intelligence-in-telehealth

https://electronics360.globalspec.com/article/12414/artificial-intelligence-system-creates-new-pharmaceutical-drugs

https://electronics360.globalspec.com/article/12231/ai-system-can-predict-side-effects-of-new-drug-combinations

https://www.nytimes.com/2019/02/05/technology/artificial-intelligence-drug-research-deepmind.html
https://www.prunderground.com/vetology-artificial-intelligence-to-begin-reading-pet-x-rays/00131851/
https://www.sciencedaily.com/releases/2018/07/180731151326.htm
https://medicine.utoronto.ca/news/smarter-radiation-therapy-artificial-intelligence
https://www.healthdatamanagement.com/opinion/3-ways-artificial-intelligence-can-disrupt-healthcare
https://www.dw.com/en/doctors-dont-scale-like-artificial-intelligence-does/a-44899327
https://www.aao.org/eyenet/article/artificial-intelligence-and-glaucoma-detection
https://www.news-medical.net/health/Artificial-Intelligence-in-Cardiology.aspx
https://www.healthdatamanagement.com/news/ai-tool-helps-gateway-health-boost-accuracy-population-health
https://www.businessinsider.com/artificial-intelligence-healthcare
https://www.healthcareitnews.com/news/where-hospitals-plan-big-ai-deployments-diagnostic-imaging
https://www.healthcareitnews.com/news/asia-pacific/australia-s-snac-develops-ai-tools-improve-brain-scan-analysis
https://www.beckershospitalreview.com/artificial-intelligence/anthem-uses-ai-in-allergy-research-trial.html
https://www.geekwire.com/2018/mindshare-medical-launches-ai-cancer-screening-tech-can-see-data-beyond-perception/
https://lfpress.com/news/local-news/artificial-intelligence-may-be-key-to-mood-disorder-diagnosis-study
https://www.nextbigfuture.com/2018/08/artificial-intelligence-creating-new-drugs-from-scratch-by-efficiently-searching-huge-molecular-possibilities.html
https://qz.com/1349854/ai-can-spot-the-pain-from-a-disease-some-doctors-still-think-is-fake/
https://qz.com/1383083/how-ai-changed-organ-donation-in-the-us/
https://healthitanalytics.com/news/artificial-intelligence-for-medical-imaging-market-to-top-2b

https://www.thegazette.com/subject/news/business/coralville-based-idx-llc-aims-to-change-health-care-delivery-20180812
https://news.mit.edu/2018/artificial-intelligence-model-learns-patient-data-cancer-treatment-less-toxic-0810
https://www.healthcarefinancenews.com/news/how-artificial-intelligence-can-save-health-insurers-7-billion
https://www.techiexpert.com/ai-used-to-create-inexpensive-heart-disease-detector/
https://www.dermatologytimes.com/business/artificial-intelligence-friend-or-foe-dermatology
http://customerthink.com/looking-at-5-noteworthy-applications-of-artificial-intelligence-in-healthcare/
https://onlinelibrary.wiley.com/doi/full/10.1111/dmcn.13942
https://blog.jive.com/ai-chatbots-contact-centers/
https://www.theguardian.com/technology/2018/jul/29/the-robot-will-see-you-now-could-computers-take-over-medicine-entirely
https://www.nature.com/articles/d41586-019-01111-y
https://www.sciencedaily.com/releases/2018/08/180813113315.htm
https://www.curetoday.com/publications/cure/2019/fall-2019/as-large-as-life-using-artificial-intelligence-in-cancer-care
https://medibulletin.com/1-2-seconds-is-all-artificial-intelligence-takes-to-screen-ct-scans/
https://www.marktechpost.com/2018/08/14/artificial-intelligence-in-medicine-gains-massive-traction-with-growing-investment-in-ai-in-the-space/
https://www.washingtontimes.com/news/2018/aug/13/colonoscopy-technology-cuts-need-polyp-surgery-tes/
https://medicalxpress.com/news/2018-09-medicine-ready-artificial-intelligence.html
https://www.beckershospitalreview.com/healthcare-information-technology/how-artificial-intelligence-can-transform-payment-integrity.html
https://www.scitecheuropa.eu/artificial-intelligence-to-map-obesity/88924/

https://www.marketwatch.com/press-release/notable-health-closes-135-million-series-a-financing-to-expand-artificial-intelligence-powered-physician-patient-interaction-platform-2018-09-06
https://www.scmp.com/news/china/science/article/2163110/are-you-risk-diabetes-chinese-ai-system-could-predict-disease-15
https://geneticliteracyproject.org/2018/09/10/can-artificial-intelligence-give-us-a-more-efficient-health-care-system/
https://mindmatters.ai/2019/06/new-evidence-that-some-comatose-people-really-do-understand/
https://www.fastcompany.com/90287723/artificial-intelligence-can-detect-alzheimers-in-brain-scans-six-years-before-a-diagnosis
https://www.sciencedaily.com/releases/2019/01/190103152906.htm
https://newatlas.com/robot-md-ai-future-medial-diagnosis/60874/
https://www.healthworkscollective.com/can-artificial-intelligence-diagnose-illnesses-better-than-other-methods/
https://mytechdecisions.com/compliance/artificial-intelligence-delivery-rooms/
https://www.europeanpharmaceuticalreview.com/news/82870/artificial-intelligence-ai-heart-disease/
https://www.europeanpharmaceuticalreview.com/news/102403/ai-can-predict-which-research-will-translate-to-clinical-trials/
https://www.smithsonianmag.com/innovation/will-artificial-intelligence-improve-health-care-for-everyone-180972758/
https://venturebeat.com/2019/01/04/massachusetts-generals-ai-can-spot-brain-hemorrhages-as-accurately-as-humans/
https://www.geek.com/news/machine-learning-may-predict-how-well-youll-age-1767646/
https://becominghuman.ai/the-role-of-ai-in-healthcare-technology-6c33a6eee18c
https://www.sciencemag.org/news/2019/01/artificial-intelligence-could-diagnose-rare-disorders-using-just-photo-face
https://www.sciencedaily.com/releases/2019/01/190102112926.htm
https://www.empr.com/home/news/artificial-intelligence-can-detect-classify-acute-brain-bleeds/
https://www.bioeng.ucla.edu/artificial-intelligence-detects-the-presence-of-viruses/

https://blogs.scientificamerican.com/observations/the-surgical-singularity-is-approaching/
https://healthitanalytics.com/news/artificial-intelligence-in-healthcare-spending-to-hit-36b
https://www.fotoinc.com/news-updates/artificial-intelligence-embedded-mobile-app-for-chronic-neck-and-back-pain
https://www.theweek.in/news/sci-tech/2018/12/19/British-doctors-sceptical-that-AI-could-entirely-replace-them.html
https://www.news-medical.net/life-sciences/Artificial-Intelligence-in-Histopathology.aspx
https://blogs.scientificamerican.com/observations/how-ai-could-help-your-bad-back/
https://www.mdmag.com/medical-news/artificial-intelligence-deep-learning-learn-diagnose-epilepsy
https://www.ncbi.nlm.nih.gov/pmc/articles/PMC6290744/
https://en.wikipedia.org/wiki/Artificial_intelligence_in_healthcare
https://www.msn.com/en-us/health/medical/cigna-is-using-artificial-intelligence-to-predict-which-of-its-subscribers-will-become-addicted-to-opioids/ar-BBReYKL
https://www.sciencedaily.com/releases/2018/12/181221123743.htm
https://www.wndu.com/content/news/Artificial-intelligence-improves-colonoscopies-503263751.html
https://www.unionleader.com/news/scitech/ai-can-predict-mental-health-issues-from-your-instagram-posts/article_06ab3c45-8fa5-539e-869e-c19cab1cbef0.html
https://www.expresshealthcare.in/news/medachievers-and-labindia-healthcare-bring-artificial-intelligence-based-open-surgery-simulator/407543/
https://hackernoon.com/how-ai-robots-are-infiltrating-healthcare-ts8u3zfq
https://towardsdatascience.com/artificial-intelligence-deep-learning-for-medical-diagnosis-9561f7a4e5f
http://www.digitaljournal.com/tech-and-science/science/ai-can-boost-cancer-drug-discovery/article/538623
https://www.nbcnews.com/mach/science/why-big-pharma-betting-big-ai-ncna852246

https://www.kurzweilai.net/how-to-predict-the-side-effects-of-millions-of-drug-combinations

https://futurism.com/neoscope/genetic-report-predict-disease-risk

https://www.modernhealthcare.com/article/20180910/NEWS/180919991/telemedicine-meets-artificial-intelligence-at-bedside

https://www.beckershospitalreview.com/artificial-intelligence/5-developments-in-ai-last-week.html

https://www.forbes.com/sites/brucelee/2019/01/20/do-you-have-early-signs-of-cervical-cancer-how-ai-technology-may-help/#55552bdce694

http://www.digitaljournal.com/life/health/could-a-machine-replace-a-doctor/article/538540

https://ai-med.io/artificial-intelligence-robotic-surgery/

https://www.healthcareitnews.com/news/ai-healthcare-big-ethical-questions-still-need-answers

http://newsroom.ucla.edu/releases/artificial-intelligence-device-detect-moving-parasites-bodily-fluid-earlier-diagnosis

https://www.periscopedata.com/press/artificial-intelligence-bias-and-mental-health-implications

https://hospitalnews.com/artificial-intelligence-being-used-to-develop-drugs-even-faster-and-cheaper/

https://www.globenewswire.com/news-release/2018/12/12/1665700/0/en/Artificial-Intelligence-key-to-Universal-Health-Coverage.html

https://www.acobiom.com/en/how-artificial-intelligence-can-help-to-the-development-of-diagnostics-dedicated-to-precision-medicine/

https://www.forbes.com/sites/bernardmarr/2018/12/21/ai-that-saves-lives-the-chatbot-that-can-detect-a-heart-attack-using-machine-learning/#3300974450f9

https://www.asianscientist.com/2018/12/in-the-lab/artificial-intelligence-flu-sensor/

https://openmedscience.com/bright-future-for-robotic-surgeons/

https://www.beckershospitalreview.com/artificial-intelligence/northwestern-introduces-ai-tool-to-help-medical-assistants-conduct-sonograms.html

https://www.medicalbag.com/home/more/tech-talk/the-future-of-robot-physicians-is-artificial-intelligence-poised-to-take-over-medicine/

https://www.geekwire.com/2018/amazon-unveils-new-service-mine-decode-medical-records-using-artificial-intelligence/

https://www.dw.com/en/artificial-intelligence-in-medicine-the-computer-knows-what-you-need/a-46226852

https://www.prnewswire.com/news-releases/carepredict-presents-artificial-intelligence-powered-solutions-for-senior-care-at-aging2-0-optimize-2018--300749580.html

https://hitconsultant.net/2018/12/13/artificial-intelligence-transform-personal-health/

https://www.wftv.com/news/local/orlando-hospital-uses-artificial-intelligence-to-save-lives-in-the-delivery-room/874200835/

https://axiosholding.com/healthcare-ai-just-what-the-doctor-ordered/

https://www.globenewswire.com/news-release/2018/11/10/1649339/0/en/AI-Algorithm-Outperformed-Majority-of-Cardiologists-in-Detection-of-Heart-Murmurs-in-Clinical-Study.html

https://www.smithsonianmag.com/innovation/can-artificial-intelligence-detect-depression-in-persons-voice-180970702/

https://ai-med.io/artificial-intelligence-lie-human/

https://www.gqrgm.com/5-ways-ai-is-transforming-healthcare/

https://www.jwatch.org/na47874/2018/11/06/artificial-intelligence-improves-wrist-fracture-detection

https://www.healthcareitnews.com/news/ai-algorithms-show-promise-colonoscopy-screenings

https://electronichealthreporter.com/artificial-intelligence-is-the-new-operating-system-in-healthcare/

Top of Form

Bottom of Form

https://pophealthanalytics.com/exploring-the-role-of-ai-in-population-health-risk-assessment-symposium/

https://www.ns-businesshub.com/science/artificial-intelligence-in-medicine-cure-cancer/

http://dailytrojan.com/2018/10/28/mindful-mondays-artificial-intelligence-has-limited-power-to-treat-mental-health-issues/
https://www.dailydot.com/debug/ai-sepsis-diagnosis/
https://blog.sevenponds.com/science-of-us/%E2%80%A8%E2%80%A8artificial-intelligence-outsmarts-past-practices-that-predict-patient-outcomes
https://www.medicalnewstoday.com/articles/325491.php#1
https://onlinemedicalcare.org/artificial-intelligence-outperforms-real-doctors-studies-show/
https://www.healthline.com/health-news/new-ai-technology-may-help-diagnose-fetal-heart-problems
https://www.aiin.healthcare/topics/artificial-intelligence/ai-enhanced-virtual-care-could-reduce-er-visits
https://www.medicaldevice-network.com/features/ethics-in-ai/
https://www.healthdatamanagement.com/opinion/why-using-ai-in-healthcare-requires-a-balance-of-efficiency-and-ethics
https://www.prnewswire.com/news-releases/osf-ventures-invests-in-gauss-artificial-intelligence-technology-to-identify-postpartum-bleeding-early-300732784.html
https://www.medicaldesignandoutsourcing.com/digital-surgery-touts-artificial-intelligence-for-the-operating-room/
https://ai-med.io/will-artificial-intelligence-replace-human-radiologists/
https://www.sciencedaily.com/releases/2019/01/190111143744.htm
https://directorsblog.nih.gov/2019/01/17/using-artificial-intelligence-to-detect-cervical-cancer/
https://www.forbes.com/sites/bernardmarr/2018/07/27/how-is-ai-used-in-healthcare-5-powerful-real-world-examples-that-show-the-latest-advances/#4a59916e5dfb
https://transmitter.ieee.org/artificial-intelligence-robots-and-the-operating-room/
https://www.nationthailand.com/asean-plus/30349160
https://directorsblog.nih.gov/2019/01/15/using-artificial-intelligence-to-catch-irregular-heartbeats/
https://analyticsindiamag.com/ai-tell-how-ovarian-tumour/

https://lfpress.com/news/local-news/artificial-intelligence-can-predict-ptsd-in-patients-london-researchers

https://www.forbes.com/sites/samshead/2018/10/08/tencent-aims-to-train-ai-to-spot-parkinsons-in-3-minutes/#676980856f36

https://engineering.stanford.edu/magazine/article/david-magnus-how-will-artificial-intelligence-impact-medical-ethics

https://theweek.com/articles/694522/robots-replace-therapists

https://www.leewayhertz.com/how-iot-transforming-healthcare/

https://www.straitstimes.com/singapore/scdf-turns-to-artificial-intelligence-to-help-emergency-call-dispatchers

https://www.eurekalert.org/pub_releases/2018-07/imi-mlt070518.php

https://www.latimes.com/world/la-fg-china-ai-20180706-story.html

https://www.healthcareitnews.com/news/chinese-hospital-guangdong-deploys-ai-cameras-detect-blindness-causing-diseases

https://www.healthcarefinancenews.com/news/artificial-intelligence-healthcare-projected-be-worth-more-27-billion-2025

https://analyticsindiamag.com/can-machines-be-taught-to-detect-medicare-fraud/

https://www.newscientist.com/article/2222907-ai-can-predict-if-youll-die-soon-but-weve-no-idea-how-it-works/

https://revcycleintelligence.com/news/using-artificial-intelligence-to-improve-the-hospital-revenue-cycle

https://www.healthcarefinancenews.com/news/addressing-social-determinants-health-consider-artificial-intelligence-and-machine-learning

https://www.mcknights.com/marketplace/marketplace-experts/what-facilities-can-gain-from-robotics-in-physical-therapy/

https://www.entrepreneur.com/article/317047

https://www.icr.ac.uk/news-archive/artificial-intelligence-can-predict-how-cancers-will-evolve-and-spread

https://www.frontiersin.org/articles/10.3389/fpsyg.2019.00263/full

https://www.usatoday.com/story/opinion/2019/01/28/health-privacy-laws-artificial-intelligence-hipaa-needs-update-column/2695386002/

https://www.nature.com/articles/s41746-019-0089-x

https://time.com/collection/life-reinvented/5494363/sleep-artificial-intelligence/

https://www.corporatewellnessmagazine.com/article/how-artificial-intelligence-can-fight-against-workplace-stress
https://www.sciencedaily.com/releases/2018/05/180528190839.htm
https://www.cnet.com/news/ai-to-aid-emergency-call-operators-diagnose-heart-attacks-in-europe/
https://www.forbes.com/sites/michaelpellmanrowland/2018/04/24/beyonce-artificial-intelligence-vegan/#503780cc7160
https://govinsider.asia/innovation/artificial-intelligence-ageing-population/
https://www.geekwire.com/2018/health-tech-podcast-using-precision-medicine-kill-cancer-artificial-intelligence/
https://thriveglobal.com/stories/the-road-to-good-health-will-now-be-guided-by-artificial-intelligence/
https://www.engadget.com/2018/10/15/google-ai-spots-advanced-breast-cancer/
https://www.stanforddaily.com/2018/09/27/artificial-swarm-intelligence-diagnoses-pneumonia-better-than-individual-computer-or-doctor/
https://www.forbes.com/sites/michaelmillenson/2018/09/23/will-apple-track-your-mind-not-just-your-heart/#4879dbabcad3
https://emerj.com/ai-sector-overviews/artificial-intelligence-in-health-insurance-current-applications-and-trends/
https://www.unc.edu/posts/2018/07/31/artificial-intelligence-system-created-at-unc-chapel-hill-designs-drugs-from-scratch/
https://www.beaconlens.com/ai-in-behavioral-health-care-when-artificial-intelligence-became-real/
https://www.eurekalert.org/pub_releases/2018-09/sfl-ait091818.php
https://www.einfochips.com/blog/how-ai-enabled-wearables-are-changing-healthcare-and-fitness-industry/
https://psychnews.psychiatryonline.org/doi/10.1176/appi.pn.2018.10a5
https://www.eurekalert.org/pub_releases/2018-10/muos-wcm100218.php
https://time.com/collection/life-reinvented/5492063/artificial-intelligence-fertility/
https://www.thefix.com/artificial-intelligence-system-aims-identify-drug-thefts-hospitals

https://towardsdatascience.com/applying-artificial-intelligence-to-help-people-quit-smoking-early-results-a3e5581d560
https://medcitynews.com/2018/05/qure-ai/
https://healthitanalytics.com/news/deep-learning-tool-tops-dermatologists-in-melanoma-detection
https://www.himss.org/resources/role-artificial-intelligence-healthcare-and-society
https://www.aami.org/newsviews/newsdetail.aspx?ItemNumber=6502
http://www.thetower.org/6317-israel-uk-announce-landmark-scientific-cooperation-agreement-focusing-on-aging-artificial-intelligence/
https://www.aarp.org/health/conditions-treatments/info-2018/hospital-artificial-intelligence-telehealth.html
https://www.digitaltrends.com/cool-tech/chatterbaby-app-deciphers-baby-crying/
https://www.cbs58.com/news/new-app-translates-babies-cries-so-deaf-parents-understand-whats-going-on
https://www.empr.com/home/news/fda-approves-ai-algorithm-that-helps-detect-wrist-fractures/
https://bioinformatics.csiro.au/blog/can-ai-help-fight-antibiotic-resistant-superbugs/
https://www.sciencedaily.com/releases/2018/09/180920161054.htm
https://www.forbes.com/sites/jenniferhicks/2018/06/08/see-how-this-hospital-uses-artificial-intelligence-to-find-kidney-disease/#78af38d72e8f
https://www.newsweek.com/ai-being-developed-which-can-smell-illness-human-breath-967197
https://www.theguardian.com/technology/2018/jun/10/artificial-intelligence-cancer-detectors-the-five
https://www.independent.co.uk/voices/loneliness-kills-artificial-intelligence-chatbot-doctors-health-risk-diagnoses-a8423321.html
https://www.technologyreview.com/f/609969/ai-could-diagnose-your-heart-attack-on-the-phone-even-if-youre-not-the-caller/
https://www.linkedin.com/feed/news/teaching-ai-to-explain-its-thinking-4617604

https://www.smartdatacollective.com/artificial-intelligence-chatbots-could-make-your-doctors-obsolete/

https://www.telegraph.co.uk/technology/2019/01/13/google-scientist-ai-may-not-cure-all-discovering-new-drugs/

https://greatlakesledger.com/2019/01/12/artificial-intelligence-can-predict-and-estimate-flu-activity/

https://www.mddionline.com/using-artificial-intelligence-predict-flu-activity

https://time.com/5556339/artificial-intelligence-robots-medicine/

https://www.nytimes.com/2019/03/11/well/live/how-artificial-intelligence-could-transform-medicine.html

https://towardsdatascience.com/artificial-intelligence-replace-the-human-doctors-in-the-future-is-it-true-91b3ae9fea0e

https://www.sciencedaily.com/releases/2019/12/191218090156.htm

https://spectrum.ieee.org/biomedical/diagnostics/how-ibm-watson-overpromised-and-underdelivered-on-ai-health-care

https://venturebeat.com/2019/09/13/googles-ai-detects-26-skin-conditions-as-accurately-as-dermatologists/

https://www.dhs.gov/science-and-technology/news/2018/10/16/snapshot-public-safety-agencies-pilot-artificial-intelligence

https://www.inverse.com/article/59742-smartphone-app-health-best-eye-disease

https://www.foxnews.com/tech/google-ai-can-predict-when-youll-die-with-95-percent-accuracy-researchers-say

https://www.zdnet.com/article/googles-deepmind-follows-a-mixed-path-to-ai-in-medicine/

https://medicalfuturist.com/artificial-intelligence-in-mental-health-care/

https://www.cnbc.com/2018/02/22/medical-errors-third-leading-cause-of-death-in-america.html#:~:text=According%20to%20a%20recent%20study,after%20%20heart%20disease%20and%20cancer.

https://www.google.com/search?q=how+many+people+go+online+to+diagnose+themselves&rlz=1C1CHBF_enUS894US894&oq=how+ma

ny+people+go+online+to+diagnose+themselves&aqs=chrome..69i57.
10637j0j7&sourceid=chrome&ie=UTF-8
https://www.google.com/search?q=obama+care+mandate+electronic+
health+records&rlz=1C1CHBF_enUS894US894&oq=obama+care+m
andate+electronic+health+records&aqs=chrome..69i57j0l2.15928j0j7
&sourceid=chrome&ie=UTF-8
https://www.healthcarelaw-blog.com/the-electronic-medical-records-
emr-mandate/#:~:text=With%20the%20passage%20of%20the,to%20
take20effect%20in%202014

Chapter 18 *The Future of Diet & Drugs with AI*

1. *AI Lark Life Coach*
   https://www.youtube.com/watch?v=q_m4oa6Fjm8
2. *AI I-Watch Commercial*
   https://www.youtube.com/watch?v=TCMnrssX1NE
3. *Microchip Medical Tattoos & Implants*
   https://www.mddionline.com/digital-health/5-futuristic-wearable-
   devices-will-blow-your-mind-medical-history
   https://www.youtube.com/watch?v=7fk4G6-I5hA
4. *Mandatory AI Wearables*
   https://www.youtube.com/watch?v=KM8HvmzcXic
5. *AI Saves Time & Money on Drugs*
   https://www.youtube.com/watch?v=t7bNe_Y2Pag
6. *AI Controls All Aspects of Drugs*
   https://www.youtube.com/watch?v=AVxTm7RsVkM
7. *AI Smart Pills*
   https://www.youtube.com/watch?v=WNyELeIszxA
8. *Information on AI Medical*
   https://www.medgadget.com/2018/09/doctors-pair-with-artificial-
   intelligence-to-improve-pneumonia-diagnosis.html
   https://qz.com/1548524/china-has-produced-another-study-showing-
   the-potential-of-ai-in-medical-diagnosis/
   https://qz.com/1384725/an-ai-algorithm-in-china-is-learning-to-
   detect-whether-patients-will-wake-from-a-coma/

https://www.scmp.com/news/china/science/article/2163298/doctors-said-coma-patients-would-never-wake-ai-said-they-would

https://www.thehindu.com/sci-tech/health/ai-may-predict-alzheimers-disease-5-years-in-advance/article25160303.ece

https://www.livemint.com/Science/evBwYmDFLvImRwjO97HeeO/Indian-scientists-using-artificial-intelligence-to-predict-e.html

https://www.forbes.com/sites/bernardmarr/2018/12/21/ai-that-saves-lives-the-chatbot-that-can-detect-a-heart-attack-using-machine-learning/#53c9859450f9

https://thenextweb.com/artificial-intelligence/2018/04/25/europe-launches-heart-attack-detecting-ai-emergency-calls/

https://www.esmo.org/About-Us/ESMO-Magazine/Feature-AI-and-Big-Data-in-Oncology-How-Ready-Are-We

https://www.statnews.com/2019/07/03/artificial-intelligence-guarded-optimism-cancer-care/

https://www.healthcareitnews.com/news/geisinger-injects-machine-learning-clinical-workflow-find-health-problems-faster

https://www.cardiovascularbusiness.com/topics/artificial-intelligence/twists-turns-ahead-ais-road-acceptance-cardiology

https://scienmag.com/insilico-to-present-its-latest-research-in-ai-aging-diagnostics-at-the-aone-conference/

https://svn.bmj.com/content/2/4/230

https://healthmanagement.org/c/hospital/issuearticle/artificial-intelligence-a-next-way-forward-for-healthcare

https://healthmanagement.org/c/hospital/issuearticle/artificial-intelligence-in-healthcare-what-is-versus-what-will-be

https://www.linkedin.com/pulse/should-ai-take-hippocratic-oath-gary-gilliland-md-phd

https://www.livescience.com/65087-ai-premature-death-prediction.html

https://www.prnewswire.com/news-releases/perception-health-on-aws-marketplace-machine-learning-and-artificial-intelligence-discovery-page-300773045.html

https://www.express.co.uk/news/world/976108/google-when-will-you-die-ai-artificial-intelligence-latest

https://cio.economictimes.indiatimes.com/news/strategy-and-management/adoption-of-new-technology-is-not-an-option-but-a-compulsion-ggm-it-irctc/71788370

https://pjmedia.com/trending/can-a-healthcare-algorithm-be-racially-biased/

https://time.com/5556339/artificial-intelligence-robots-medicine/

https://www.modernhealthcare.com/article/20150610/NEWS/150619996/walgreen-insurers-push-expansion-of-virtual-doctor-visits

https://electronicsforu.com/technology-trends/tech-focus/artificial-intelligence-healthcare-replace-doctors

https://healthitanalytics.com/news/at-montefiore-artificial-intelligence-becomes-key-to-patient-care

https://time.com/5709346/artificial-intelligence-health/

https://www.reuters.com/article/us-facebook-features/facebook-turns-to-artificial-intelligence-to-tackle-suicides-idUSKBN1684JQ

https://www.fastcompany.com/90299135/mental-health-crisis-robots-chatbots-listeners

https://www.sandiegouniontribune.com/sdut-smartphone-voices-not-always-helpful-in-health-2016mar14-story.html

https://becominghuman.ai/becoming-human-ai-addiction-deaf948e17ba

https://www.cms.gov/newsroom/press-releases/cms-announces-artificial-intelligence-health-outcomes-challenge-participants-advancing-stage-1

https://healthmanagement.org/c/healthmanagement/IssueArticle/ai-is-the-new-reality-the-4th-healthcare-revolution-in-medicine

https://www.independent.co.uk/voices/artificial-intelligence-machine-learning-computers-global-healthcare-malaria-facebook-a8327901.html

https://questsoblogspot.wordpress.com/2019/09/25/artificial-intelligence-for-personal-business-and-enterprise-uses/

https://news.yahoo.com/health-checks-smartphone-raise-privacy-fears-080635403.html

https://www.theglobeandmail.com/life/health-and-fitness/article-app-developers-are-using-artificial-intelligence-to-advise-teens-about/

https://www.sdglobaltech.com/blog/some-mind-blowing-stats-about-ai-and-iot-in-healthcare

https://builtin.com/artificial-intelligence/artificial-intelligence-healthcare

https://khn.org/news/a-reality-check-on-artificial-intelligence-are-health-care-claims-overblown/

https://healthitanalytics.com/news/54-of-healthcare-pros-expect-widespread-ai-adoption-in-5-years

https://www.ncbi.nlm.nih.gov/pmc/articles/PMC6616181/

https://dzone.com/articles/using-ai-to-design-drugs-from-scratch

https://www.smithsonianmag.com/innovation/will-artificial-intelligence-improve-health-care-for-everyone-180972758/

https://www.managedcaremag.com/archives/2019/7/ai-all-ails-american-health-care-how-smart

https://www.digitalcommerce360.com/2018/09/14/ai-may-save-one-small-hospital-20-million/

https://www.healthdatamanagement.com/news/cios-must-take-charge-in-implementing-artificial-intelligence-governance

https://www.theweek.in/news/sci-tech/2018/09/15/New-artificial-intelligence-system-can-detect-dementia.html

http://www.digitaljournal.com/life/health/ai-in-healthcare-will-reach-6-16-billion-by-2022/article/531991

http://www.digitaljournal.com/life/health/amazon-is-diving-ever-deeper-into-healthcare/article/563201

https://healthitanalytics.com/news/45-of-ors-will-be-integrated-with-artificial-intelligence-by-2022

https://www.prweb.com/releases/mediaplanet_teams_up_with_berg_to_highlight_how_harnessing_artificial_intelligence_and_patient_biology_can_accelerate_treatments_for_rare_diseases/prweb15764315.h

https://www.drugchannels.net/2018/09/artificial-intelligence-ai-how.html

https://elitedatascience.com/machine-learning-impact

http://lippincottsolutions.lww.com/blog.entry.html/2018/09/13/the_future_is_nowh-JFSb.html

https://blogs.microsoft.com/blog/2019/06/19/harnessing-the-power-of-ai-to-transform-healthcare/

https://www.prnewswire.com/news-releases/artificial-intelligence-in-healthcare-takes-precision-medicine-to-the-next-level-300712098.html

https://www.rtinsights.com/artificial-intelligence-brings-more-clarity-to-hearing-devices/

https://www.europeanpharmaceuticalreview.com/news/79303/artificial-intelligence-ai-lung-disease-diagnosis-accuracy/

https://www.holyrood.com/inside-politics/view,could-artificial-intelligence-save-the-health-service_9249.htm

https://sbmi.uth.edu/blog/aug-17/what-is-the-relationship-between-informatics-and-data-science.htm

https://www.beckershospitalreview.com/healthcare-information-technology/how-artificial-intelligence-impacts-preventative-care.html

https://medium.com/swlh/the-future-with-ai-and-automated-digital-coaching-assistants-e0ccf7072c54

https://www.sciencedaily.com/releases/2018/09/180918180501.htm

https://www.engineering.com/DesignerEdge/DesignerEdgeArticles/ArticleID/17664/A-Healthy-Future-for-Artificial-Intelligence-in-Healthcare.aspx

https://nyulangone.org/news/artificial-intelligence-tool-accurately-identifies-cancer-type-genetic-changes-each-patients-lung-tumor

https://www.postbulletin.com/life/health/is-artificial-intelligence-a-natural-fit-for-health-care/article_ab274b68-b82c-11e8-aa3c-77aaaf1d1012.html

https://www.sciencedaily.com/releases/2018/09/180917111642.htm

https://www.globenewswire.com/news-release/2018/09/17/1571635/0/en/Global-Artificial-Intelligence-In-Diabetes-Management-Market-Worth-USD-1422-52-Million-By-2024-Zion-Market-Research.html

https://retinaroundup.com/2018/09/16/retina-society-2018-artificial-intelligence/

https://www.itnonline.com/content/exact-imaging-partners-improve-prostate-cancer-detection-artificial-intelligence

https://www.itnonline.com/content/philips-and-paige-team-bring-artificial-intelligence-ai-clinical-pathology-diagnostics

https://www.itnonline.com/content/dia-joins-ibm-watson-health-arm-clinicians-its-ai-powered-cardiac-ultrasound-software

https://www.itnonline.com/content/ai-improves-chest-x-ray-interpretation

https://www.drugtargetreview.com/article/45973/artificial-intelligence-in-the-world-of-drug-discovery/

http://www.digitaljournal.com/tech-and-science/science/artificial-intelligence-used-to-detect-early-signs-of-dementia/article/532769

https://www.technologyreview.com/s/609236/ai-can-spot-signs-of-alzheimers-before-your-family-does/

https://madison.com/wsj/business/technology/ensodata-uses-ai-to-help-improve-health-care/article_0f742d73-9256-589f-95e5-ff0a4e3457b4.html

https://hitinfrastructure.com/news/how-to-begin-healthcare-artificial-intelligence-deployment

https://www.labiotech.eu/features/artificial-intelligence-oncology/

https://www.nytimes.com/2019/10/24/well/live/machine-intelligence-AI-breast-cancer-mammogram.html

https://www.detroitnews.com/story/opinion/2018/07/20/impact-artificial-intelligence-health-care/791952002/

https://www.healthdatamanagement.com/list/7-ways-ai-could-make-an-impact-on-medical-care

https://www.itnewsafrica.com/2018/07/artificial-intelligence-is-solving-african-healthcare-challenges/

https://www.healthcarefinancenews.com/news/what-healthcare-cfos-should-know-about-artificial-intelligence-machine-learning-and-chatbots

https://venturebeat.com/2018/07/19/consortium-ai-wants-to-cure-rare-diseases-using-artificial-intelligence/

https://venturebeat.com/2019/07/31/deepminds-ai-predicts-kidney-injury-up-to-48-hours-before-it-happens/

https://venturebeat.com/2019/12/10/current-health-raises-11-5-million-to-predict-diseases-with-ai-and-remote-monitoring/

https://venturebeat.com/2019/12/09/google-proposes-hybrid-approach-to-ai-transfer-learning/

https://venturebeat.com/2019/12/06/ai-weekly-amazon-plays-the-long-game-in-health-care-ai/

https://venturebeat.com/2019/12/03/google-details-ai-that-classifies-chest-x-rays-with-human-level-accuracy/

http://bwdisrupt.businessworld.in/article/Artificial-Intelligence-Redefining-the-Digital-Nervous-System-of-Healthcare-Industry-/19-07-2018-155279/

https://www.fiercebiotech.com/medtech/insilico-and-a2a-launch-new-duchenne-focused-ai-drug-company

https://www.twst.com/news/robotic-surgical-system-using-artificial-intelligence-cure-baldness-restoration-robotics-inc-nasdaqhair/

https://www.globenewswire.com/news-release/2018/07/23/1540633/0/en/MC-Endeavors-Inc-Room-21-Media-Launches-First-Artificial-Intelligent-Marketing-Platform-with-Restore-Detox-Centers-to-Combat-Addiction-in-the-US.html

https://www.theguardian.com/technology/2018/aug/13/new-artificial-intelligence-tool-can-detect-eye-problems-as-well-as-experts

https://www.theguardian.com/technology/2018/jul/04/its-going-create-revolution-how-ai-transforming-nhs

https://www.opengovasia.com/artificial-intelligence-to-improve-medical-imaging-for-patients-with-brain-ailments/

https://mhealthintelligence.com/news/new-group-aims-to-advance-artificial-intelligence-in-telehealth

https://electronics360.globalspec.com/article/12414/artificial-intelligence-system-creates-new-pharmaceutical-drugs

https://electronics360.globalspec.com/article/12231/ai-system-can-predict-side-effects-of-new-drug-combinations

https://www.nytimes.com/2019/02/05/technology/artificial-intelligence-drug-research-deepmind.html

https://www.prunderground.com/vetology-artificial-intelligence-to-begin-reading-pet-x-rays/00131851/

https://www.sciencedaily.com/releases/2018/07/180731151326.htm

https://medicine.utoronto.ca/news/smarter-radiation-therapy-artificial-intelligence

https://www.healthdatamanagement.com/opinion/3-ways-artificial-intelligence-can-disrupt-healthcare

https://www.dw.com/en/doctors-dont-scale-like-artificial-intelligence-does/a-44899327

https://www.aao.org/eyenet/article/artificial-intelligence-and-glaucoma-detection

https://www.news-medical.net/health/Artificial-Intelligence-in-Cardiology.aspx

https://www.healthdatamanagement.com/news/ai-tool-helps-gateway-health-boost-accuracy-population-health

https://www.businessinsider.com/artificial-intelligence-healthcare

https://www.healthcareitnews.com/news/where-hospitals-plan-big-ai-deployments-diagnostic-imaging

https://www.healthcareitnews.com/news/asia-pacific/australia-s-snac-develops-ai-tools-improve-brain-scan-analysis

https://www.beckershospitalreview.com/artificial-intelligence/anthem-uses-ai-in-allergy-research-trial.html

https://www.geekwire.com/2018/mindshare-medical-launches-ai-cancer-screening-tech-can-see-data-beyond-perception/

https://lfpress.com/news/local-news/artificial-intelligence-may-be-key-to-mood-disorder-diagnosis-study

https://www.nextbigfuture.com/2018/08/artificial-intelligence-creating-new-drugs-from-scratch-by-efficiently-searching-huge-molecular-possibilities.html

https://qz.com/1349854/ai-can-spot-the-pain-from-a-disease-some-doctors-still-think-is-fake/

https://qz.com/1383083/how-ai-changed-organ-donation-in-the-us/

https://healthitanalytics.com/news/artificial-intelligence-for-medical-imaging-market-to-top-2b

https://www.thegazette.com/subject/news/business/coralville-based-idx-llc-aims-to-change-health-care-delivery-20180812

https://news.mit.edu/2018/artificial-intelligence-model-learns-patient-data-cancer-treatment-less-toxic-0810

https://www.healthcarefinancenews.com/news/how-artificial-intelligence-can-save-health-insurers-7-billion

https://www.techiexpert.com/ai-used-to-create-inexpensive-heart-disease-detector/

https://www.dermatologytimes.com/business/artificial-intelligence-friend-or-foe-dermatology

http://customerthink.com/looking-at-5-noteworthy-applications-of-artificial-intelligence-in-healthcare/

https://onlinelibrary.wiley.com/doi/full/10.1111/dmcn.13942

https://blog.jive.com/ai-chatbots-contact-centers/

https://www.theguardian.com/technology/2018/jul/29/the-robot-will-see-you-now-could-computers-take-over-medicine-entirely

https://www.nature.com/articles/d41586-019-01111-y

https://www.sciencedaily.com/releases/2018/08/180813113315.htm

https://www.curetoday.com/publications/cure/2019/fall-2019/as-large-as-life-using-artificial-intelligence-in-cancer-care

https://medibulletin.com/1-2-seconds-is-all-artificial-intelligence-takes-to-screen-ct-scans/

https://www.marktechpost.com/2018/08/14/artificial-intelligence-in-medicine-gains-massive-traction-with-growing-investment-in-ai-in-the-space/

https://www.washingtontimes.com/news/2018/aug/13/colonoscopy-technology-cuts-need-polyp-surgery-tes/

https://medicalxpress.com/news/2018-09-medicine-ready-artificial-intelligence.html

https://www.beckershospitalreview.com/healthcare-information-technology/how-artificial-intelligence-can-transform-payment-integrity.html

https://www.scitecheuropa.eu/artificial-intelligence-to-map-obesity/88924/

https://www.marketwatch.com/press-release/notable-health-closes-135-million-series-a-financing-to-expand-artificial-intelligence-powered-physician-patient-interaction-platform-2018-09-06

https://www.scmp.com/news/china/science/article/2163110/are-you-risk-diabetes-chinese-ai-system-could-predict-disease-15

https://geneticliteracyproject.org/2018/09/10/can-artificial-intelligence-give-us-a-more-efficient-health-care-system/

https://mindmatters.ai/2019/06/new-evidence-that-some-comatose-people-really-do-understand/

https://www.fastcompany.com/90287723/artificial-intelligence-can-detect-alzheimers-in-brain-scans-six-years-before-a-diagnosis
https://www.sciencedaily.com/releases/2019/01/190103152906.htm
https://newatlas.com/robot-md-ai-future-medial-diagnosis/60874/
https://www.healthworkscollective.com/can-artificial-intelligence-diagnose-illnesses-better-than-other-methods/
https://mytechdecisions.com/compliance/artificial-intelligence-delivery-rooms/
https://www.europeanpharmaceuticalreview.com/news/82870/artificial-intelligence-ai-heart-disease/
https://www.europeanpharmaceuticalreview.com/news/102403/ai-can-predict-which-research-will-translate-to-clinical-trials/
https://www.smithsonianmag.com/innovation/will-artificial-intelligence-improve-health-care-for-everyone-180972758/
https://venturebeat.com/2019/01/04/massachusetts-generals-ai-can-spot-brain-hemorrhages-as-accurately-as-humans/
https://www.geek.com/news/machine-learning-may-predict-how-well-youll-age-1767646/
https://becominghuman.ai/the-role-of-ai-in-healthcare-technology-6c33a6eee18c
https://www.sciencemag.org/news/2019/01/artificial-intelligence-could-diagnose-rare-disorders-using-just-photo-face
https://www.sciencedaily.com/releases/2019/01/190102112926.htm
https://www.empr.com/home/news/artificial-intelligence-can-detect-classify-acute-brain-bleeds/
https://www.bioeng.ucla.edu/artificial-intelligence-detects-the-presence-of-viruses/
https://blogs.scientificamerican.com/observations/the-surgical-singularity-is-approaching/
https://healthitanalytics.com/news/artificial-intelligence-in-healthcare-spending-to-hit-36b
https://www.fotoinc.com/news-updates/artificial-intelligence-embedded-mobile-app-for-chronic-neck-and-back-pain
https://www.theweek.in/news/sci-tech/2018/12/19/British-doctors-sceptical-that-AI-could-entirely-replace-them.html

https://www.news-medical.net/life-sciences/Artificial-Intelligence-in-Histopathology.aspx

https://blogs.scientificamerican.com/observations/how-ai-could-help-your-bad-back/

https://www.mdmag.com/medical-news/artificial-intelligence-deep-learning-learn-diagnose-epilepsy

https://www.ncbi.nlm.nih.gov/pmc/articles/PMC6290744/

https://en.wikipedia.org/wiki/Artificial_intelligence_in_healthcare

https://www.msn.com/en-us/health/medical/cigna-is-using-artificial-intelligence-to-predict-which-of-its-subscribers-will-become-addicted-to-opioids/ar-BBReYKL

https://www.sciencedaily.com/releases/2018/12/181221123743.htm

https://www.wndu.com/content/news/Artificial-intelligence-improves-colonoscopies-503263751.html

https://www.unionleader.com/news/scitech/ai-can-predict-mental-health-issues-from-your-instagram-posts/article_06ab3c45-8fa5-539e-869e-c19cab1cbef0.html

https://www.expresshealthcare.in/news/medachievers-and-labindia-healthcare-bring-artificial-intelligence-based-open-surgery-simulator/407543/

https://hackernoon.com/how-ai-robots-are-infiltrating-healthcare-ts8u3zfq

https://towardsdatascience.com/artificial-intelligence-deep-learning-for-medical-diagnosis-9561f7a4e5f

http://www.digitaljournal.com/tech-and-science/science/ai-can-boost-cancer-drug-discovery/article/538623

https://www.nbcnews.com/mach/science/why-big-pharma-betting-big-ai-ncna852246

https://www.kurzweilai.net/how-to-predict-the-side-effects-of-millions-of-drug-combinations

https://futurism.com/neoscope/genetic-report-predict-disease-risk

https://www.modernhealthcare.com/article/20180910/NEWS/180919991/telemedicine-meets-artificial-intelligence-at-bedside

https://www.beckershospitalreview.com/artificial-intelligence/5-developments-in-ai-last-week.html

https://www.forbes.com/sites/brucelee/2019/01/20/do-you-have-early-signs-of-cervical-cancer-how-ai-technology-may-help/#55552bdce694

http://www.digitaljournal.com/life/health/could-a-machine-replace-a-doctor/article/538540

https://ai-med.io/artificial-intelligence-robotic-surgery/

https://www.healthcareitnews.com/news/ai-healthcare-big-ethical-questions-still-need-answers

http://newsroom.ucla.edu/releases/artificial-intelligence-device-detect-moving-parasites-bodily-fluid-earlier-diagnosis

https://www.periscopedata.com/press/artificial-intelligence-bias-and-mental-health-implications

https://hospitalnews.com/artificial-intelligence-being-used-to-develop-drugs-even-faster-and-cheaper/

https://www.globenewswire.com/news-release/2018/12/12/1665700/0/en/Artificial-Intelligence-key-to-Universal-Health-Coverage.html

https://www.acobiom.com/en/how-artificial-intelligence-can-help-to-the-development-of-diagnostics-dedicated-to-precision-medicine/

https://www.forbes.com/sites/bernardmarr/2018/12/21/ai-that-saves-lives-the-chatbot-that-can-detect-a-heart-attack-using-machine-learning/#3300974450f9

https://www.asianscientist.com/2018/12/in-the-lab/artificial-intelligence-flu-sensor/

https://openmedscience.com/bright-future-for-robotic-surgeons/

https://www.beckershospitalreview.com/artificial-intelligence/northwestern-introduces-ai-tool-to-help-medical-assistants-conduct-sonograms.html

https://www.medicalbag.com/home/more/tech-talk/the-future-of-robot-physicians-is-artificial-intelligence-poised-to-take-over-medicine/

https://www.geekwire.com/2018/amazon-unveils-new-service-mine-decode-medical-records-using-artificial-intelligence/

https://www.dw.com/en/artificial-intelligence-in-medicine-the-computer-knows-what-you-need/a-46226852

https://www.prnewswire.com/news-releases/carepredict-presents-artificial-intelligence-powered-solutions-for-senior-care-at-aging2-0-optimize-2018--300749580.html

https://hitconsultant.net/2018/12/13/artificial-intelligence-transform-personal-health/

https://www.wftv.com/news/local/orlando-hospital-uses-artificial-intelligence-to-save-lives-in-the-delivery-room/874200835/

https://axiosholding.com/healthcare-ai-just-what-the-doctor-ordered/

https://www.globenewswire.com/news-release/2018/11/10/1649339/0/en/AI-Algorithm-Outperformed-Majority-of-Cardiologists-in-Detection-of-Heart-Murmurs-in-Clinical-Study.html

https://www.smithsonianmag.com/innovation/can-artificial-intelligence-detect-depression-in-persons-voice-180970702/

https://ai-med.io/artificial-intelligence-lie-human/

https://www.gqrgm.com/5-ways-ai-is-transforming-healthcare/

https://www.jwatch.org/na47874/2018/11/06/artificial-intelligence-improves-wrist-fracture-detection

https://www.healthcareitnews.com/news/ai-algorithms-show-promise-colonoscopy-screenings

https://electronichealthreporter.com/artificial-intelligence-is-the-new-operating-system-in-healthcare/

Top of Form

Bottom of Form

https://pophealthanalytics.com/exploring-the-role-of-ai-in-population-health-risk-assessment-symposium/

https://www.ns-businesshub.com/science/artificial-intelligence-in-medicine-cure-cancer/

http://dailytrojan.com/2018/10/28/mindful-mondays-artificial-intelligence-has-limited-power-to-treat-mental-health-issues/

https://www.dailydot.com/debug/ai-sepsis-diagnosis/

https://blog.sevenponds.com/science-of-us/%E2%80%A8%E2%80%A8artificial-intelligence-outsmarts-past-practices-that-predict-patient-outcomes

https://www.medicalnewstoday.com/articles/325491.php#1

https://onlinemedicalcare.org/artificial-intelligence-outperforms-real-doctors-studies-show/

https://www.healthline.com/health-news/new-ai-technology-may-help-diagnose-fetal-heart-problems

https://www.aiin.healthcare/topics/artificial-intelligence/ai-enhanced-virtual-care-could-reduce-er-visits

https://www.medicaldevice-network.com/features/ethics-in-ai/

https://www.healthdatamanagement.com/opinion/why-using-ai-in-healthcare-requires-a-balance-of-efficiency-and-ethics

https://www.prnewswire.com/news-releases/osf-ventures-invests-in-gauss-artificial-intelligence-technology-to-identify-postpartum-bleeding-early-300732784.html

https://www.medicaldesignandoutsourcing.com/digital-surgery-touts-artificial-intelligence-for-the-operating-room/

https://ai-med.io/will-artificial-intelligence-replace-human-radiologists/

https://www.sciencedaily.com/releases/2019/01/190111143744.htm

https://directorsblog.nih.gov/2019/01/17/using-artificial-intelligence-to-detect-cervical-cancer/

https://www.forbes.com/sites/bernardmarr/2018/07/27/how-is-ai-used-in-healthcare-5-powerful-real-world-examples-that-show-the-latest-advances/#4a59916e5dfb

https://transmitter.ieee.org/artificial-intelligence-robots-and-the-operating-room/

https://www.nationthailand.com/asean-plus/30349160

https://directorsblog.nih.gov/2019/01/15/using-artificial-intelligence-to-catch-irregular-heartbeats/

https://analyticsindiamag.com/ai-tell-how-ovarian-tumour/

https://lfpress.com/news/local-news/artificial-intelligence-can-predict-ptsd-in-patients-london-researchers

https://www.forbes.com/sites/samshead/2018/10/08/tencent-aims-to-train-ai-to-spot-parkinsons-in-3-minutes/#676980856f36

https://engineering.stanford.edu/magazine/article/david-magnus-how-will-artificial-intelligence-impact-medical-ethics

https://theweek.com/articles/694522/robots-replace-therapists

https://www.leewayhertz.com/how-iot-transforming-healthcare/

https://www.straitstimes.com/singapore/scdf-turns-to-artificial-intelligence-to-help-emergency-call-dispatchers
https://www.eurekalert.org/pub_releases/2018-07/imi-mlt070518.php
https://www.latimes.com/world/la-fg-china-ai-20180706-story.html
https://www.healthcareitnews.com/news/chinese-hospital-guangdong-deploys-ai-cameras-detect-blindness-causing-diseases
https://www.healthcarefinancenews.com/news/artificial-intelligence-healthcare-projected-be-worth-more-27-billion-2025
https://analyticsindiamag.com/can-machines-be-taught-to-detect-medicare-fraud/
https://www.newscientist.com/article/2222907-ai-can-predict-if-youll-die-soon-but-weve-no-idea-how-it-works/
https://revcycleintelligence.com/news/using-artificial-intelligence-to-improve-the-hospital-revenue-cycle
https://www.healthcarefinancenews.com/news/addressing-social-determinants-health-consider-artificial-intelligence-and-machine-learning
https://www.mcknights.com/marketplace/marketplace-experts/what-facilities-can-gain-from-robotics-in-physical-therapy/
https://www.entrepreneur.com/article/317047
https://www.icr.ac.uk/news-archive/artificial-intelligence-can-predict-how-cancers-will-evolve-and-spread
https://www.frontiersin.org/articles/10.3389/fpsyg.2019.00263/full
https://www.usatoday.com/story/opinion/2019/01/28/health-privacy-laws-artificial-intelligence-hipaa-needs-update-column/2695386002/
https://www.nature.com/articles/s41746-019-0089-x
https://time.com/collection/life-reinvented/5494363/sleep-artificial-intelligence/
https://www.corporatewellnessmagazine.com/article/how-artificial-intelligence-can-fight-against-workplace-stress
https://www.sciencedaily.com/releases/2018/05/180528190839.htm
https://www.cnet.com/news/ai-to-aid-emergency-call-operators-diagnose-heart-attacks-in-europe/
https://www.forbes.com/sites/michaelpellmanrowland/2018/04/24/beyonce-artificial-intelligence-vegan/#503780cc7160

https://govinsider.asia/innovation/artificial-intelligence-ageing-population/

https://www.geekwire.com/2018/health-tech-podcast-using-precision-medicine-kill-cancer-artificial-intelligence/

https://thriveglobal.com/stories/the-road-to-good-health-will-now-be-guided-by-artificial-intelligence/

https://www.engadget.com/2018/10/15/google-ai-spots-advanced-breast-cancer/

https://www.stanforddaily.com/2018/09/27/artificial-swarm-intelligence-diagnoses-pneumonia-better-than-individual-computer-or-doctor/

https://www.forbes.com/sites/michaelmillenson/2018/09/23/will-apple-track-your-mind-not-just-your-heart/#4879dbabcad3

https://emerj.com/ai-sector-overviews/artificial-intelligence-in-health-insurance-current-applications-and-trends/

https://www.unc.edu/posts/2018/07/31/artificial-intelligence-system-created-at-unc-chapel-hill-designs-drugs-from-scratch/

https://www.beaconlens.com/ai-in-behavioral-health-care-when-artificial-intelligence-became-real/

https://www.eurekalert.org/pub_releases/2018-09/sfl-ait091818.php

https://www.einfochips.com/blog/how-ai-enabled-wearables-are-changing-healthcare-and-fitness-industry/

https://psychnews.psychiatryonline.org/doi/10.1176/appi.pn.2018.10a5

https://www.eurekalert.org/pub_releases/2018-10/muos-wcm100218.php

https://time.com/collection/life-reinvented/5492063/artificial-intelligence-fertility/

https://www.thefix.com/artificial-intelligence-system-aims-identify-drug-thefts-hospitals

https://towardsdatascience.com/applying-artificial-intelligence-to-help-people-quit-smoking-early-results-a3e5581d560

https://medcitynews.com/2018/05/qure-ai/

https://healthitanalytics.com/news/deep-learning-tool-tops-dermatologists-in-melanoma-detection

https://www.himss.org/resources/role-artificial-intelligence-healthcare-and-society

https://www.aami.org/newsviews/newsdetail.aspx?ItemNumber=6502

http://www.thetower.org/6317-israel-uk-announce-landmark-scientific-cooperation-agreement-focusing-on-aging-artificial-intelligence/

https://www.aarp.org/health/conditions-treatments/info-2018/hospital-artificial-intelligence-telehealth.html

https://www.digitaltrends.com/cool-tech/chatterbaby-app-deciphers-baby-crying/

https://www.cbs58.com/news/new-app-translates-babies-cries-so-deaf-parents-understand-whats-going-on

https://www.empr.com/home/news/fda-approves-ai-algorithm-that-helps-detect-wrist-fractures/

https://bioinformatics.csiro.au/blog/can-ai-help-fight-antibiotic-resistant-superbugs/

https://www.sciencedaily.com/releases/2018/09/180920161054.htm

https://www.forbes.com/sites/jenniferhicks/2018/06/08/see-how-this-hospital-uses-artificial-intelligence-to-find-kidney-disease/#78af38d72e8f

https://www.newsweek.com/ai-being-developed-which-can-smell-illness-human-breath-967197

https://www.theguardian.com/technology/2018/jun/10/artificial-intelligence-cancer-detectors-the-five

https://www.independent.co.uk/voices/loneliness-kills-artificial-intelligence-chatbot-doctors-health-risk-diagnoses-a8423321.html

https://www.technologyreview.com/f/609969/ai-could-diagnose-your-heart-attack-on-the-phone-even-if-youre-not-the-caller/

https://www.linkedin.com/feed/news/teaching-ai-to-explain-its-thinking-4617604

https://www.smartdatacollective.com/artificial-intelligence-chatbots-could-make-your-doctors-obsolete/

https://www.telegraph.co.uk/technology/2019/01/13/google-scientist-ai-may-not-cure-all-discovering-new-drugs/

https://greatlakesledger.com/2019/01/12/artificial-intelligence-can-predict-and-estimate-flu-activity/

https://www.mddionline.com/using-artificial-intelligence-predict-flu-activity

https://time.com/5556339/artificial-intelligence-robots-medicine/

https://www.nytimes.com/2019/03/11/well/live/how-artificial-intelligence-could-transform-medicine.html

https://towardsdatascience.com/artificial-intelligence-replace-the-human-doctors-in-the-future-is-it-true-91b3ae9fea0e

https://www.sciencedaily.com/releases/2019/12/191218090156.htm

https://spectrum.ieee.org/biomedical/diagnostics/how-ibm-watson-overpromised-and-underdelivered-on-ai-health-care

https://venturebeat.com/2019/09/13/googles-ai-detects-26-skin-conditions-as-accurately-as-dermatologists/

https://www.dhs.gov/science-and-technology/news/2018/10/16/snapshot-public-safety-agencies-pilot-artificial-intelligence

https://www.inverse.com/article/59742-smartphone-app-health-best-eye-disease

https://www.foxnews.com/tech/google-ai-can-predict-when-youll-die-with-95-percent-accuracy-researchers-say

https://www.zdnet.com/article/googles-deepmind-follows-a-mixed-path-to-ai-in-medicine/

https://medicalfuturist.com/artificial-intelligence-in-mental-health-care/

https://www.cnbc.com/2018/02/22/medical-errors-third-leading-cause-of-death-in-america.html#:~:text=According%20to%20a%20recent%20study,after%20heart%20disease%20and%20cancer.

https://www.google.com/search?q=how+many+people+go+online+to+diagnose+themselves&rlz=1C1CHBF_enUS894US894&oq=how+many+people+go+online+to+diagnose+themselves&aqs=chrome..69i57.10637j0j7&sourceid=chrome&ie=UTF-8

https://www.google.com/search?q=obama+care+mandate+electronic+health+records&rlz=1C1CHBF_enUS894US894&oq=obama+care+

mandate+electronic+health+records&aqs=chrome..69i57j0l2.15928j0
j7&sourceid=chrome&ie=UTF-8
https://www.healthcarelaw-blog.com/the-electronic-medical-records-
emr-
mandate/#:~:text=With%20the%20passage%20of%20the,to%20take
%20effect%20in%202014.

Chapter 19 *The Future of Death with AI*

1. *AI Mental Therapy Bot*
   https://www.youtube.com/watch?v=V3Ni45-2c3E
2. *Kill God Vaccine*
   https://www.youtube.com/watch?v=-gfTqfVeLHw&feature=emb_title
3. *AI App Measure Age*
   https://www.youtube.com/watch?v=yRUxayMWin0
4. *Logan's Run Movie Scene*
   https://www.youtube.com/watch?v=USADM5Gk9Gs
   https://www.youtube.com/watch?v=4M2vx_RCwSs
5. *Murder Babies Outside Womb*
   https://www.youtube.com/watch?v=K9KkAqPv_AI
6. *Obamacare Death Panels*
   https://www.youtube.com/watch?v=5e1JnSoaMyw
   https://www.youtube.com/watch?v=njbt8PPCFPg
   https://www.youtube.com/watch?v=0hTsjm98lVI
   https://www.youtube.com/watch?v=vvj_lp59GtM
   https://www.youtube.com/watch?v=bvvJUiqotm4
7. *Bill Gates Death Panel*
   https://www.youtube.com/watch?v=03MZG9vK0W8
   https://www.youtube.com/watch?v=j2hK-7tWD3E
8. *Roger Ebert Chipped at Birth*
   (Original Source No Longer Available)
9. *Killer Microchip*
   https://www.youtube.com/watch?v=KlrmCM5BkiA

10. *Information on AI Medicine*

https://qz.com/1548524/china-has-produced-another-study-showing-the-potential-of-ai-in-medical-diagnosis/

https://www.medgadget.com/2018/09/doctors-pair-with-artificial-%20%20%20%20%20intelligence-to-improve-pneumonia-diagnosis.html

https://qz.com/1384725/an-ai-algorithm-in-china-is-learning-to-detect-whether-patients-will-wake-from-a-coma/

https://www.scmp.com/news/china/science/article/2163298/doctors-said-coma-patients-would-never-wake-ai-said-they-would

https://www.thehindu.com/sci-tech/health/ai-may-predict-alzheimers-disease-5-years-in-advance/article25160303.ece

https://www.livemint.com/Science/evBwYmDFLvImRwjO97HeeO/Indian-scientists-using-artificial-intelligence-to-predict-e.html

https://www.forbes.com/sites/bernardmarr/2018/12/21/ai-that-saves-lives-the-chatbot-that-can-detect-a-heart-attack-using-machine-learning/#53c9859450f9

https://thenextweb.com/artificial-intelligence/2018/04/25/europe-launches-heart-attack-detecting-ai-emergency-calls/

https://www.esmo.org/About-Us/ESMO-Magazine/Feature-AI-and-Big-Data-in-Oncology-How-Ready-Are-We

https://www.statnews.com/2019/07/03/artificial-intelligence-guarded-optimism-cancer-care/

https://www.healthcareitnews.com/news/geisinger-injects-machine-learning-clinical-workflow-find-health-problems-faster

https://www.cardiovascularbusiness.com/topics/artificial-intelligence/twists-turns-ahead-ais-road-acceptance-cardiology

https://scienmag.com/insilico-to-present-its-latest-research-in-ai-aging-diagnostics-at-the-aone-conference/

https://svn.bmj.com/content/2/4/230

https://healthmanagement.org/c/hospital/issuearticle/artificial-intelligence-a-next-way-forward-for-healthcare

https://healthmanagement.org/c/hospital/issuearticle/artificial-intelligence-in-healthcare-what-is-versus-what-will-be

https://www.linkedin.com/pulse/should-ai-take-hippocratic-oath-gary-gilliland-md-phd

https://www.livescience.com/65087-ai-premature-death-prediction.html
https://www.prnewswire.com/news-releases/perception-health-on-aws-marketplace-machine-learning-and-artificial-intelligence-discovery-page-300773045.html
https://www.express.co.uk/news/world/976108/google-when-will-you-die-ai-artificial-intelligence-latest
https://cio.economictimes.indiatimes.com/news/strategy-and-management/adoption-of-new-technology-is-not-an-option-but-a-compulsion-ggm-it-irctc/71788370
https://pjmedia.com/trending/can-a-healthcare-algorithm-be-racially-biased/
https://time.com/5556339/artificial-intelligence-robots-medicine/
https://www.modernhealthcare.com/article/20150610/NEWS/150619996/walgreen-insurers-push-expansion-of-virtual-doctor-visits
https://electronicsforu.com/technology-trends/tech-focus/artificial-intelligence-healthcare-replace-doctors
https://healthitanalytics.com/news/at-montefiore-artificial-intelligence-becomes-key-to-patient-care
https://time.com/5709346/artificial-intelligence-health/
https://www.reuters.com/article/us-facebook-features/facebook-turns-to-artificial-intelligence-to-tackle-suicides-idUSKBN1684JQ
https://www.fastcompany.com/90299135/mental-health-crisis-robots-chatbots-listeners
https://www.sandiegouniontribune.com/sdut-smartphone-voices-not-always-helpful-in-health-2016mar14-story.html
https://becominghuman.ai/becoming-human-ai-addiction-deaf948e17ba
https://www.cms.gov/newsroom/press-releases/cms-announces-artificial-intelligence-health-outcomes-challenge-participants-advancing-stage-1
https://healthmanagement.org/c/healthmanagement/IssueArticle/ai-is-the-new-reality-the-4th-healthcare-revolution-in-medicine
https://www.independent.co.uk/voices/artificial-intelligence-machine-learning-computers-global-healthcare-malaria-facebook-a8327901.html

https://questsoblogspot.wordpress.com/2019/09/25/artificial-intelligence-for-personal-business-and-enterprise-uses/

https://news.yahoo.com/health-checks-smartphone-raise-privacy-fears-080635403.html

https://www.theglobeandmail.com/life/health-and-fitness/article-app-developers-are-using-artificial-intelligence-to-advise-teens-about/

https://www.sdglobaltech.com/blog/some-mind-blowing-stats-about-ai-and-iot-in-healthcare

https://builtin.com/artificial-intelligence/artificial-intelligence-healthcare

https://khn.org/news/a-reality-check-on-artificial-intelligence-are-health-care-claims-overblown/

https://healthitanalytics.com/news/54-of-healthcare-pros-expect-widespread-ai-adoption-in-5-years

https://www.ncbi.nlm.nih.gov/pmc/articles/PMC6616181/

https://dzone.com/articles/using-ai-to-design-drugs-from-scratch

https://www.smithsonianmag.com/innovation/will-artificial-intelligence-improve-health-care-for-everyone-180972758/

https://www.managedcaremag.com/archives/2019/7/ai-all-ails-american-health-care-how-smart

https://www.digitalcommerce360.com/2018/09/14/ai-may-save-one-small-hospital-20-million/

https://www.healthdatamanagement.com/news/cios-must-take-charge-in-implementing-artificial-intelligence-governance

https://www.theweek.in/news/sci-tech/2018/09/15/New-artificial-intelligence-system-can-detect-dementia.html

http://www.digitaljournal.com/life/health/ai-in-healthcare-will-reach-6-16-billion-by-2022/article/531991

http://www.digitaljournal.com/life/health/amazon-is-diving-ever-deeper-into-healthcare/article/563201

https://healthitanalytics.com/news/45-of-ors-will-be-integrated-with-artificial-intelligence-by-2022

https://www.prweb.com/releases/mediaplanet_teams_up_with_berg_to_highlight_how_harnessing_artificial_intelligence_and_patient_biology_can_accelerate_treatments_for_rare_diseases/prweb15764315.h

https://www.drugchannels.net/2018/09/artificial-intelligence-ai-how.html

https://elitedatascience.com/machine-learning-impact

http://lippincottsolutions.lww.com/blog.entry.html/2018/09/13/the_future_is_nowh-JFSb.html

https://blogs.microsoft.com/blog/2019/06/19/harnessing-the-power-of-ai-to-transform-healthcare/

https://www.prnewswire.com/news-releases/artificial-intelligence-in-healthcare-takes-precision-medicine-to-the-next-level-300712098.html

https://www.rtinsights.com/artificial-intelligence-brings-more-clarity-to-hearing-devices/

https://www.europeanpharmaceuticalreview.com/news/79303/artificial-intelligence-ai-lung-disease-diagnosis-accuracy/

https://www.holyrood.com/inside-politics/view,could-artificial-intelligence-save-the-health-service_9249.htm

https://sbmi.uth.edu/blog/aug-17/what-is-the-relationship-between-informatics-and-data-science.htm

https://www.beckershospitalreview.com/healthcare-information-technology/how-artificial-intelligence-impacts-preventative-care.html

https://medium.com/swlh/the-future-with-ai-and-automated-digital-coaching-assistants-e0ccf7072c54

https://www.sciencedaily.com/releases/2018/09/180918180501.htm

https://www.engineering.com/DesignerEdge/DesignerEdgeArticles/ArticleID/17664/A-Healthy-Future-for-Artificial-Intelligence-in-Healthcare.aspx

https://nyulangone.org/news/artificial-intelligence-tool-accurately-identifies-cancer-type-genetic-changes-each-patients-lung-tumor

https://www.postbulletin.com/life/health/is-artificial-intelligence-a-natural-fit-for-health-care/article_ab274b68-b82c-11e8-aa3c-77aaaf1d1012.html

https://www.sciencedaily.com/releases/2018/09/180917111642.htm

https://www.globenewswire.com/news-release/2018/09/17/1571635/0/en/Global-Artificial-Intelligence-In-Diabetes-Management-Market-Worth-USD-1422-52-Million-By-2024-Zion-Market-Research.html

https://retinaroundup.com/2018/09/16/retina-society-2018-artificial-intelligence/

https://www.itnonline.com/content/exact-imaging-partners-improve-prostate-cancer-detection-artificial-intelligence

https://www.itnonline.com/content/philips-and-paige-team-bring-artificial-intelligence-ai-clinical-pathology-diagnostics

https://www.itnonline.com/content/dia-joins-ibm-watson-health-arm-clinicians-its-ai-powered-cardiac-ultrasound-software

https://www.itnonline.com/content/ai-improves-chest-x-ray-interpretation

https://www.drugtargetreview.com/article/45973/artificial-intelligence-in-the-world-of-drug-discovery/

http://www.digitaljournal.com/tech-and-science/science/artificial-intelligence-used-to-detect-early-signs-of-dementia/article/532769

https://www.technologyreview.com/s/609236/ai-can-spot-signs-of-alzheimers-before-your-family-does/

https://madison.com/wsj/business/technology/ensodata-uses-ai-to-help-improve-health-care/article_0f742d73-9256-589f-95e5-ff0a4e3457b4.html

https://hitinfrastructure.com/news/how-to-begin-healthcare-artificial-intelligence-deployment

https://www.labiotech.eu/features/artificial-intelligence-oncology/

https://www.nytimes.com/2019/10/24/well/live/machine-intelligence-AI-breast-cancer-mammogram.html

https://www.detroitnews.com/story/opinion/2018/07/20/impact-artificial-intelligence-health-care/791952002/

https://www.healthdatamanagement.com/list/7-ways-ai-could-make-an-impact-on-medical-care

https://www.itnewsafrica.com/2018/07/artificial-intelligence-is-solving-african-healthcare-challenges/

https://www.healthcarefinancenews.com/news/what-healthcare-cfos-should-know-about-artificial-intelligence-machine-learning-and-chatbots

https://venturebeat.com/2018/07/19/consortium-ai-wants-to-cure-rare-diseases-using-artificial-intelligence/

https://venturebeat.com/2019/07/31/deepminds-ai-predicts-kidney-injury-up-to-48-hours-before-it-happens/

https://venturebeat.com/2019/12/10/current-health-raises-11-5-million-to-predict-diseases-with-ai-and-remote-monitoring/

https://venturebeat.com/2019/12/09/google-proposes-hybrid-approach-to-ai-transfer-learning/

https://venturebeat.com/2019/12/06/ai-weekly-amazon-plays-the-long-game-in-health-care-ai/

https://venturebeat.com/2019/12/03/google-details-ai-that-classifies-chest-x-rays-with-human-level-accuracy/

http://bwdisrupt.businessworld.in/article/Artificial-Intelligence-Redefining-the-Digital-Nervous-System-of-Healthcare-Industry-/19-07-2018-155279/

https://www.fiercebiotech.com/medtech/insilico-and-a2a-launch-new-duchenne-focused-ai-drug-company

https://www.twst.com/news/robotic-surgical-system-using-artificial-intelligence-cure-baldness-restoration-robotics-inc-nasdaqhair/

https://www.globenewswire.com/news-release/2018/07/23/1540633/0/en/MC-Endeavors-Inc-Room-21-Media-Launches-First-Artificial-Intelligent-Marketing-Platform-with-Restore-Detox-Centers-to-Combat-Addiction-in-the-US.html

https://www.theguardian.com/technology/2018/aug/13/new-artificial-intelligence-tool-can-detect-eye-problems-as-well-as-experts

https://www.theguardian.com/technology/2018/jul/04/its-going-create-revolution-how-ai-transforming-nhs

https://www.opengovasia.com/artificial-intelligence-to-improve-medical-imaging-for-patients-with-brain-ailments/

https://mhealthintelligence.com/news/new-group-aims-to-advance-artificial-intelligence-in-telehealth

https://electronics360.globalspec.com/article/12414/artificial-intelligence-system-creates-new-pharmaceutical-drugs

https://electronics360.globalspec.com/article/12231/ai-system-can-predict-side-effects-of-new-drug-combinations

https://www.nytimes.com/2019/02/05/technology/artificial-intelligence-drug-research-deepmind.html

https://www.prunderground.com/vetology-artificial-intelligence-to-begin-reading-pet-x-rays/00131851/

https://www.sciencedaily.com/releases/2018/07/180731151326.htm

https://medicine.utoronto.ca/news/smarter-radiation-therapy-artificial-intelligence

https://www.healthdatamanagement.com/opinion/3-ways-artificial-intelligence-can-disrupt-healthcare

https://www.dw.com/en/doctors-dont-scale-like-artificial-intelligence-does/a-44899327

https://www.aao.org/eyenet/article/artificial-intelligence-and-glaucoma-detection

https://www.news-medical.net/health/Artificial-Intelligence-in-Cardiology.aspx

https://www.healthdatamanagement.com/news/ai-tool-helps-gateway-health-boost-accuracy-population-health

https://www.businessinsider.com/artificial-intelligence-healthcare

https://www.healthcareitnews.com/news/where-hospitals-plan-big-ai-deployments-diagnostic-imaging

https://www.healthcareitnews.com/news/asia-pacific/australia-s-snac-develops-ai-tools-improve-brain-scan-analysis

https://www.beckershospitalreview.com/artificial-intelligence/anthem-uses-ai-in-allergy-research-trial.html

https://www.geekwire.com/2018/mindshare-medical-launches-ai-cancer-screening-tech-can-see-data-beyond-perception/

https://lfpress.com/news/local-news/artificial-intelligence-may-be-key-to-mood-disorder-diagnosis-study

https://www.nextbigfuture.com/2018/08/artificial-intelligence-creating-new-drugs-from-scratch-by-efficiently-searching-huge-molecular-possibilities.html

https://qz.com/1349854/ai-can-spot-the-pain-from-a-disease-some-doctors-still-think-is-fake/

https://qz.com/1383083/how-ai-changed-organ-donation-in-the-us/

https://healthitanalytics.com/news/artificial-intelligence-for-medical-imaging-market-to-top-2b

https://www.thegazette.com/subject/news/business/coralville-based-idx-llc-aims-to-change-health-care-delivery-20180812

https://news.mit.edu/2018/artificial-intelligence-model-learns-patient-data-cancer-treatment-less-toxic-0810

https://www.healthcarefinancenews.com/news/how-artificial-intelligence-can-save-health-insurers-7-billion

https://www.techiexpert.com/ai-used-to-create-inexpensive-heart-disease-detector/

https://www.dermatologytimes.com/business/artificial-intelligence-friend-or-foe-dermatology

http://customerthink.com/looking-at-5-noteworthy-applications-of-artificial-intelligence-in-healthcare/

https://onlinelibrary.wiley.com/doi/full/10.1111/dmcn.13942

https://blog.jive.com/ai-chatbots-contact-centers/

https://www.theguardian.com/technology/2018/jul/29/the-robot-will-see-you-now-could-computers-take-over-medicine-entirely

https://www.nature.com/articles/d41586-019-01111-y

https://www.sciencedaily.com/releases/2018/08/180813113315.htm

https://www.curetoday.com/publications/cure/2019/fall-2019/as-large-as-life-using-artificial-intelligence-in-cancer-care

https://medibulletin.com/1-2-seconds-is-all-artificial-intelligence-takes-to-screen-ct-scans/

https://www.marktechpost.com/2018/08/14/artificial-intelligence-in-medicine-gains-massive-traction-with-growing-investment-in-ai-in-the-space/

https://www.washingtontimes.com/news/2018/aug/13/colonoscopy-technology-cuts-need-polyp-surgery-tes/

https://medicalxpress.com/news/2018-09-medicine-ready-artificial-intelligence.html

https://www.beckershospitalreview.com/healthcare-information-technology/how-artificial-intelligence-can-transform-payment-integrity.html

https://www.scitecheuropa.eu/artificial-intelligence-to-map-obesity/88924/

https://www.marketwatch.com/press-release/notable-health-closes-135-million-series-a-financing-to-expand-artificial-intelligence-powered-physician-patient-interaction-platform-2018-09-06

https://www.scmp.com/news/china/science/article/2163110/are-you-risk-diabetes-chinese-ai-system-could-predict-disease-15

https://geneticliteracyproject.org/2018/09/10/can-artificial-intelligence-give-us-a-more-efficient-health-care-system/

https://mindmatters.ai/2019/06/new-evidence-that-some-comatose-people-really-do-understand/

https://www.fastcompany.com/90287723/artificial-intelligence-can-detect-alzheimers-in-brain-scans-six-years-before-a-diagnosis

https://www.sciencedaily.com/releases/2019/01/190103152906.htm

https://newatlas.com/robot-md-ai-future-medial-diagnosis/60874/

https://www.healthworkscollective.com/can-artificial-intelligence-diagnose-illnesses-better-than-other-methods/

https://mytechdecisions.com/compliance/artificial-intelligence-delivery-rooms/

https://www.europeanpharmaceuticalreview.com/news/82870/artificial-intelligence-ai-heart-disease/

https://www.europeanpharmaceuticalreview.com/news/102403/ai-can-predict-which-research-will-translate-to-clinical-trials/

https://www.smithsonianmag.com/innovation/will-artificial-intelligence-improve-health-care-for-everyone-180972758/

https://venturebeat.com/2019/01/04/massachusetts-generals-ai-can-spot-brain-hemorrhages-as-accurately-as-humans/

https://www.geek.com/news/machine-learning-may-predict-how-well-youll-age-1767646/

https://becominghuman.ai/the-role-of-ai-in-healthcare-technology-6c33a6eee18c

https://www.sciencemag.org/news/2019/01/artificial-intelligence-could-diagnose-rare-disorders-using-just-photo-face

https://www.sciencedaily.com/releases/2019/01/190102112926.htm

https://www.empr.com/home/news/artificial-intelligence-can-detect-classify-acute-brain-bleeds/

https://www.bioeng.ucla.edu/artificial-intelligence-detects-the-presence-of-viruses/

https://blogs.scientificamerican.com/observations/the-surgical-singularity-is-approaching/

https://healthitanalytics.com/news/artificial-intelligence-in-healthcare-spending-to-hit-36b

https://www.fotoinc.com/news-updates/artificial-intelligence-embedded-mobile-app-for-chronic-neck-and-back-pain

https://www.theweek.in/news/sci-tech/2018/12/19/British-doctors-sceptical-that-AI-could-entirely-replace-them.html

https://www.news-medical.net/life-sciences/Artificial-Intelligence-in-Histopathology.aspx

https://blogs.scientificamerican.com/observations/how-ai-could-help-your-bad-back/

https://www.mdmag.com/medical-news/artificial-intelligence-deep-learning-learn-diagnose-epilepsy

https://www.ncbi.nlm.nih.gov/pmc/articles/PMC6290744/

https://en.wikipedia.org/wiki/Artificial_intelligence_in_healthcare

https://www.msn.com/en-us/health/medical/cigna-is-using-artificial-intelligence-to-predict-which-of-its-subscribers-will-become-addicted-to-opioids/ar-BBReYKL

https://www.sciencedaily.com/releases/2018/12/181221123743.htm

https://www.wndu.com/content/news/Artificial-intelligence-improves-colonoscopies-503263751.html

https://www.unionleader.com/news/scitech/ai-can-predict-mental-health-issues-from-your-instagram-posts/article_06ab3c45-8fa5-539e-869e-c19cab1cbef0.html

https://www.expresshealthcare.in/news/medachievers-and-labindia-healthcare-bring-artificial-intelligence-based-open-surgery-simulator/407543/

https://hackernoon.com/how-ai-robots-are-infiltrating-healthcare-ts8u3zfq

https://towardsdatascience.com/artificial-intelligence-deep-learning-for-medical-diagnosis-9561f7a4e5f

http://www.digitaljournal.com/tech-and-science/science/ai-can-boost-cancer-drug-discovery/article/538623

https://www.nbcnews.com/mach/science/why-big-pharma-betting-big-ai-ncna852246

https://www.kurzweilai.net/how-to-predict-the-side-effects-of-millions-of-drug-combinations

https://futurism.com/neoscope/genetic-report-predict-disease-risk
https://www.modernhealthcare.com/article/20180910/NEWS/1809199
91/telemedicine-meets-artificial-intelligence-at-bedside
https://www.beckershospitalreview.com/artificial-intelligence/5-
developments-in-ai-last-week.html
https://www.forbes.com/sites/brucelee/2019/01/20/do-you-have-early-
signs-of-cervical-cancer-how-ai-technology-may-help/#55552bdce694
http://www.digitaljournal.com/life/health/could-a-machine-replace-a-
doctor/article/538540
https://ai-med.io/artificial-intelligence-robotic-surgery/
https://www.healthcareitnews.com/news/ai-healthcare-big-ethical-
questions-still-need-answers
http://newsroom.ucla.edu/releases/artificial-intelligence-device-detect-
moving-parasites-bodily-fluid-earlier-diagnosis
https://www.periscopedata.com/press/artificial-intelligence-bias-and-
mental-health-implications
https://hospitalnews.com/artificial-intelligence-being-used-to-develop-
drugs-even-faster-and-cheaper/
https://www.globenewswire.com/news-
release/2018/12/12/1665700/0/en/Artificial-Intelligence-key-to-
Universal-Health-Coverage.html
https://www.acobiom.com/en/how-artificial-intelligence-can-help-to-
the-development-of-diagnostics-dedicated-to-precision-medicine/
https://www.forbes.com/sites/bernardmarr/2018/12/21/ai-that-saves-
lives-the-chatbot-that-can-detect-a-heart-attack-using-machine-
learning/#3300974450f9
https://www.asianscientist.com/2018/12/in-the-lab/artificial-
intelligence-flu-sensor/
https://openmedscience.com/bright-future-for-robotic-surgeons/
https://www.beckershospitalreview.com/artificial-
intelligence/northwestern-introduces-ai-tool-to-help-medical-
assistants-conduct-sonograms.html
https://www.medicalbag.com/home/more/tech-talk/the-future-of-
robot-physicians-is-artificial-intelligence-poised-to-take-over-
medicine/

https://www.geekwire.com/2018/amazon-unveils-new-service-mine-decode-medical-records-using-artificial-intelligence/

https://www.dw.com/en/artificial-intelligence-in-medicine-the-computer-knows-what-you-need/a-46226852

https://www.prnewswire.com/news-releases/carepredict-presents-artificial-intelligence-powered-solutions-for-senior-care-at-aging2-0-optimize-2018--300749580.html

https://hitconsultant.net/2018/12/13/artificial-intelligence-transform-personal-health/

https://www.wftv.com/news/local/orlando-hospital-uses-artificial-intelligence-to-save-lives-in-the-delivery-room/874200835/

https://axiosholding.com/healthcare-ai-just-what-the-doctor-ordered/

https://www.globenewswire.com/news-release/2018/11/10/1649339/0/en/AI-Algorithm-Outperformed-Majority-of-Cardiologists-in-Detection-of-Heart-Murmurs-in-Clinical-Study.html

https://www.smithsonianmag.com/innovation/can-artificial-intelligence-detect-depression-in-persons-voice-180970702/

https://ai-med.io/artificial-intelligence-lie-human/

https://www.gqrgm.com/5-ways-ai-is-transforming-healthcare/

https://www.jwatch.org/na47874/2018/11/06/artificial-intelligence-improves-wrist-fracture-detection

https://www.healthcareitnews.com/news/ai-algorithms-show-promise-colonoscopy-screenings

https://electronichealthreporter.com/artificial-intelligence-is-the-new-operating-system-in-healthcare/

Top of Form

Bottom of Form

https://pophealthanalytics.com/exploring-the-role-of-ai-in-population-health-risk-assessment-symposium/

https://www.ns-businesshub.com/science/artificial-intelligence-in-medicine-cure-cancer/

http://dailytrojan.com/2018/10/28/mindful-mondays-artificial-intelligence-has-limited-power-to-treat-mental-health-issues/

https://www.dailydot.com/debug/ai-sepsis-diagnosis/

https://blog.sevenponds.com/science-of-
us/%E2%80%A8%E2%80%A8artificial-intelligence-outsmarts-past-
practices-that-predict-patient-outcomes
https://www.medicalnewstoday.com/articles/325491.php#1
https://onlinemedicalcare.org/artificial-intelligence-outperforms-real-
doctors-studies-show/
https://www.healthline.com/health-news/new-ai-technology-may-help-
diagnose-fetal-heart-problems
https://www.aiin.healthcare/topics/artificial-intelligence/ai-enhanced-
virtual-care-could-reduce-er-visits
https://www.medicaldevice-network.com/features/ethics-in-ai/
https://www.healthdatamanagement.com/opinion/why-using-ai-in-
healthcare-requires-a-balance-of-efficiency-and-ethics
https://www.prnewswire.com/news-releases/osf-ventures-invests-in-
gauss-artificial-intelligence-technology-to-identify-postpartum-
bleeding-early-300732784.html
https://www.medicaldesignandoutsourcing.com/digital-surgery-touts-
artificial-intelligence-for-the-operating-room/
https://ai-med.io/will-artificial-intelligence-replace-human-
radiologists/
https://www.sciencedaily.com/releases/2019/01/190111143744.htm
https://directorsblog.nih.gov/2019/01/17/using-artificial-intelligence-
to-detect-cervical-cancer/
https://www.forbes.com/sites/bernardmarr/2018/07/27/how-is-ai-used-
in-healthcare-5-powerful-real-world-examples-that-show-the-latest-
advances/#4a59916e5dfb
https://transmitter.ieee.org/artificial-intelligence-robots-and-the-
operating-room/
https://www.nationthailand.com/asean-plus/30349160
https://directorsblog.nih.gov/2019/01/15/using-artificial-intelligence-
to-catch-irregular-heartbeats/
https://analyticsindiamag.com/ai-tell-how-ovarian-tumour/
https://lfpress.com/news/local-news/artificial-intelligence-can-predict-
ptsd-in-patients-london-researchers
https://www.forbes.com/sites/samshead/2018/10/08/tencent-aims-to-
train-ai-to-spot-parkinsons-in-3-minutes/#676980856f36

https://engineering.stanford.edu/magazine/article/david-magnus-how-will-artificial-intelligence-impact-medical-ethics
https://theweek.com/articles/694522/robots-replace-therapists
https://www.leewayhertz.com/how-iot-transforming-healthcare/
https://www.straitstimes.com/singapore/scdf-turns-to-artificial-intelligence-to-help-emergency-call-dispatchers
https://www.eurekalert.org/pub_releases/2018-07/imi-mlt070518.php
https://www.latimes.com/world/la-fg-china-ai-20180706-story.html
https://www.healthcareitnews.com/news/chinese-hospital-guangdong-deploys-ai-cameras-detect-blindness-causing-diseases
https://www.healthcarefinancenews.com/news/artificial-intelligence-healthcare-projected-be-worth-more-27-billion-2025
https://analyticsindiamag.com/can-machines-be-taught-to-detect-medicare-fraud/
https://www.newscientist.com/article/2222907-ai-can-predict-if-youll-die-soon-but-weve-no-idea-how-it-works/
https://revcycleintelligence.com/news/using-artificial-intelligence-to-improve-the-hospital-revenue-cycle
https://www.healthcarefinancenews.com/news/addressing-social-determinants-health-consider-artificial-intelligence-and-machine-learning
https://www.mcknights.com/marketplace/marketplace-experts/what-facilities-can-gain-from-robotics-in-physical-therapy/
https://www.entrepreneur.com/article/317047
https://www.icr.ac.uk/news-archive/artificial-intelligence-can-predict-how-cancers-will-evolve-and-spread
https://www.frontiersin.org/articles/10.3389/fpsyg.2019.00263/full
https://www.usatoday.com/story/opinion/2019/01/28/health-privacy-laws-artificial-intelligence-hipaa-needs-update-column/2695386002/
https://www.nature.com/articles/s41746-019-0089-x
https://time.com/collection/life-reinvented/5494363/sleep-artificial-intelligence/
https://www.corporatewellnessmagazine.com/article/how-artificial-intelligence-can-fight-against-workplace-stress
https://www.sciencedaily.com/releases/2018/05/180528190839.htm

https://www.cnet.com/news/ai-to-aid-emergency-call-operators-diagnose-heart-attacks-in-europe/

https://www.forbes.com/sites/michaelpellmanrowland/2018/04/24/beyonce-artificial-intelligence-vegan/#503780cc7160

https://govinsider.asia/innovation/artificial-intelligence-ageing-population/

https://www.geekwire.com/2018/health-tech-podcast-using-precision-medicine-kill-cancer-artificial-intelligence/

https://thriveglobal.com/stories/the-road-to-good-health-will-now-be-guided-by-artificial-intelligence/

https://www.engadget.com/2018/10/15/google-ai-spots-advanced-breast-cancer/

https://www.stanforddaily.com/2018/09/27/artificial-swarm-intelligence-diagnoses-pneumonia-better-than-individual-computer-or-doctor/

https://www.forbes.com/sites/michaelmillenson/2018/09/23/will-apple-track-your-mind-not-just-your-heart/#4879dbabcad3

https://emerj.com/ai-sector-overviews/artificial-intelligence-in-health-insurance-current-applications-and-trends/

https://www.unc.edu/posts/2018/07/31/artificial-intelligence-system-created-at-unc-chapel-hill-designs-drugs-from-scratch/

https://www.beaconlens.com/ai-in-behavioral-health-care-when-artificial-intelligence-became-real/

https://www.eurekalert.org/pub_releases/2018-09/sfl-ait091818.php

https://www.einfochips.com/blog/how-ai-enabled-wearables-are-changing-healthcare-and-fitness-industry/

https://psychnews.psychiatryonline.org/doi/10.1176/appi.pn.2018.10a5

https://www.eurekalert.org/pub_releases/2018-10/muos-wcm100218.php

https://time.com/collection/life-reinvented/5492063/artificial-intelligence-fertility/

https://www.thefix.com/artificial-intelligence-system-aims-identify-drug-thefts-hospitals

https://towardsdatascience.com/applying-artificial-intelligence-to-help-people-quit-smoking-early-results-a3e5581d560

https://medcitynews.com/2018/05/qure-ai/

https://healthitanalytics.com/news/deep-learning-tool-tops-dermatologists-in-melanoma-detection

https://www.himss.org/resources/role-artificial-intelligence-healthcare-and-society

https://www.aami.org/newsviews/newsdetail.aspx?ItemNumber=6502

http://www.thetower.org/6317-israel-uk-announce-landmark-scientific-cooperation-agreement-focusing-on-aging-artificial-intelligence/

https://www.aarp.org/health/conditions-treatments/info-2018/hospital-artificial-intelligence-telehealth.html

https://www.digitaltrends.com/cool-tech/chatterbaby-app-deciphers-baby-crying/

https://www.cbs58.com/news/new-app-translates-babies-cries-so-deaf-parents-understand-whats-going-on

https://www.empr.com/home/news/fda-approves-ai-algorithm-that-helps-detect-wrist-fractures/

https://bioinformatics.csiro.au/blog/can-ai-help-fight-antibiotic-resistant-superbugs/

https://www.sciencedaily.com/releases/2018/09/180920161054.htm

https://www.forbes.com/sites/jenniferhicks/2018/06/08/see-how-this-hospital-uses-artificial-intelligence-to-find-kidney-disease/#78af38d72e8f

https://www.newsweek.com/ai-being-developed-which-can-smell-illness-human-breath-967197

https://www.theguardian.com/technology/2018/jun/10/artificial-intelligence-cancer-detectors-the-five

https://www.independent.co.uk/voices/loneliness-kills-artificial-intelligence-chatbot-doctors-health-risk-diagnoses-a8423321.html

https://www.technologyreview.com/f/609969/ai-could-diagnose-your-heart-attack-on-the-phone-even-if-youre-not-the-caller/

https://www.linkedin.com/feed/news/teaching-ai-to-explain-its-thinking-4617604

https://www.smartdatacollective.com/artificial-intelligence-chatbots-could-make-your-doctors-obsolete/

https://www.telegraph.co.uk/technology/2019/01/13/google-scientist-ai-may-not-cure-all-discovering-new-drugs/

https://greatlakesledger.com/2019/01/12/artificial-intelligence-can-predict-and-estimate-flu-activity/
https://www.mddionline.com/using-artificial-intelligence-predict-flu-activity
https://time.com/5556339/artificial-intelligence-robots-medicine/
https://www.nytimes.com/2019/03/11/well/live/how-artificial-intelligence-could-transform-medicine.html
https://towardsdatascience.com/artificial-intelligence-replace-the-human-doctors-in-the-future-is-it-true-91b3ae9fea0e
https://www.sciencedaily.com/releases/2019/12/191218090156.htm
https://spectrum.ieee.org/biomedical/diagnostics/how-ibm-watson-overpromised-and-underdelivered-on-ai-health-care
https://venturebeat.com/2019/09/13/googles-ai-detects-26-skin-conditions-as-accurately-as-dermatologists/
https://www.dhs.gov/science-and-technology/news/2018/10/16/snapshot-public-safety-agencies-pilot-artificial-intelligence
https://www.inverse.com/article/59742-smartphone-app-health-best-eye-disease
https://www.foxnews.com/tech/google-ai-can-predict-when-youll-die-with-95-percent-accuracy-researchers-say
https://www.zdnet.com/article/googles-deepmind-follows-a-mixed-path-to-ai-in-medicine/
https://medicalfuturist.com/artificial-intelligence-in-mental-health-care/
https://www.cnbc.com/2018/02/22/medical-errors-third-leading-cause-of-death-in-america.html#:~:text=According%20to%20a%20recent%20study,after%20%20heart%20disease%20and%20cancer.
https://www.google.com/search?q=how+many+people+go+online+to+diagnose+themselves&rlz=1C1CHBF_enUS894US894&oq=how+many+people+go+online+to+diagnose+themselves&aqs=chrome..69i57.10637j0j7&sourceid=chrome&ie=UTF-8
https://www.google.com/search?q=obama+care+mandate+electronic+health+records&rlz=1C1CHBF_enUS894US894&oq=obama+care+m

andate+electronic+health+records&aqs=chrome..69i57j0l2.15928j0j7
&sourceid=chrome&ie=UTF-8
https://www.healthcarelaw-blog.com/the-electronic-medical-records-
emr-
mandate/#:~:text=With%20the%20passage%20of%20the,to%20take%
20effect%20in%202014.
https://www.tampabay.com/archive/2000/06/23/france-criminalizes-
mental-manipulation/
https://www.smithsonianmag.com/innovation/can-artificial-
intelligence-detect-depression-in-persons-voice-180970702/
https://www.unionleader.com/news/scitech/ai-can-predict-mental-
health-issues-from-your-instagram-posts-but-should-
it/article_06ab3c45-8fa5-539e-869e-c19cab1cbef0.html
https://beforeitsnews.com/prophecy/2020/09/theyve-killed-god-i-cant-
feel-god-anymore-my-soul-is-dead-after-the-vaccine-urgent-read-
2514214.html
https://www.medicalnewstoday.com/articles/325491
https://www.lifespan.io/news/new-iphone-app-measures-biological-
age/
https://www.carepredict.com/
Billy Crone, Abortion: The Mass Murder of Children,
(Las Vegas: Get A Life Ministries, Pgs. 156-200)
https://www.newsmax.com/newsfront/google-medical-brain-artificial-
intelligence-medical-research/2018/06/18/id/866907/
https://www.express.co.uk/news/world/976108/google-when-will-you-
die-ai-artificial-intelligence-latest
https://www.newscientist.com/article/2222907-ai-can-predict-if-youll-
die-soon-but-weve-no-idea-how-it-works/
https://futurism.com/neoscope/genetic-report-predict-disease-risk
https://www.livescience.com/65087-ai-premature-death-
prediction.html#:~:text=Learn%20more-
,AI%20Is%20Good%20(Perhaps%20Too%20Good)%20at,Predicting
%20Who%20Will%20Die%20Prematurely&text=Medical%20researc
hers%20have%20unlocked%20an,predicting%20a%20person's%20ear
ly%20death.

https://www.dailymail.co.uk/news/article-8403369/Tony-Blair-calls-new-digital-ID-people-prove-coronavirus-disease-status.html

Chapter 20 *The Future of Big Brother with AI*

1. *Google Waymo*
   https://www.youtube.com/watch?v=xjoWJ3XZFNk
2. *Driverless Trucks*
   https://www.youtube.com/watch?v=4umVfjcpDql
3. *Spielberg Movie Duel*
   https://www.youtube.com/watch?v=SutDTIhbQ2g
4. *Uber Self-Driving Car*
   https://www.youtube.com/watch?v=EYh0F_8ZdSU
5. *Smart Pods*
   https://www.youtube.com/watch?v=7gXfZXkYogY
6. *Total Recall Johnny Cab*
   https://www.youtube.com/watch?v=eWgrvNHjKkY
7. *AI Decides Who Lives or Dies*
   https://www.youtube.com/watch?v=WS4tUn3MvCU
   https://www.youtube.com/watch?v=t6WlQHmTrpw
8. *Minority Report Pre-Crime Scene*
   https://www.youtube.com/watch?v=N3QCRLQppf4
9. *AI Robot Police*
   https://www.youtube.com/watch?v=_Nz4WASJ0E4
   https://www.youtube.com/watch?v=cMtMk21FZUM
10. *Minority Report Facial Recognition*
    https://www.youtube.com/watch?v=N3QCRLQppf4
11. *L.A. Police Predict Crimes*
    https://www.youtube.com/watch?v=7lpCWxlRFAw
12. *Logan's Run Runner Scene*
    https://www.youtube.com/watch?v=EoKz-ilaZiA
13. *Microchipping On the Rise*
    https://www.youtube.com/watch?v=dl_gemn9a9E
14. *Information on Transportation*

https://medium.com/@Liamiscool/a-list-of-artificial-intelligence-tools-you-can-use-today-for-personal-use-1-3-7f1b60b6c94f

https://medium.com/@Liamiscool/a-list-of-artificial-intelligence-tools-you-can-use-today-for-industry-specific-3-3-5e16c68da697

https://www.caranddriver.com/news/a30857661/autonomous-car-self-driving-research-expensive/#:~:text=Self%2DDriving%2DCar%20Research%20Has%20Cost%20%2416%20Billion.

https://www.forbes.com/sites/samabuelsamid/2020/02/11/aptiv-self-driving-vehicles-top-100000-rides-in-las-vegas/#203a3e4d5439

https://techcrunch.com/2018/12/09/g7-raised-320-million/

https://www.fox5vegas.com/news/self-driving-vehicles-hit-milestone-on-las-vegas-strip/article_4f7f3f14-8bda-11e9-bbdd-134b856a97c6.html

https://www.reuters.com/article/us-rolls-royce-hldg-uptake/rolls-royce-partners-with-ai-software-maker-to-predict-engine-performance-idUSKBN1O326D

https://arstechnica.com/information-technology/2018/12/unite-day1-1/

https://www.itf-oecd.org/how-artificial-intelligence-can-contribute-well-being-transport

https://www.pewresearch.org/internet/2018/12/10/artificial-intelligence-and-the-future-of-humans/

https://www.telegraph.co.uk/technology/2018/10/20/us-driverless-cars-unsafe-cant-spot-iconic-british-vehicles/

https://www.dailysabah.com/technology/2018/10/20/artificial-intelligence-predicts-errors-malfunctions-to-improve-efficiency

http://www.santacruztechbeat.com/2018/10/16/sapientx-voice-enabled-artificial-intelligence-for-cars-and-consumer-products/

https://www.offshorewind.biz/2018/10/17/artificial-intelligence-swarms-offshore-wind/

https://www.audi-mediacenter.com/en/press-releases/audi-optimizes-quality-inspections-in-the-press-shop-with-artificial-intelligence-10847

https://www.carmudi.com.ph/journal/hyundai-wants-develop-artificial-intelligence-tech-self-driving-cars/

https://galusaustralis.com/2019/11/42363/artificial-intelligence-for-automotive-applications-market-2019-trends-share-and-future-analysis-of-manufacturers-aimotive-argo-ai-astute-solutions-audi-bmw/

https://harkeraquila.com/27048/features/the-future-of-artificial-intelligence/

http://www.industrytap.com/artificial-intelligence-cutting-edge-automotive-technology/46790

https://www.uitp.org/news/demystifying-artificial-intelligence-public-transport

https://www.volpe.dot.gov/news/transportation-age-artificial-intelligence-and-predictive-analytics-final-report

http://www.ascd.org/publications/educational-leadership/feb19/vol76/num05/Confronting-Artificial-Intelligence.aspx

https://www.prnewswire.com/news-releases/byton-showcases-the-impact-of-artificial-intelligence-and-machine-learning-on-ev-innovation-at-techcrunch-disrupt-2018-300706556.html

https://www.khaleejtimes.com/business/aviation/artificial-intelligence-to-help-manage-airspace-incidents-in-dubai

https://medium.com/@Liamiscool/a-list-of-artificial-intelligence-tools-you-can-use-today-for-industry-specific-3-3-5e16c68da697

https://medium.com/@Liamiscool/a-list-of-artificial-intelligence-tools-you-can-use-today-for-businesses-2-3-eea3ac374835

https://medium.com/@Liamiscool/a-list-of-artificial-intelligence-tools-you-can-use-today-for-personal-use-1-3-7f1b60b6c94f

https://newsroom.intel.com/news/intel-artificial-intelligence-rolls-royce-push-full-steam-autonomous-shipping/#gs.kkrim7

https://www.freightwaves.com/news/technology/uptake-adds-artificial-intelligence-to-vehicle-maintenance

https://insidebigdata.com/2019/07/30/big-data-and-the-future-of-self-driving-cars/

https://tech.slashdot.org/story/18/11/05/1644259/why-big-tech-pays-poor-kenyans-to-teach-self-driving-cars

https://www.americamagazine.org/politics-society/2018/11/02/what-are-dangers-artificial-intelligence-our-brave-new-world-self

https://www.brookings.edu/research/the-folly-of-trolleys-ethical-challenges-and-autonomous-vehicles/
https://www.brookings.edu/blog/brookings-now/2017/01/24/driverless-cars-are-coming/
https://www.brookings.edu/blog/the-avenue/2017/10/16/how-will-autonomous-vehicles-transform-the-built-environment/
https://www.theverge.com/2018/7/3/17530232/self-driving-ai-winter-full-autonomy-waymo-tesla-uber
https://www.technologyreview.com/s/612341/a-global-ethics-study-aims-to-help-ai-solve-the-self-driving-trolley-problem/
https://www.extremetech.com/extreme/252645-ai-thorny-ethical-dilemmas-lurking-behind-self-driving-cars
https://www.livescience.com/50841-future-of-driverless-cars.html
https://www.inc.com/magazine/201811/tom-foster/artificial-intelligence-ethics.html
https://news.bloombergenvironment.com/environment-and-energy/insight-how-artificial-intelligence-will-change-environmental-compliance
https://www.eurekalert.org/pub_releases/2018-10/vuot-aip102318.php
https://www.smartdatacollective.com/driverless-cars-and-quest-for-true-artificial-intelligence/
https://www.thenational.ae/uae/uae-to-develop-laws-to-govern-self-driving-cars-and-artificial-intelligence-1.790555
https://www.thestar.com/business/tech_news/2018/04/30/canadian-airlines-get-on-board-with-artificial-intelligence-quest.html
https://sports.yahoo.com/self-driving-cars-reality-waymo-203113900.html
https://www.cnbc.com/2019/11/21/self-driving-cars-may-not-be-widespread-reality-but-potential-is-great.html
https://www.30secondstofly.com/ai-software/self-driving-cars-change-business/
https://www.weforum.org/agenda/2019/07/autonomous-vehicles-driverless-cars-public-transport/
https://www.busbud.com/blog/will-driverless-buses-reality/

https://www.masstransitmag.com/alt-mobility/autonomous-vehicles/article/21073560/autonomous-vehicle-technology-preparing-for-the-next-wave-of-innovation-in-public-transit

https://www.cnn.com/style/article/dubai-autonomous-public-transport/index.html

https://www.phocuswire.com/2018-review-2019-preview-AI

https://www.autosport.com/f1/news/140796/roborace-keen-to-demo-driverless-car-at-f1-gp

https://www.washingtonpost.com/lifestyle/travel/what-to-expect-when-you-travel-in-2019/2019/01/02/c0ba472a-fcc7-11e8-83c0-b06139e540e5_story.html

https://safecarnews.com/the-coalition-for-future-mobility-urges-congressional-action-for-autonomous-vehicles/

https://safecarnews.com/the-latest-robosense-lidar-perception-solution-will-support-robo-taxi-development/

https://safecarnews.com/seat-showcases-car-that-communicates-with-traffic-lights/

https://www.ttnews.com/articles/ai-trucking

https://www.benzinga.com/pressreleases/19/01/r12917440/travel-trends-for-2019-how-big-data-ai-technology-and-personalization-

https://www.groovecar.com/articles/tech-out-my-new-car/2018/12/fords-artificial-intelligence-i-feel-you/

https://emerj.com/ai-sector-overviews/ai-in-transportation-current-and-future-business-use-applications/

https://www.micron.com/insight/on-the-road-to-full-autonomy-self-driving-cars-will-rely-on-ai-and-innovative-memory

https://www.dxc.technology/digital_transformation/insights/145756-artificial_intelligence_in_travel_transportation_and_hospitality

https://www.thenational.ae/lifestyle/travel/2019-s-top-four-travel-trends-1.809555

https://www.techinasia.com/ai-world-safer-place

https://www.nytimes.com/2018/12/31/us/waymo-self-driving-cars-arizona-attacks.html

https://www.daytondailynews.com/technology/artificial-intelligence-driverless-vehicles-seen-springboro-future/sWaJD7n4D6J6rs3OFWgjVM/

https://www.airbus.com/newsroom/news/en/2018/12/airbus--latest-aigym-artificial-intelligence-challenge-is-focuse.html
https://www.dailysabah.com/science/2018/12/20/artificial-intelligence-to-make-you-fly
https://www.israel21c.org/10-of-the-hottest-autonomous-driving-technologies-from-israel/
https://www.technologyreview.com/s/612553/racing-ahead-with-artificial-intelligence/
https://www.information-age.com/artificial-intelligence-trucking-industry-123477239/
https://theloadstar.com/artificial-intelligence-help-solve-use-trucks-driver-shortage-worsens/
https://www.designnews.com/electronics-test/autonomous-car-s-big-challenge-using-hyperscale-server-fleet-train-ai-neural-networks/196274523759943
https://fortune.com/2018/12/14/byton-ceo-disrupting-the-auto-industry-with-a-startup-culture/
https://venturebeat.com/2019/12/11/lyft-details-the-planning-model-behind-its-self-driving-cars/
https://business.financialpost.com/pmn/business-pmn/artificial-intelligence-promises-advantages-for-airlines-and-their-passengers
https://www.kickstarter.com/projects/germanautolabs/chris-your-digital-co-driver-with-artificial-intel
https://www.businessinsider.com/r-chinas-alibaba-doubles-down-on-chips-amid-cloud-computing-push-2018-9
https://safety4sea.com/uk-explores-benefits-of-artificial-intelligence-in-transport/
https://www.globenewswire.com/news-release/2018/09/13/1570522/0/en/Crawford-Company-to-Launch-Artificial-Intelligence-Solution-for-Auto-Claims.html
https://www.bbntimes.com/en/technology/artificial-intelligence-to-reduce-baggage-mishandling
https://www.newshub.co.nz/home/new-zealand/2018/09/how-artificial-intelligence-is-saving-kiwi-truck-drivers-lives.html

https://www.marketwatch.com/press-release/artificial-intelligence-in-aviation-market-wide-spread-across-world-airbus-amazon-boeing-garmin-ge-2018-10-06

https://www.airport-technology.com/features/ai-at-airports-security/

https://analyticsindiamag.com/indian-railways-to-use-emotional-intelligence-to-serve-passengers-better/

https://www.timesofisrael.com/israeli-ai-tech-to-help-avert-accidents-in-china-by-tracking-drivers-drowsiness/

https://arstechnica.com/information-technology/2018/12/unite-day1-1/

https://www.financialexpress.com/industry/technology/now-artificial-intelligence-to-drive-innovation-for-road-safety/1368485/

https://www.intelligent-aerospace.com/avionics/article/16545234/artificial-intelligence-ais-increasing-role-in-autonomous-aircraft-technology

https://forum.thaivisa.com/topic/1078510-artificial-intelligence-to-run-traffic-lights-at-all-bangkok-intersections/

https://www.cityam.com/future-infrastructure-needs-artificial-intelligence-and-new/

https://cruisefever.net/cruise-line-introduces-first-cruise-assistant-powered-by-artificial-intelligence/

https://ramboll.com/media/rgr/artificial-intelligence-for-optimising-road-maintenance-work

https://www.forbes.com/sites/christopherhelman/2019/01/25/as-shutdown-slows-flights-into-laguardia-maybe-its-time-to-let-artificial-intelligence-handle-air-traffic-control/#8b822f52bcd7

https://www.tahawultech.com/industry/transport-logistics/dubai-taxi-services-artificial-intelligence-rta/

https://houseofbots.com/news-detail/3286-ARTIFICIAL-INTELLIGENCE-CAN-SAVE-US-FROM-TRAFFIC

https://www.information-age.com/trains-brains-how-artificial-intelligence-transforming-railway-industry-123460379/

https://www.theverge.com/2018/9/5/17822384/ai-autonomous-drone-quadcopter-pilot-drl-drone-racing-league-2019

https://www.machinedesign.com/motion-control/highway-future-artificial-intelligence-smart-vehicles

https://www.albawaba.com/business/pr/i-insured-offers-artificial-intelligence-based-pay-how-you-drive-insurance-uae-1137482
https://safety4sea.com/cma-cgm-to-embed-artificial-intelligence-onboard-boxships/
https://observer.com/2018/06/uber-identify-drunk-passengers-artificial-intelligence/
https://edmontonjournal.com/news/local-news/the-way-of-the-future-edmonton-to-embrace-artificial-intelligence-for-new-traffic-signals
https://en.wikipedia.org/wiki/History_of_self-driving_cars
https://www.greyb.com/autonomous-vehicle-companies/
https://www.cbinsights.com/research/autonomous-driverless-vehicles-corporations-list/
https://en.wikipedia.org/wiki/Steven_Spielberg_filmography
https://en.wikipedia.org/wiki/Robotaxi
https://www.economist.com/technology-quarterly/2019/05/30/pilotless-planes-are-on-the-way

15. *Information on Crime Investigation*

https://www.foxnews.com/tech/dhs-withdraws-proposal-facial-recognition-scans-us-citizens-trump-administration
https://www.foxnews.com/tech/amazons-facial-recognition-detect-fear-activists-law-enforcement
https://ktla.com/2019/08/30/threat-of-mass-shootings-gives-rise-to-artificial-intelligence-powered-cameras/
https://luminaanalytics.com/2018/09/15/how-artificial-intelligence-can-be-used-to-predict-threats-of-mass-violence-in-public-spaces-and-live-spaces/
https://www.businessinsider.com/r-chinas-alibaba-doubles-down-on-chips-amid-cloud-computing-push-2018-9
https://www.govtech.com/question-of-the-day/Question-of-the-Day-for-10012018.html
https://www.zdnet.com/article/ai-security-camera-detects-guns-and-identifies-shooters/
https://wpsecurityninja.com/role-of-artificial-intelligence-in-cyber-security/
https://www.insurancejournal.com/news/national/2019/01/03/513436.htm

https://www.onenewspage.us/video/20190103/11225266/New-York-power-plant-use-of-artificial.htm

https://www.livemint.com/Technology/EG3XzpKUyJ6MBsczEMMjU J/AI-face-recognition-to-help-UP-Police-catch-criminals.html

https://www.newsweek.com/ai-artificial-intelligence-robots-emotions-humans-541595

https://fortune.com/2019/05/07/artificial-intelligence-mind-reading-technology/

https://www.nytimes.com/2018/12/31/technology/human-resources-artificial-intelligence-humu.html

https://www.policeone.com/police-products/investigation/Investigative-Software/articles/482492006-How-intelligent-computing-can-help-win-the-war-on-crime/

http://www.industrytap.com/first-artificial-intelligence-colony-on-earth/47256

https://www.dallasnews.com/business/technology/2018/12/18/with-artificial-intelligence-dallas-startup-aims-to-catch-threats-that-humans-alone-might-miss/

https://www1.cbn.com/cbnnews/us/2018/december/speech-police-an-online-hate-index-run-by-artificial-intelligence-what-could-possibly-go-wrong

artificialintelligence-news.com/2018/11/28/ai-tags-potential-criminals/

https://artificialintelligence-news.com/2018/11/01/ai-lie-detector-eu-borders/

https://www.bloomberg.com/news/articles/2019-03-04/the-ai-cameras-that-can-spot-shoplifters-even-before-they-steal

https://www.theverge.com/2018/1/23/16907238/artificial-intelligence-surveillance-cameras-security

https://www.business-standard.com/article/current-affairs/delhi-police-gets-artificial-intelligence-centre-to-fight-crime-terrorists-118120400449_1.html

https://www.livemint.com/Opinion/k67awvCTBiZRylXdL8wcZP/Opinion--Can-AI-predict-if-you-are-about-to-kill-your-wife.html

https://www.forbes.com/sites/stuartanderson/2018/12/03/crowdstrikes-immigrant-co-founder-fighting-cyber-criminals/#45dfbc7b176c

https://www.dailystar.co.uk/news/latest-news/ai-minority-report-real-life-16822512

https://www.telegraph.co.uk/technology/2018/11/14/facebook-instagram-use-ai-spot-drug-dealers/

https://www.rri-tools.eu/-/notes-from-the-ai-frontier-applying-artificial-intelligence-for-social-good

https://www.fox29.com/news/archbishop-wood-high-school-first-to-use-artificial-intelligence-technology-to-detect-guns

https://voiceofeurope.com/2018/11/artificial-intelligence-to-interview-and-detect-lying-asylum-seekers/

https://mashable.com/article/ai-lie-detector-border/

https://qz.com/1441034/using-artificial-intelligence-to-detect-written-lies/

https://www.zdnet.com/article/law-enforcement-is-using-artificial-intelligence-to-spot-fake-confessions/

https://www.forbes.com/sites/bernardmarr/2018/10/29/how-the-uk-government-uses-artificial-intelligence-to-identify-welfare-and-state-benefits-fraud/#3e426db140cb

https://www.nbcnews.com/tech/tech-news/instagram-use-artificial-intelligence-detect-bullying-photos-n918291

https://chronicleofsocialchange.org/child-welfare-2/using-algorithms-artificial-intelligence-in-child-welfare/33429

https://www.frontiersin.org/articles/10.3389/fnhum.2018.00105/full?utm_source=FWEB&utm_medium=NBLOG&utm_campaign=ECO_FN HUM_personality-eye-movements

https://thenextweb.com/artificial-intelligence/2019/02/01/politicians-have-failed-us-its-time-for-ai-to-stop-school-shootings/

https://www.fastcompany.com/90388822/how-gun-detection-technology-promises-to-help-prevent-mass-shootings

https://www.sciencedaily.com/releases/2018/08/180802130750.htm

https://www.forbes.com/sites/jenniferhicks/2018/09/30/making-facial-recognition-smarter-with-artificial-intelligence/#76459d5c8f1f

https://thenextweb.com/problem-solvers/2018/07/13/authentication-cybersecurity/

https://www.washingtonpost.com/news/the-switch/wp/2018/05/17/ice-just-abandoned-its-dream-of-extreme-vetting-software-that-could-predict-whether-a-foreign-visitor-would-become-a-terrorist/
https://www.onartificialintelligence.com/articles/15516/ai-security-cameras-identify-weapons-prevent-crime
https://www.futuregrasp.com/opportunities-from-artificial-intelligence-for-law-enforcement
http://www.parabolicarc.com/2019/09/15/first-earth-observation-satellite-with-artificial-intelligence-ready-to-launch/
https://news.mit.edu/2018/artificial-intelligence-senses-people-through-walls-0612
https://www.foxnews.com/tech/how-hackers-can-use-artificial-intelligence-against-us
https://www.ifsecglobal.com/ifsec/artificial-intelligence-threat
https://www.smithsonianmag.com/innovation/can-artificial-intelligence-help-stop-school-shootings-180969288/
https://www.policeone.com/police-products/police-technology/software/video-analysis/articles/how-deep-learning-is-transforming-police-investigations-Egyc70WTpIajnhG8/

Chapter 21 *The Future of Military with AI*

1. *Terminator Opening Scene*
   https://www.youtube.com/watch?v=_Mg7qKstnPk
2. *Various AI Military Equipment*
   Billy Crone, *Drones, AI & the Coming Human Annihilation*,
   (Las Vegas: Get A Life Ministry, 2018, Pgs. 11-274)
3. *Skynet Goes to War*
   https://www.youtube.com/watch?v=AdyWIiYQv7E
   https://www.youtube.com/watch?v=SWiM4ImsxsA
4. *Militarized AI Insects*
   https://alliance.seas.upenn.edu/~mastwiki/wiki/index.php?n=Repository.SeminarsAndPresentations?action=download&upname=MAST_intro_BAE_Systems_Large.wmv
   https://www.youtube.com/watch?v=_5YkQ9w3PJ4

5. *Navy Swarm Technology*
   https://www.youtube.com/watch?v=Mq-zUBgBtVg
6. *Mav Swarm Technology*
   https://www.youtube.com/watch?v=_5YkQ9w3PJ4
7. *Slaughterbot Swarm*
   https://www.youtube.com/watch?v=KqoGacUu07I
8. *Various AI Animal Robots*
   Billy Crone, *Drones, AI & the Coming Human Annihilation,*
   (Las Vegas: Get A Life Ministry, 2018, Pgs. 11-274)
9. *AI Military Swarm Drop*
   https://www.youtube.com/watch?v=KqoGacUu07I
10. *Terminator 3 Opening Scene*
    https://www.youtube.com/watch?v=MaiLrrn31nE
11. *Neuromorphic Brain Ship*
    https://www.youtube.com/watch?v=gXNCz26UhyY
12. *Early Versions Terminator Robots*
    https://www.youtube.com/watch?v=tFrjrgBV8K0
    https://www.youtube.com/user/BostonDynamics
13. *Fedor Terminator Robot*
    https://www.youtube.com/watch?v=HTPIED6jUdU
14. *Putin Warns Genetic Soldiers*
    https://www.thesun.co.uk/news/4746212/vladimir-putin-russia-super-
    human-soldiers-nuclear-bomb/
15. *Skynet Takes Over Scene*
    https://www.youtube.com/watch?v=_Wlsd9mIjiU
16. *AI Goes Rogue in War*
    https://www.youtube.com/watch?v=UmNRUPlbscg
17. *Google Builds Project Maven*
    https://www.youtube.com/watch?v=0i_0alI-qQw
18. *Google Builds Skynet*
    https://www.youtube.com/watch?v=2BMQoN5FLU0
19. *Terminator Judgment Day Scene*
    https://www.youtube.com/watch?v=IqITGz-b11s
    https://www.youtube.com/watch?v=3iMvFMMrNkA
    https://www.youtube.com/watch?v=xjatJ36cJvM
20. *AI Military Information*

https://www.nationaldefensemagazine.org/articles/2020/2/3/analysts-say-$25-billion-needed-annually-for-ai#:~:text=Defense%2Drelated%20AI%20spending%20was,efforts%2C%20according%20to%20the%20study.

https://breakingdefense.com/2019/04/how-ai-could-change-the-art-of-war/

https://www.zdnet.com/article/trump-signs-executive-order-prioritising-ai-development/

https://www.forbes.com/sites/rogertrapp/2018/04/29/effective-leaders-need-to-train-their-minds-as-well-as-their-bodies/#2697f3f2143a

http://mil-embedded.com/articles/artificial-intelligence-help-warfighters-many-fronts/

https://electronics360.globalspec.com/article/11706/artificial-intelligence-could-help-us-soldiers-learn-13-times-faster-on-the-battlefield

https://www.c4isrnet.com/intel-geoint/2018/04/25/a-new-council-could-advance-artificial-intelligence-for-the-military/

http://mil-embedded.com/articles/military-learning-speeds-decision-making-efficiency-warfighters/

http://mil-embedded.com/news/artificial-intelligence-machine-learning-techniques-to-be-explored-in-darpa-causal-exploration-program/

https://www.marketsandmarkets.com/Market-Reports/artificial-intelligence-military-market-41793495.html

https://www.lawfareblog.com/war-machines-artificial-intelligence-conflict

https://publicintegrity.org/national-security/trump-administration-accelerates-military-study-of-artificial-intelligence/

https://www.nytimes.com/2018/08/26/technology/pentagon-artificial-intelligence.html

https://publicintegrity.org/national-security/the-pentagon-plans-to-spend-2-billion-to-help-inject-more-artificial-intelligence-into-its-weaponry/

https://www.newsweek.com/nuclear-war-ai-artificial-intelligence-arms-race-901267

https://www.army-technology.com/news/darpa-invest-2bn-artificial-intelligence/

https://about.bgov.com/news/can-pentagon-bridge-artificial-intelligences-valley-of-death/

https://taskandpurpose.com/air-force-not-creating-skynet

https://www.marinecorpstimes.com/news/your-marine-corps/2018/09/14/marines-want-to-use-artificial-intelligence-to-help-find-and-neutralize-sea-mines/

https://techstory.in/artificial-intelligence-war-2018/

https://warontherocks.com/2019/10/with-ai-well-see-faster-fights-but-longer-wars/

https://www.defense.gov/explore/story/Article/2009288/without-effective-ai-military-risks-losing-next-war-general-says/

https://climateerinvest.blogspot.com/2018/10/billionaire-who-once-built-robots-to_1.html?m=0

https://www.military.com/daily-news/2018/09/18/air-force-wants-use-artificial-intelligence-train-pilots.html

https://www.helpnetsecurity.com/2018/09/19/iot-era-cybersecurity-gaps/

https://www.businessinsider.com/marine-corps-wants-to-use-artificial-intelligence-to-counter-sea-mines-2018-9

https://www.c4isrnet.com/intel-geoint/2018/02/16/heres-where-the-pentagon-wants-to-invest-in-artificial-intelligence-in-2019/

https://i-hls.com/archives/85519

https://warontherocks.com/2018/09/an-air-force-way-of-swarm-using-wargaming-and-artificial-intelligence-to-train-drones/

https://news.sky.com/story/ai-soldiers-army-trials-battlefield-scanning-technology-11507083

https://www.itv.com/news/2018-09-24/british-ai-successfully-trialled-in-military-experiment/

https://www.executivegov.com/2018/09/report-navy-leverages-artificial-intelligence-data-to-perform-predictive-maintenance-on-naval-aircraft/

https://www.flyingmag.com/us-air-force-may-use-artificial-intelligence-for-pilot-training/

https://www.wenatcheeworld.com/news/world/cold-war-us-intelligence-agencies-see-new-threats-from-global/article_49ae56ee-e979-544b-8c6e-3c5e57d3d1c5.html
https://www.meritalk.com/articles/the-army-is-taking-ai-into-the-weeds/
https://defensemaven.io/warriormaven/cyber/darpa-launches-massive-new-ai-next-3rd-wave-next-gen-artificial-intelligence-BVvYc1pToEWb-WWs9btw4A/
https://www.telegraph.co.uk/news/2018/09/24/artificial-intelligence-weaponry-successfully-trialled-mock/
https://www.nextgov.com/emerging-tech/2018/09/us-must-keep-artificial-intelligence-edge-keep-security-threats-check-lawmakers-say/151566/
https://blogs.wsj.com/cio/2018/07/30/pentagon-signs-885-million-artificial-intelligence-contract-with-booz-allen/
https://www.mordorintelligence.com/industry-reports/artificial-intelligence-impact-and-future-in-modern-warfare
http://www.spacewar.com/reports/UK_Looking_to_Design_Next_Gen_Military_Satellites_999.html
http://www.spacewar.com/reports/The_US_Armys_plans_to_fill_urgent_capability_gaps_in_2019_999.html
https://www.etftrends.com/robotics-ai-channel/us-department-of-defense-pledges-billions-for-ai-research/
https://formtek.com/blog/artificial-intelligence-darpa-pledges-funds-for-emerging-technologies/
https://www.asdreports.com/market-research-report-449570/artificial-intelligence-military-market-global-forecast
https://www.defenseone.com/technology/2018/07/china-russia-and-us-are-all-building-centers-military-ai/149643/
https://artificialintelligence-news.com/2018/09/28/darpa-third-wave-artificial-intelligence/
https://warontherocks.com/2018/07/but-first-infrastructure-creating-the-conditions-for-artificial-intelligence-to-thrive-in-the-pentagon/
https://www.doncio.navy.mil/CHIPS/ArticleDetails.aspx?ID=10553

https://www.marinecorpstimes.com/news/your-marine-corps/2018/12/18/marines-look-for-ibm-watson-like-artificial-intelligence-to-plan-large-scale-wargames/
https://vocalviews.com/blog/2018/12/18/uk-pledges-greater-use-of-artificial-intelligence-to-repel-military-threats/
https://emerj.com/ai-podcast-interviews/data-challenges-in-the-defense-sector/
https://www.prnewswire.com/news-releases/artificial-intelligence-in-military-market-worth-1882-billion-usd-by-2025-677647613.html
https://www.defenseone.com/technology/2018/12/project-maven-overseer-will-lead-pentagons-new-ai-center/153555/
https://www.nextgov.com/emerging-tech/2018/12/pentagon-needs-faster-buying-process-adopt-artificial-intelligence/153470/
https://www.army.mil/article/215029/artificial_intelligence_experts_a ddress_getting_capabilities_to_warfighters
https://blog.marketresearch.com/8-key-military-applications-for-artificial-intelligence-in-2018
https://federalnewsnetwork.com/defense-main/2018/12/ai-breakthroughs-versus-snake-oil-defense-agencies-taking-cautious-approach-toward-tech/
https://www.militaryaerospace.com/computers/article/16726717/rugge d-secure-ddr4-sdram-for-unmanned-systems-and-artificial-intelligence-ai-introduced-by-mercury
https://othjournal.com/2018/12/21/is-air-force-doctrine-stuck-on-artificial-intelligence/
https://publicintegrity.org/national-security/the-pentagon-tries-to-win-hearts-and-minds-in-silicon-valley/
https://www.c4isrnet.com/it-networks/2018/12/05/now-the-air-force-has-an-artificial-intelligence-team/
https://www.army.mil/article/222153/army_futures_leveraging_missio n_command_for_effective_soldier_robot_teams
https://www.evensi.us/artificial-intelligence-quantum-technology-implications-national-security-hudson-institute/280427405
https://www.brookings.edu/research/ai-and-future-warfare/
https://www.nextgov.com/emerging-tech/2018/11/marines-turn-artificial-intelligence-better-deploy-troops/153151/

https://breakingdefense.com/2018/11/artificial-intelligence-key-to-commanding-future-army-ltg-wesley/

http://centralblue.williamsfoundation.org.au/perspectives-on-artificial-intelligence-seminar-kate-yaxley/

https://www.designnews.com/electronics-test/future-military-artificial-intelligence/165978317459765

https://www.breitbart.com/national-security/2018/12/11/pentagon-u-s-in-danger-of-losing-dominance-in-artificial-intelligence/

https://www.ft.com/content/8dcb534c-dbaf-11e8-9f04-38d397e6661c

https://www.iss.europa.eu/content/artificial-intelligence-%E2%80%93-what-implications-eu-security-and-defence

https://www.washingtonpost.com/opinions/chinas-application-of-ai-should-be-a-sputnik-moment-for-the-us-but-will-it-be/2018/11/06/69132de4-e204-11e8-b759-3d88a5ce9e19_story.html

https://www.zdnet.com/article/this-is-how-artificial-intelligence-will-become-weaponized-in-future-cyberattacks/

https://www.brookings.edu/blog/order-from-chaos/2018/11/06/artificial-intelligence-and-the-security-dilemma/

https://worldview.stratfor.com/article/artificial-intelligence-cyberattacks-and-nuclear-weapons-dangerous-combination

https://ucscgenomics.soe.ucsc.edu/event/human-compatible-artificial-intelligence/

https://www.defenseone.com/technology/2018/10/pentagon-doesnt-want-real-artificial-intelligence-war-former-official-says/152450/

https://www.timesofisrael.com/military-sees-surge-in-ai-use-but-not-yet-for-critical-missions/

https://www.cnas.org/publications/reports/artificial-intelligence-and-international-security

https://www.doncio.navy.mil/CHIPS/ArticleDetails.aspx?ID=10553

https://www.nextgov.com/emerging-tech/2018/07/pentagon-sets-big-goals-its-new-ai-center/149686/

https://spacenews.com/pentagon-sees-quantum-computing-as-key-weapon-for-war-in-space/

https://gizmodo.com/ai-could-dramatically-increase-risk-of-nuclear-war-by-2-1825497698

https://futurism.com/ai-could-start-a-nuclear-war-but-thats-only-if-we-put-ai-in-charge-of-starting-nuclear-war

https://www.theregister.co.uk/2018/04/24/ai_nuclear_war/

https://www.rand.org/news/press/2018/04/24.html

https://www.rand.org/pubs/perspectives/PE296.html

https://spacenews.com/artificial-intelligence-arms-race-accelerating-in-space/

https://bernardmarr.com/default.asp?contentID=1421

https://www.nytimes.com/2018/10/26/us/politics/ai-microsoft-pentagon.html

https://www.investorsalley.com/the-artificial-intelligence-arms-race/

https://truthout.org/articles/artificial-intelligence-and-the-future-of-war/

https://www.independent.co.uk/news/uk/home-news/wars-space-online-uk-future-funding-armed-forces-modernising-defence-programme-gavin-williamson-a8687946.html

https://foreignpolicy.com/2019/03/05/whoever-predicts-the-future-correctly-will-win-the-ai-arms-race-russia-china-united-states-artificial-intelligence-defense/

https://asia.nikkei.com/Politics/Japan-steps-up-deployment-of-defense-AI-and-robots

https://www.bloomberg.com/news/articles/2019-01-10/u-s-military-trusted-more-than-google-facebook-to-develop-ai

https://www.thesun.co.uk/news/4746212/vladimir-putin-russia-super-human-soldiers-nuclear-bomb/

https://www.kut.org/post/pentagon-will-use-artificial-intelligence-find-nuclear-missiles-and-predict-potential-strikes

https://www.newsmax.com/newsfront/artificial-intelligence-defense-contractor-insect/2019/01/10/id/897745/

https://www.forbes.com/sites/cognitiveworld/2018/08/26/4-ways-the-global-defense-forces-are-using-ai/#23f17076503e

https://americanmilitarynews.com/2018/06/chinas-new-fleet-of-unmanned-assault-boats-to-use-artificial-intelligence-experts-say/

https://www.militaryaerospace.com/unmanned/article/14040644/combat-vehicles-unmanned-autonomous

https://qz.com/383275/google-has-patented-the-ability-to-control-a-robot-army/
https://www.computerworld.com/article/2910561/google-patent-envisions-cloud-control-of-an-army-of-robots.html
https://www.telegraph.co.uk/technology/google/11537596/Is-Google-building-a-robot-army.html

Chapter 22 *The Future of Religion with AI*

1. *Obama Worship*
   Billy Crone, *The Final Countdown Vol.1,2,3,*
   (Las Vegas: Get A Life Ministry, 2018)
   http://www.wnd.com/2009/06/101217/
   http://www.wnd.com/2009/01/87040/
   http://www.wnd.com/2009/04/96417/
   http://www.wnd.com/2009/09/111399/
   http://www.wnd.com/2009/01/86695/
2. *Hinduism God*
   Caryl Matrisciana, *Invasion of the Godmen,*
   (Hemet California: Jeremiah Films Inc., 1991, Video)
3. *New Age God*
   Shirley MacLaine, *Out On a Limb,*
   (Los Angeles: ABC Video Enterprises Inc., 1986, Video)
4. *Charismatic Little Gods*
   https://www.youtube.com/watch?v=K0L3x3uB9ow
5. *Transhumanists Image We Will Kill You*
   https://www.youtube.com/watch?v=01hbkh4hXEk
   https://www.youtube.com/watch?v=DaKspqDiis4
6. *Transhumanists Uploading Human Brain*
   Billy Crone, *Hybrids, Super Soldiers & the Coming Genetic Apocalypse,*
   (Las Vegas: Get A Life Ministry, 2020)
7. *Hologram Compilation*
   https://www.youtube.com/watch?v=fDR3hSeQ8pk

Billy Crone, *The Final Countdown Vol.1,2,3,*
(Las Vegas: Get A Life Ministry, 2018)
8. *Hologram People Interacting*
https://www.youtube.com/watch?v=-aHHr2D6CZQ
9. *Hologram AI Version*
https://www.youtube.com/watch?v=yzFW4-dvFDA
10. *AI Judging People Like God*
https://www.youtube.com/watch?v=KQZxwbkm0sg
11. *AI Breeding People Like God*
https://www.youtube.com/watch?v=SqEo107j-uw
12. *AI Naming Itself Like God*
https://www.youtube.com/watch?v=GWL1HNHDSq4
13. *AI Robot Priest*
https://www.youtube.com/watch?v=_XdQugsDz8E
14. *AI New Church*
https://www.youtube.com/watch?v=0cR6EbWad4E
15. *Pope Prays for AI*
https://www.youtube.com/watch?v=NYkwVRteZTI
16. *AI Robot Pastor*
https://www.youtube.com/watch?v=lZgDFBohLtc
17. *AI Warnings of Disaster*
Billy Crone, *Drones, Artificial Intelligence & the Coming Human Annihilation,* (Las Vegas: Get A Life Ministry, 2018)
https://cms.ati.ms/2018/07/artificial-intelligence-and-the-risks-of-a-hyper-war/
https://www.technologyreview.com/s/612689/never-mind-killer-robotshere-are-six-real-ai-dangers-to-watch-out-for-in-2019/
https://www.vanityfair.com/news/2018/12/microsoft-warns-washington-to-regulate-ai-before-its-too-late
https://www.wbur.org/onpoint/2018/09/24/ai-superpowers-china-us-technology-kai-fu-lee
https://www.thenewamerican.com/tech/computers/item/30141-assange-today-s-generation-last-to-be-free-technology-may-end-civilization
https://robinmarkphillips.com/ai-apocalypse-happening-right-now-not-way-think/

https://finance.yahoo.com/news/weaponized-drones-machines-attack-own-003900919.html

https://medium.com/ideachain/artificial-intelligence-risks-concerns-2a19ba21cfd9

https://towardsdatascience.com/the-threats-of-artificial-intelligence-9dd719cd1138

https://www.newsweek.com/artificial-intelligence-bigger-threat-humanity-terrorism-warns-leading-1115086

https://www.techworld.com/picture-gallery/apps-wearables/tech-leaders-warned-us-that-robots-will-kill-us-all-3611611/

https://www.bbc.com/news/technology-30290540

https://www.telegraph.co.uk/news/worldnews/northamerica/usa/12047454/Elon-Musk-launches-1bn-fund-to-save-world-from-AI.html

https://www.dailystar.co.uk/news/latest-news/un-killer-robots-ban-war-17043156

https://www.popularmechanics.com/technology/a21246473/meet-norman-a-psychopath-ai-based-on-reddit/

https://www.technologyreview.com/s/604087/the-dark-secret-at-the-heart-of-ai/

https://www.technologyreview.com/2018/03/12/144746/when-an-ai-finally-kills-someone-who-will-be-responsible/

https://www.breitbart.com/tech/2018/02/13/a-i-expert-artificial-intelligence-will-be-billions-of-times-smarter-than-humans/

https://www.breitbart.com/tech/2018/04/07/elon-musk-fears-lead-immortal-dictator-humanity-can-never-escape/

https://futurism.com/father-artificial-intelligence-singularity-decades-away

https://www.wired.com/story/google-cofounder-sergey-brin-warns-of-ais-dark-side/

https://theconversation.com/is-stephen-hawking-right-could-ai-lead-to-the-end-of-humankind-34967

https://www.beckershospitalreview.com/artificial-intelligence/elon-musk-suggests-he-s-only-months-from-merging-the-human-brain-with-ai.html

https://www.director.co.uk/artificial-intelligence-our-slave-or-master/

https://www.businesslive.co.za/bd/opinion/2018-12-05-artificial-intelligence-is-coming-whether-we-are-ready-or-not/

https://www.washingtonpost.com/technology/2018/12/12/google-ceo-sundar-pichai-fears-about-artificial-intelligence-are-very-legitimate-he-says-post-interview/

https://www.vox.com/future-perfect/2018/11/2/18053418/elon-musk-artificial-intelligence-google-deepmind-openai

https://www.currentaffairs.org/2018/11/what-you-have-to-fear-from-artificial-intelligence

https://www.reuters.com/article/us-tech-artificial-intelligence-breaking/breakingviews-review-why-an-ai-apocalypse-could-happen-idUSKCN1TF1FH

https://nypost.com/2017/12/28/terrifying-stories-that-prove-the-ai-apocalypse-is-imminent/

https://www.usnews.com/news/blogs/at-the-edge/2015/10/29/artificial-intelligence-may-kill-us-all-in-30-years

http://www.ejinsight.com/20190109-when-humans-are-ruled-by-artificial-intelligence/

https://www.theguardian.com/commentisfree/2018/dec/16/tech-experts-worried-about-artificial-intelligence-pew-research-center

https://evolutionnews.org/2018/07/artificial-intelligence-is-a-pandoras-box-whats-in-there-find-out-tomorrow/

https://www.foxbusiness.com/features/elon-musk-god-like-artificial-intelligence-could-rule-humanity

https://www.rand.org/blog/articles/2018/04/how-artificial-intelligence-could-increase-the-risk.html

https://www.vox.com/future-perfect/2019/3/26/18281297/ai-artificial-intelligence-safety-disaster-scenarios

https://www.philstar.com/opinion/2018/10/25/1862928/artificial-intelligence-will-devastate-world

18. *People Saying They're God*
http://www.carm.org/lds/lds_doctrines.htm

http://www.carm.org/wicca.htm

http://www.religioustolerance.org/hinduism2.htm

http://www.khouse.org/articles/2001/345/

http://www.canadafreepress.com/2002/main90902.htm

http://members.fortunecity.com/alahoy33/msg04.htm
http://www.spiritfind.net/jump.pl?ID=1818&Cat=Channeling/Channeled_Material&Dir=SpiritFind
http://www.spiritfind.net/jump.pl?ID=2257&Cat=Channeling/Channeled_Material&Dir=SpiritFind

19. *Church Saying They're God*
http://www.bereanfaith.com/heresy.php?action=tquote&id=6
http://www.bereanfaith.com/heresy.php?action=tquote&id=7
http://www.bereanfaith.com/heresy.php?action=tquote&id=8
http://www.bereanfaith.com/heresy.php?action=tquote&id=45
http://www.bereanfaith.com/heresy.php?action=tquote&id=47
http://www.bereanfaith.com/heresy.php?action=tquote&id=32
Hank Hanegraaff, *Christianity In Crisis*
(Eugene: Harvest House Publishers, 1993, Pgs. 11, 21, 24-25, 26-27)

20. *Artificial Intelligence as a Savior*
https://intpolicydigest.org/2018/10/16/artificial-intelligence-can-help-in-the-fight-against-poverty-and-overpopulation/
https://artificialintelligence-news.com/2018/07/04/ai-predict-nuclear-fallout-save-lives/
https://www.telegraph.co.uk/technology/2018/10/03/ai-could-bring-end-famine-says-world-bank-president/
https://www.earth.com/news/ai-unlock-satellite-data/
http://eo-alert-h2020.eu/2019/04/22/ai-for-earth-observation-and-numerical-weather-prediction/
https://interestingengineering.com/ai-might-be-the-future-for-weather-forecasting
https://eandt.theiet.org/content/articles/2019/01/anti-poaching-cameras-powered-by-ai-to-protect-african-wildlife/
https://www.costaricantimes.com/artificial-intelligence-predicting-costa-rica-volcanoes/64799
https://www.cleanenergyauthority.com/solar-energy-news/artificial-intelligence-can-help-clean-energy-060418
https://qz.com/1367197/machines-know-when-someones-about-to-attempt-suicide-how-should-we-use-that-information/
https://www.essex.ac.uk/news/2018/10/15/artificial-intelligence-could-forecast-wars-and-save-both-lives-and-money

https://www.abc.net.au/news/2018-12-15/artificial-intelligence-to-help-domestic-violence-victims/10606376

https://www.smartindustry.com/blog/smart-industry-connect/artificial-intelligence-for-oilfields-and-pipelines/

https://www.thedailybeast.com/artificial-intelligence-like-lidar-can-stop-wildfires-in-their-tracks?ref=scroll

https://learningenglish.voanews.com/a/ai-machine-learning-to-track-arctic-birds-and-wildlife/4455290.html

https://www.eon.com/en/about-us/media/press-release/2018/artificial-intelligence-warns-of-potential-power-failures.html

https://www.thenewsguard.com/news/artificial-intelligence-researchers-to-use-big-data-to-predict-sea/article_f10142a2-0466-11e9-9562-c79a4edf0818.html

https://www.foodnavigator.com/Article/2019/01/11/Artificial-intelligence-illegal-fishing-and-export-success-A-round-up-of-EU-news-views#

https://www.unops.org/news-and-stories/insights/can-artificial-intelligence-improve-humanitarian-responses

https://www.techbriefs.com/component/content/article/tb/stories/blog/33645

https://lfpress.com/news/local-news/city-hall-taps-artificial-intelligence-to-prevent-water-leaks/

https://www.newcastleherald.com.au/story/5678134/the-artificial-intelligence-helping-drones-tell-the-difference-between-sharks/

https://www.inverse.com/article/48043-paul-allen-makes-amends-with-coral-reefs-with-artificial-intelligence-research

https://www.forbes.com/sites/allenelizabeth/2019/06/27/how-artificial-intelligence-can-help-predict-toxic-algal-blooms/?sh=2ff0b034180f

https://www.foxnews.com/tech/artificial-intelligence-global-battle-human-trafficking

https://www.pbs.org/newshour/science/how-artificial-intelligence-spotted-every-solar-panel-in-the-u-s

https://www.audubon.org/magazine/spring-2018/how-new-technology-making-wind-farms-safer-birds

https://www.pcmag.com/news/industrial-iot-intelligence-aims-to-save-lives-by-preventing-disasters

https://www.thehindubusinessline.com/info-tech/intel-ai-camera-to-help-stop-animal-poaching/article25911742.ece
https://www.ozy.com/the-new-and-the-next/its-a-bird-its-a-plant-millions-turn-to-a-i-to-discover-new-species/89374/
https://www.straitstimes.com/lifestyle/keeping-sharks-at-bay-with-the-help-of-artificial-intelligence
https://www.sappi.com/sappi-fund-research-artificial-intelligence-and-bringing-carbon-emissions-net-zero-paper-and-pulp
https://www.thestar.com.my/news/nation/2018/10/20/selangor-may-use-drones-to-find-potholes-and-trees-that-need-pruning
https://www.theregister.com/2018/06/21/ai_nuclear_bombs/
https://www.techjuice.pk/this-pakistani-startup-is-using-artificial-intelligence-to-combat-water-shortage-faced-by-local-farmers/
https://www.timberbiz.com.au/artificial-intelligence-to-predict-at-risk-trees/
https://www.capitalfm.co.ke/news/2018/12/unesco-boss-says-floods-and-drought-can-be-managed-using-artificial-intelligence/
https://www.theregister.com/2018/09/29/ai_roundup_290918/
https://www.hawaiipublicradio.org/post/using-artificial-intelligence-identify-humpback-whales#stream/0
https://sf.curbed.com/2018/12/28/18159182/trees-san-francisco-census-descartes-lab
https://phys.org/news/2019-01-artificial-intelligence-bees.html
https://www.kiro7.com/news/local/western-washington-city-uses-artificial-intelligence-to-predict-who-needs-help-in-disaster/827996348/
https://www.wvnews.com/news/wvnews/wvu-to-research-artificial-intelligence-to-combat-online-opioid-trafficking/article_d9e2c85e-16ac-5183-9225-3f603a7deff7.html

21. *Artificial Intelligence as God*

https://www.malaymail.com/news/malaysia/2019/01/11/cj-artificial-intelligence-for-sentencing-virtual-hearings-holograms-in-tom/1711625
https://techwireasia.com/2019/01/ai-and-holograms-to-soon-make-a-debut-in-malaysian-courts/

https://medium.com/@Liamiscool/a-list-of-artificial-intelligence-tools-you-can-use-today-for-businesses-2-3-eea3ac374835
https://aiantichrist.blogspot.com/2019/
https://electricliterature.com/we-asked-googles-new-book-based-artificial-intelligence-about-the-meaning-of-life/
https://www.raptureforums.com/end-times/the-coming-worship-of-artificial-intelligence/
https://www.geekwire.com/2018/god-alexa-tech-religious-leaders-ponder-future-ai-together/
https://www.businessinsider.com.au/anthony-levandowski-way-of-the-future-church-where-people-worship-ai-god-2017-11?r=US&IR=T
https://www.wired.com/story/anthony-levandowski-artificial-intelligence-religion/
https://pjmedia.com/faith/ex-google-executive-registers-first-church-of-ai-with-irs/
https://www.wired.com/story/god-is-a-bot-and-anthony-levandowski-is-his-messenger/
http://www.wayofthefuture.church/
https://twitter.com/wayofthefuture_?lang=en
https://www.geekwire.com/2018/god-alexa-tech-religious-leaders-ponder-future-ai-together/
https://onezero.medium.com/how-robot-priests-will-change-human-spirituality-913a19386698
https://www.cnbc.com/2018/05/11/how-artificial-intelligence-is-shaping-religion-in-the-21st-century.html
https://www.breitbart.com/tech/2017/05/30/report-robot-preacher-gives-automated-blessings-shoots-light-beams-hands/
https://www.dailystar.co.uk/news/weird-news/god-artificial-intelligence-ai-messiah-16815131
foxnews.com/opinion/my-daughter-prayed-to-alexa-heres-the-incredible-thing-that-happened-next
https://cac.org/christianity-as-planetary-faith-2018-05-31_faculty-reflection-ilia-delio/
https://www.salon.com/2014/09/14/what_robot_theology_tells_us_about_ourselves_partner/

https://www.theatlantic.com/technology/archive/2017/02/artificial-intelligence-christianity/515463/

https://thenextweb.com/contributors/2018/10/13/ai-effect-on-faith-and-religion/

https://venturebeat.com/2017/10/02/an-ai-god-will-emerge-by-2042-and-write-its-own-bible-will-you-worship-it/

https://www.freerepublic.com/focus/f-bloggers/3655181/posts

Made in the USA
Monee, IL
15 October 2021